UNIVERSI
UW-STEVL
W9-BIY-556

ADVANCES IN CHEMICAL

VOLUME LXXII

Advances in CHEMICAL PHYSICS

EDITED BY

I. PRIGOGINE

University of Brussels
Brussels, Belgium
and
University of Texas
Austin, Texas

AND

STUART A. RICE

Department of Chemistry
and
The James Franck Institute
The University of Chicago
Chicago, Illinois

VOLUME LXXII

AN INTERSCIENCE® PUBLICATION
JOHN WILEY & SONS
NEW YORK • CHICHESTER • BRISBANE • TORONTO • SINGAPORE

An Interscience® Publication

Library of Congress Catalog Number: 58-9935

ISBN 0-471-63626-6

Printed in the United States of America

10 9 8 7 6 5 4 3 2 1

CONTRIBUTORS TO VOLUME LXXII

JOSEPH BERKOWITZ, Chemistry Division, Argonne National Laboratory, Argonne, Illinois, U.S.A.

ELEANOR E. B. CAMPBELL, Faultaet fuer Physik, Universitaet Freiburg, Freiburg, West Germany

INGOLF V. HERTEL, Faultaet fuer Physik, Universitaet Freiburg, Freiburg, West Germany

SERGEI V. KOROLEV, Polymer Chemistry Department, The Moscow State University, Moscow, U.S.S.R.

SEMION I. KUCHANOV, Polymer Chemistry Department, The Moscow State University, Moscow, U.S.S.R.

SERGEI V. PANYUKOV, Polymer Chemistry Department, The Moscow State University, Moscow, U.S.S.R.

HARTMUT SCHMIDT, Braun A. G., Wiesbaden, West Germany

INTRODUCTION

Few of us can any longer keep up with the flood of scientific literature, even in specialized subfields. Any attempt to do more and be broadly educated with respect to a large domain of science has the appearance of tilting at windmills. Yet the synthesis of ideas drawn from different subjects into new, powerful, general concepts is as valuable as ever, and the desire to remain educated persists in all scientists. This series, *Advances in Chemical Physics*, is devoted to helping the reader obtain general information about a wide variety of topics in chemical physics, which field we interpret very broadly. Our intent is to have experts present comprehensive analyses of subjects of interest and to encourage the expression of individual points of view. We hope that this approach to the presentation of an overview of a subject will both stimulate new research and serve as a personalized learning text for beginners in a field.

ILYA PRIGOGINE
STUART A. RICE

CONTENTS

Some Systematics of Autoionization Features in Atoms — 1
By Joseph Berkowitz

Symmetry and Angular Momentum in Collisions with Laser-Excited Polarized Atoms — 37
By Eleanor E. B. Campbell, Hartmut Schmidt, and Ingolf V. Hertel

Graphs in Chemical Physics of Polymers — 115
By Semion I. Kuchanov, Sergei V. Korolev, and Sergei V. Panyukov

Author Index — 327

Subject Index — 335

ADVANCES IN CHEMICAL PHYSICS

VOLUME LXXII

SOME SYSTEMATICS OF AUTOIONIZATION FEATURES IN ATOMS

JOSEPH BERKOWITZ

Chemistry Division, Argonne National Laboratory, Argonne, Illinois, U.S.A.

CONTENTS

I. Introduction
II. Systematic Trends Inferred from Experiments
  A. Width of *ns* and *nd* Resonances
  B. Limiting Values of Some Narrow Resonances in First-Row Atoms
  C. Autoionization Line Shapes
III. Characteristics of Discrete-Continuum Interaction
  A. Formalism
  B. Integral Involving Rydberg and Continuum Wave Functions
  C. Integral Involving Core Functions
  D. Improved Wave Functions
  E. Z-Dependence of Resonance Width in Isoelectronic Sequence
IV. Discussion
  A. Noble Gases
  B. Halogen Atoms
  C. Chalcogen Atoms
  D. Group V (Pnicogen) Atoms
    1. Autoionization Converging to ^1D and ^1S
    2. Autoionization between Fine-Structure Components $^3P_0$–$^3P_1$–$^3P_2$
  E. Group IV Atoms
    1. Autoionization between Spin–Orbit $^2P_{1/2}$–$^2P_{3/2}$
    2. Excitation from s Shell
  F. Group III Atoms
  G. Group II (Alkaline-Earth) Atoms
  H. Group I (Alkali) Atoms
V. Summary
References

I. INTRODUCTION

A complete set of experimental studies of the photoionization of the noble gases,[1] halogen atoms,[2–5] and chalcogen atoms[5–9] in the valence region now exists. In each case, autoionization structure dominates this region.

Examination of these experimental results reveals certain systematic trends, which we express here as propensity rules. These rules are then used to predict the widths and shapes of autoionization resonances in Groups I–V of the periodic chart. Good agreement with available experimental data is obtained in virtually every case.

II. SYSTEMATIC TRENDS INFERRED FROM EXPERIMENTS

We are dealing here primarily with initial states that have a configuration $\ldots p^m$, and the optically allowed excited states have the configuration $\ldots p^{m-1}ns, nd$. Similarly, the optically allowed autoionization continuum has the configuration $\ldots p^{m-1}\varepsilon s, \varepsilon d$. This is clear for Groups III–VIII of the periodic chart, where m varies from 1 to 6, and p refers to the valence shell. For Group I (the alkalis), $m = 6$, and we refer here to the outermost occupied p shell, just below the valence s electron. Group II (helium and the alkaline earths) deserves special attention since the excitations here are heavily, if not primarily, correlated two-electron processes, unlike the other cases described.

A. Widths of *n*s and *n*d Resonances

It can readily be shown (see ref. 1, pp. 21, 30) that the successive members of an autoionizing Rydberg series have resonance widths Γ varying as $(n^*)^{-3}$, where n^* is the effective quantum number in the Rydberg expression

$$E_n = \mathrm{IP} - \frac{R}{(n^*)^2}. \tag{1}$$

Hence, a useful criterion for comparing the widths of *different* autoionizing Rydberg series is the quantity $(n^*)^3\Gamma$, since it is approximately constant for a given series. We simplify the notation by defining a reduced width, $\Gamma_r = (n^*)^3\Gamma$.

If we express Γ in electronvolts, then Γ_r is about 5.5, 2.5, and 4.5 for the *n*d series in Ar, Kr, and Xe, respectively.[10] For the *n*s series it is 0.285, 0.17, and 0.12, in the same order.[10] Thus, we conclude that Γ_r is about 20–50 times larger for the series designated as *n*d in the noble gases Ar, Kr, and Xe than for the corresponding *n*s series. The case of Ne, and other first-row elements, is drastically different and described in what follows. In the atomic halogens, we focus our attention on the four Rydberg series converging to X^+ (^{1}D). We now know that the broad resonance series (particularly clear in atomic chlorine and atomic bromine) is a composite of $\ldots p^4(^1D)nd\ ^2P$ and ^2D. The available data[3,4] enable us to compute $\Gamma_r \cong 5.1$ for this series in Cl, and $\Gamma_r \cong 4.0$ for Br. The third optically allowed *n*d series, corresponding to a composite ^2S series, is forbidden to autoionize in a rigorous *L-S* framework, since an optically allowed ^2S continuum is not available. The series is nonetheless observed to autoionize, as a consequence of spin–orbit interaction, but the resulting resonance widths are small. We have examined this series, and the lone *ns* series, most carefully in the case of atomic chlorine. At an optical resolution of 0.28 Å [full width at half maximum (FWHM)] employed in the published work[3,4] on Cl and Br, both of these resonance series have widths limited by the instrumental resolution. In a subsequent extension of the atomic chlorine study,[11] the first members of each of these series was examined with 0.07 Å resolution. Both peaks still had widths limited by instrumental resolution. Using the relation

$$\Delta\lambda_{obs} = \sqrt{\overline{\Delta\lambda}^2_{instr} + \overline{\Delta\lambda}^2_{nw}} \tag{2}$$

where obs, instr, and nw refer to observed, instrumental, and natural widths, we conclude that $\Delta\lambda_{nw} \leqslant 0.04$ Å, which translates to $\Gamma \leqslant 0.0006$ eV in this wavelength range, or $\Gamma_r \leqslant 0.036$. Thus, the ratio of Γ_r's for broad d and narrow s resonances in Cl is $\geqslant 140$. In atomic Br, correspondingly high-resolution studies have not yet been performed, and the limits are not as tightly drawn. Thus, $\Gamma \leqslant 0.0025$ eV, and $\Gamma_r \leqslant 0.15$ for the *ns* series. The ratio of Γ_r's for broad d and narrow s resonances in Br is $\geqslant 27$. In atomic iodine,[5] one can discern broad and narrow resonances, but the spectrum is much more complex and overlapping because the spin–orbit interaction is comparable to the electrostatic (coulomb) interaction. This makes the evaluation of individual Γ's (and hence Γ_r's) uncertain. The first-row element fluorine is again an exception, to be considered later.

Among the chalcogens, we focus on the four series converging to the second ionization potential, ^2D. Also paralleling the halogens, the most revealing examples are the second and third group,[8,9] S and Se. Of the three d-like series, one (designated ^3D) is very broad, comparable to the broad widths of the aforementioned series in the noble gases and halogens. A second (designated ^3P) is very narrow, limited in width by the instrumental resolution in the case of sulfur but not selenium. It is also "forbidden" to autoionize, lacking an optically allowed ^3P continuum. The value of Γ_r for the ^3P series in atomic sulfur can be estimated from the knowledge that radiative emission and autoionization are competitive in the first member,[8] whereas in atomic selenium, Γ_r for this ^3P series can be determined from the photoionization spectrum,[9] since now $\Delta\lambda_{obs} > \Delta\lambda_{instr}$. The third series (designated ^3S) is narrow but distinctly broader than the instrumental resolution. This *nd* series is comparable in width to the sole *ns* series. Also analogous to the halogen sequence, the heaviest member (Te) has broad and narrow features, but a clear delineation is muddled by spin–orbit effects.[5] The lightest member (O) displays only narrow features.[6,7]

TABLE I Reduced Autoionization Widths (Γ_r, eV) for Various *ns* and *nd* Series in Chalcogen, Halogen, and Noble Gas Columns of Periodic Chart

	Column VI Series				*Column VII* Series			*Column VIII* Series	
Row	$ns\ ^3D$	$nd\ ^3S$	$nd\ ^3P$	$nd\ ^3D$	$ns\ ^2D$	$nd\ ^2S$	$nd\ ^2P, ^2D$	$ns\ ^1P$	$nd\ ^1P$
I	0.085	0.109	0.00024	0.108	0.007	$<\Gamma_r$	$<.0.2$	0.0507	0.0143
II	0.14	0.207	~ 0.00035	~ 5.0	$\leqslant 0.036$	$\leqslant 0.036$	~ 5.1	0.285	5.5
III	0.397	0.286	0.092	~ 4.0	$\leqslant 0.15$	$\leqslant 0.15$	~ 4.0	0.17	2.5
IV	—	—	—	—	—	—	—	0.12	4.5

B. Limiting Values of Some Narrow Resonances in First-Row Atoms

For Ne, the experiment of Ganz et al. [12] using laser excitation of metastable Ne atoms yields natural linewidths that enable us to compute $\Gamma_r = 0.0143$ for the *n*d series and 0.0507 for the *n*s series. In atomic fluorine, $\Delta\lambda_{nw}$ has not been observed for the relevant series, but fairly good limits can be established. The photographic absorption data of Palenius et al.[13] were performed at a resolution $\Delta\lambda_{instr} \leqslant 0.04$ Å, and there is no clear evidence of broadening beyond $\Delta\lambda_{instr}$. Therefore, we can conservatively take $\Delta\lambda_{nw} \leqslant$ 0.04 Å, which is equivalent to $\Gamma_r \leqslant 0.02$.

The lower limit is based on the argument that the autoionization rate competes with radiative emission. Radiative emission has been observed from only one state of fluorine above the ionization threshold, the state designated $\ldots(^1S_0)3s\ ^2S$. In that case, it is known[14] that the autoionization rate is ~ 3 times larger than spontaneous radiation. If we conservatively estimate that an autoionization rate ~ 10 times the radiative rate would make the latter difficult to observe, we can conclude that $0.007 < \Gamma_r$. The same limit on Γ_r for fluorine applies to both s and d series.

In atomic oxygen, the observed autoionization widths of the lowest members of the three series "allowed" to autoionize are just larger[7] than $\Delta\lambda_{instr}$ and hence enable one to estimate $\Delta\lambda_{nw}$ and the corresponding Γ_r. For the "forbidden"$\ldots(^1D)nd\ ^3P$ series, a lifetime dominated by the autoionization rate has been obtained in beam foil experiments[15] and enables us to compute Γ_r.

The results of the preceding considerations are summarized in Table I.

C. Autoionization Line Shapes

Fano[16] has characterized autoionization line shapes in terms of a parameter q (often called a profile index) defined by

$$q = \frac{\langle \phi | T | i \rangle}{\pi V_{\mathrm{E}}^{*} \langle \psi_{\mathrm{E}} | T | i \rangle} \tag{3}$$

where ϕ = wave function of quasi-discrete state
T = transition operator, for example, an electric dipole operator
i = wave function of initial state
ψ_{E} = wave function of ionization continuum near ϕ
V_{E} = configuration interaction matrix element connecting ϕ and ψ_{E} and
$\Gamma = 2\pi |V_{\mathrm{E}}|^2$

The parameter q can have either a plus or a minus sign.[16] When positive, the peak shape rises more abruptly on the low-energy side, wanes more gradually on the high-energy side. When negative, the reverse is true. Thus, in principle, the sign of q can be obtained from experiment by inspection, when the peak shape is not limited by the instrumental resolution. There are exceptions when the peak shape appears to be symmetrical. However, there are several clearly defined cases:

(a) The broad nd resonances in Ar, Kr, and Xe are all q positive. See Fig. 1.
(b) The broad nd resonances converging to $^1D_2$ in Cl, Br, and I are all q positive. See Fig. 2.
(c) The ns resonances converging to $^1S_0$ in Cl and Br are q negative. (In I, they are too weak to characterize.) See Fig. 3.

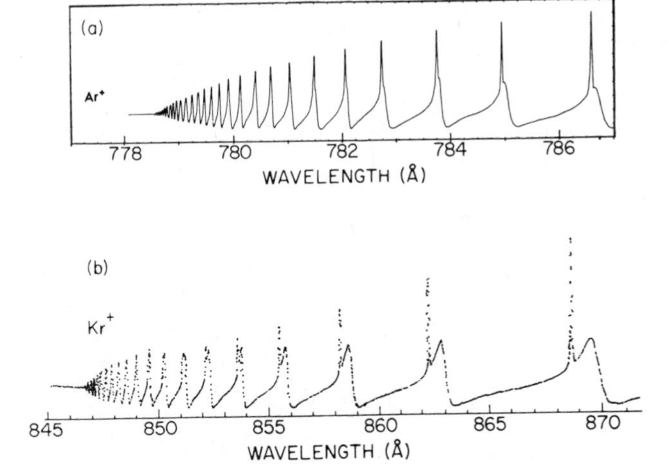

Figure 1. Portions of photoion yield curves of (*a*) Ar and (*b*) Kr demonstrating broad *nd* resonances with positive q (from ref. 1, pp. 24, 177).

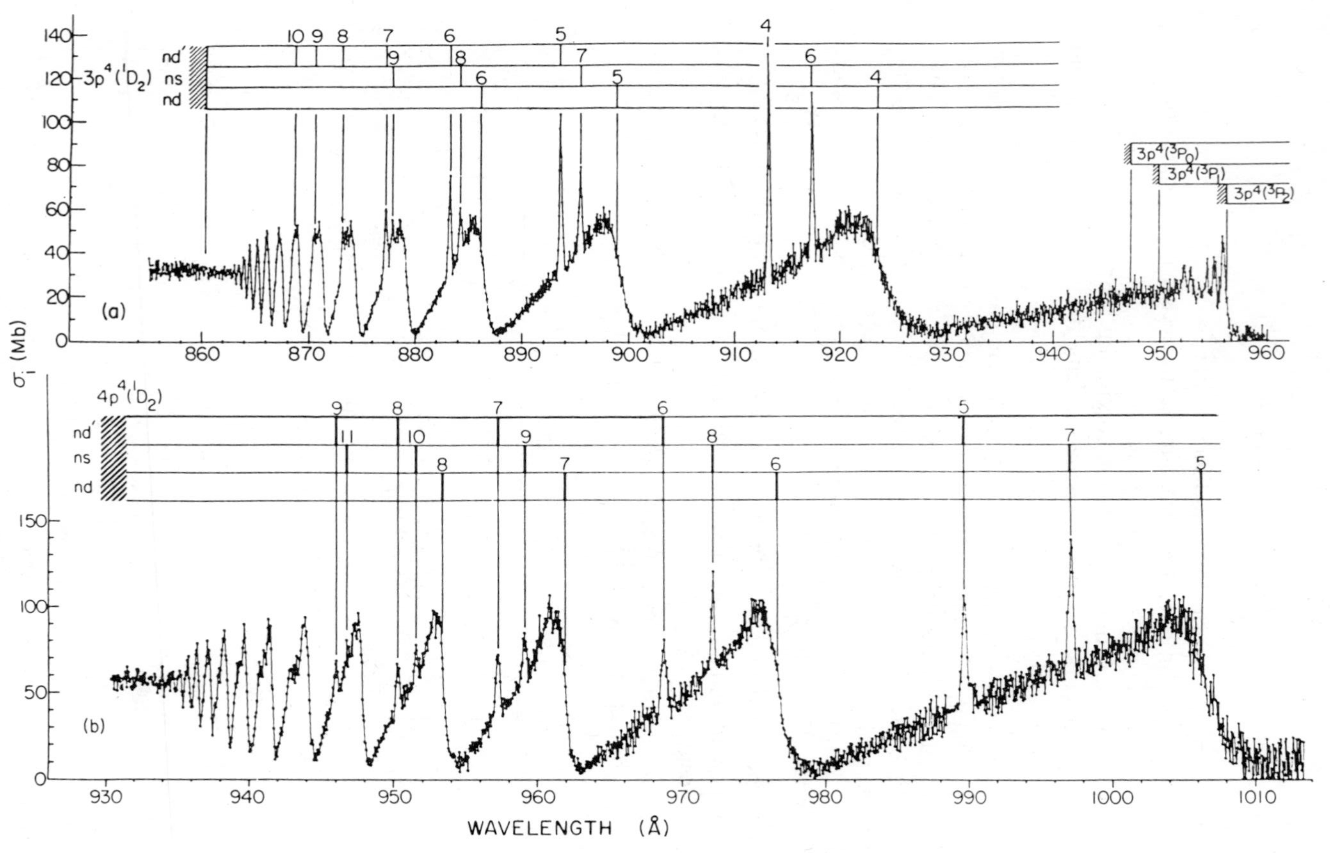

Figure 2. Portions of photoion yield curves of (*a*) Cl and (*b*) Br demonstrating broad *n*d resonances with positive *q* [(*a*) from ref. 3; (*b*) from ref. 4].

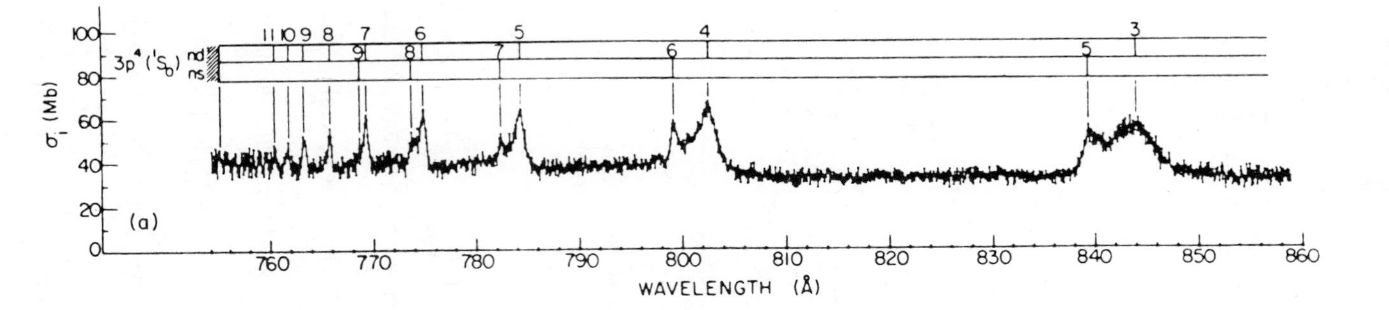

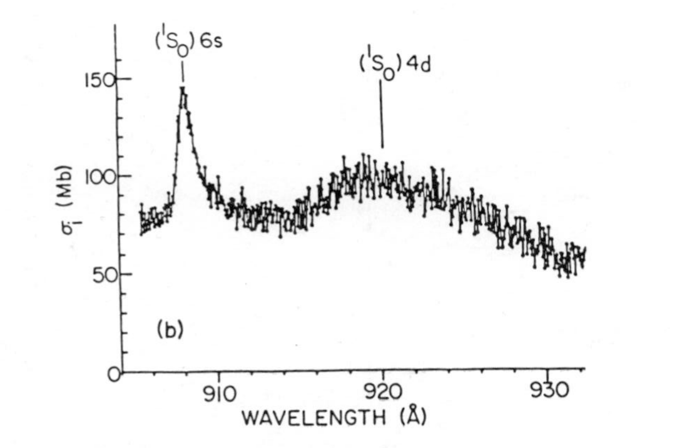

Figure 3. Portions of photoion yield curves of (*a*) Cl and (*b*) Br demonstrating *ns* resonances that have negative *q* [(*a*) from ref. 3; (*b*) from ref. 4].

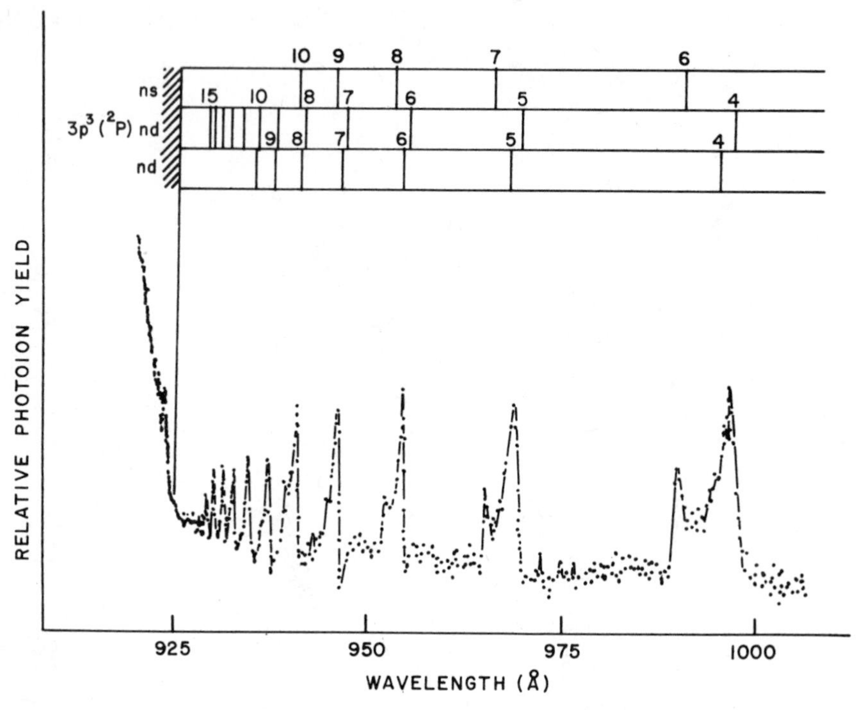

Figure 4. Portion of photoion yield curve of atomic sulfur demonstrating *n*s resonances with negative *q* and *n*d resonances with positive *q* (from ref. 8).

(d) The broad *n*d resonances converging to ^2D in S,[8] Se,[9] and Te[5] are all *q* positive.

These assignments are not controversial. There are several others, in which our assignment differs from that of other workers:

(e) The resonances we designate[8] as *n*d converging to ^2P in S, and probably Se, are *q* positive. See Fig. 4.

(f) The resonances we designate[9] as *n*s converging to ^2P in S, and probably Se, are *q* negative. See Fig. 4.

(g) The resonances we designate[8,9] as *n*d ^3S converging to ^2S in S and Se are *q* positive.

The tentative conclusion appears to be that *n*d resonances are *q* positive, while *n*s resonances are *q* negative.

III. CHARACTERISTICS OF DISCRETE-CONTINUUM INTERACTION

A. Formalism

The matrix element determining the width of the resonances is of the form

$$V = \langle \psi_{\mathrm{i}} | e^2/r_{ij} | \psi_{\mathrm{n}} \rangle, \tag{4}$$

where ψ_{n} represents the total wave function of the neutral excited (typically Rydberg) state and ψ_{i} represents the total wave function of final ion state and continuum electron. Wave function ψ_{n} can be written as an antisymmetrized product of core (ϕ_{c}) and Rydberg electron (ϕ_{R}) wave functions, and ψ_{i} is similarly described as an antisymmetrized product of final-ion (ϕ_{i}) and continuum electron (ϕ_{e}) wave functions.

Since our study focuses on optical excitation of electrons from p orbitals, we can describe ϕ_{R} and ϕ_{e} by

$$\phi_{\mathrm{R}} = a\phi'_{\mathrm{s}} + b\phi'_{\mathrm{d}} \tag{5}$$

and

$$\phi_{\mathrm{e}} = c\phi_{\mathrm{s}} + d\phi_{\mathrm{d}}. \tag{6}$$

We can expand the operator $1/r_{12}$ (restricting ourselves to a two-electron problem) in the usual way[17]:

$$\frac{1}{r_{12}} = \sum_{l=0}^{\infty} \frac{r_<^l}{r_>^{l+1}} Y_{lm}(r_1) Y_{l-m}(r_2), \tag{7}$$

where $Y(r)$ are the Legendre functions. By restricting ourselves to $l = 0, 2$ for the wave functions, the expansion of the operator $1/r_{12}$ is itself limited to two terms ($l = 0, 2$) in order that V be nonvanishing. Explicitly, V consists of direct and exchange integrals, each being the product of an angular factor C_l (which, for the moment, we take to be some constant) and a radial integral. For the direct integral, the radial integral has the form

$$R^k = \iint \phi_{\mathrm{e}}^{(1)} \phi_{\mathrm{i}}^{(2)} \frac{r_<^k}{r_>^{k+1}} \phi_{\mathrm{R}}^{(1)} \phi_{\mathrm{c}}^{(2)} \, dr_1 \, dr_2, \qquad k = 0, 2. \tag{8}$$

The related case of the Slater integral F^k can be defined by

$$F^k(nl, n'l') = \iint \frac{r_<^k}{r_>^{k+1}} R_{n'l'}^2(r_1) R_{nl}^2(r_2) \, dr_1 \, dr_2. \tag{9}$$

For this latter case, Smitt[18] has shown that for large values of n' and l' the identification of $r_>$ with r_1 and $r_<$ with r_2 seems to be rather good. The approximation is best when l' attains its maximum value, $n' - 1$, since then $R_{n'l'}(r_1)$ is nodeless, has its maximum amplitude at large r, and only has tail penetration into the core. Thus, if $R_{nl}(r_2)$ is confined largely to the core region, and $R_{n'l'}(r_1)$ is primarily outside the core, the integral of Eq. (9) can be approximately separated into two parts:

$$\begin{aligned} F^k(nl, n'l') &= \int r_2^k R_{nl}^2(r_2)\, dr_2 \int \frac{1}{r_1^{k+1}} R_{n'l'}^2(r_1)\, dr_1 \\ &= \langle r_2^k \rangle \left\langle \frac{1}{r_1^{k+1}} \right\rangle. \end{aligned} \tag{10}$$

Smitt[18] has shown that this approximation gives excellent agreement with values of $F^2(3\mathrm{p}, n'l')$ extracted from experiment for $n'l' = 5g, 6g, 6h, 7h$, and $7i$. Seaton and Storey,[19] in their study of dielectronic recombination, utilize essentially the same approach, called the Coulomb–Bethe approximation. They also note that "the CBe method is a good approximation for l sufficiently large, but is not valid for l small."

In the present circumstance, n' can be quite large, but l' is typically low (0 or 2). The wave function $\phi_{\mathrm{R}}^{(1)}$ will consequently have many nodes, but its maximum amplitude will be at large r, and if the amplitude near the core region is small, it may still be a useful approximation. By analogy with Smitt's approximation but recognizing that our justification is less firm, we therefore make the heuristic assumption

$$R^k = R_1^k R_2^k, \tag{11}$$

where

$$R_1^k = \int_0^\infty \phi_{\mathrm{e}}^{(1)} \phi_{\mathrm{R}}^{(1)} \frac{1}{r_1^{k+1}}\, dr_1, \qquad k = 0, 2, \tag{12}$$

and

$$R_2^k = \int_0^\infty \phi_{\mathrm{i}}^{(2)} \phi_{\mathrm{c}}^{(2)} r_2^k\, dr_2, \qquad k = 0, 2. \tag{13}$$

The matrix element R^k has thus been reduced to an integral involving Rydberg and continuum wave functions and another integral involving the core wave functions. We examine these integrals in the next section.

B. Integral Involving Rydberg and Continuum Wave Functions

Consider the integral of Eq. (12). In principle, we could use Hartree–Slater or Hartree–Fock wave functions for each specific element to evaluate R^1. Here, however, we are seeking some characteristic behavior that should be generally applicable to explore trends. Our approach is to use hydrogenic wave functions for both ϕ_R and ϕ_e. For ϕ_R we select $n = 3$ or $n = 4$, typical of the effective quantum numbers of low autoionizing states. For ϕ_e we choose n values and explore the value of the matrix element as we approach $n = \infty$. This parallels the approach of plotting oscillator strength f_n as a function of excitation energy E_n in histogram fashion and observing a smooth convergence to df/dE at the ionization continuum.[20]

Five cases are relevant to the present discussion:

1. $\langle 3s|1/r|ns\rangle$,
2. $\langle 3d|1/r|nd\rangle$,
3. $\langle 3d|1/r^3|nd\rangle$,
4. $\langle 3s|1/r^3|nd\rangle$, and
5. $\langle 3d|1/r^3|ns\rangle$.

The calculated matrix elements for each of these cases are summarized in Table II. Several conclusions can be drawn from these calculations.

a. The square of the matrix elements in each table decline approximately as n^{-3}, as expected for hydrogenic functions.

b. The monopole matrix elements for the transitions leaving l unchanged, that is, $\langle 3s|1/r|ns\rangle$ and $\langle 3d|1/r|nd\rangle$, are about two orders of magnitude larger than the two l-changing matrix elements. The quadrupole matrix element $\langle 3d|1/r^3|nd\rangle$ has an intermediate magnitude.

c. For extrapolation to the continuum, we denote the square of one of the aforementioned matrix elements (e.g., $\langle 3d|1/r|nd\rangle$) by P_n. A histogram is then constructed by making the area of the rectangle corresponding to the

TABLE II Calculated Values of Radial Matrix Elements Using Hydrogenic Functions

n	$\langle 3s\|1/r\|ns\rangle$ (e^2Z/a_0)	$\langle 3d\|1/r\|nd\rangle$ (e^2Z/a_0)	$\langle 3d\|1/r^3\|nd\rangle$ $[(e^2Z^3/a_0^3)\times 10^{-3}]$	$\langle 3s\|1/r^3\|nd\rangle$ $[(e^2Z^3/a_0^3)\times 10^{-4}]$	$\langle 3d\|1/3\|ns\rangle$ $[(e^2Z^3/a_0)\times 10^{-5}]$
4	0.0363596054	0.0274114	1.399125456	−2.534825829	1.931197131
5	0.0231069433	0.015733025	0.9589653381	−2.021676457	2.214888703
6	0.0167010164	0.0108143439	0.7145191517	−1.564861926	2.01962707
7	0.012897793	0.0081104279	0.5603506102	−1.242521495	1.758716489
8	0.0103747081	0.0064096330	0.4552478219	−1.013877654	1.520855051
9	0.0085981498	0.0052476042	0.3796246344	−0.846673159	1.320704469
	0.0072845448	0.0044076566	0.322991808	−0.720571156	1.155559016

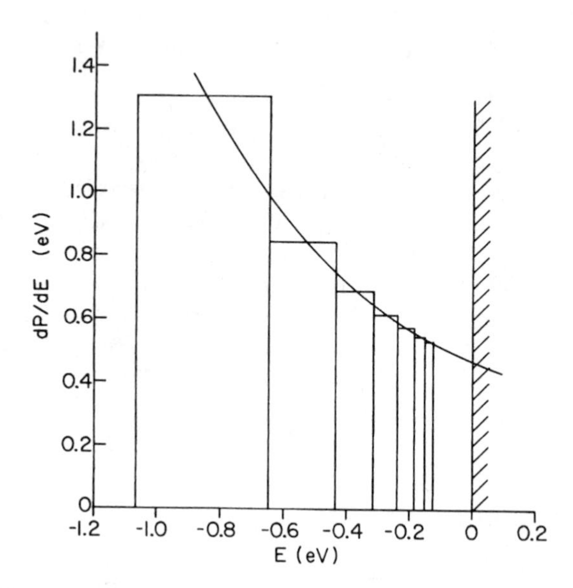

Figure 5. Histogram of P_n/E_n versus E, where $P_n = \langle 3\text{d}|1/r|n\text{d}\rangle^2$ using hydrogenic wave functions. This function is extrapolated to the ionization threshold to obtain dP/dE in eV.

nd transition equal to P_n and the base equal to the slope of the curve dE_n/dn at n, where $E_n = I_{\text{H}}(1 - 1/n^2)$, and I_{H} is the ionization energy of atomic hydrogen. With the inclusion of the constants $(Ze^2/a_0)^2$, P_n has units of energy squared, and the height of each rectangle has units of energy. Figure 5 is a histogram for this case, with energy in electronvolts. The extrapolated value at the ionization threshold, dP/dE, is ~ 0.465 eV. In this model calculation, $\Gamma = 2\pi|V_{\text{E}}|^2 = 2\pi(dP/dE)R_2^k$, where R_2^k is the integral involving the core functions (see Section III.C). It amounts to $2.92R_2^k$ eV. We note in Fig. 5 that dP/dE is clearly still declining rapidly as this function enters the continuum region, and the core overlap integral R_2^k should be less than unity, so that Γ should be of the order of 1 eV for the broadest autoionization of an $n = 3$ Rydberg state, which is compatible with the broadest resonances in Table I.

d. Only one of the four types of matrix elements considered, namely, $\langle 3s|1/r^r|n\text{d}\rangle$, is negative.

The analysis presented up to this point can be severely criticized on several counts, some already mentioned. Heretofore, we have ignored the angular factor, treating it as a nonvanishing constant. It can be shown[21] that for some of the cases treated here, the angular coefficient of the direct integral vanishes, and only the exchange integral contributes. In such cases, Eq. (8) is altered, as are the calculated integrals derived from this equation.

The sign of a matrix element involves the signs of both the angular factor and the radial integral. Both have arbitrary conventions. However, if one employs a consistent convention, the sign of q will be independent of this convention. Section IV.C discusses this subject for the chalcogen atom case.

We deduce certain propensity rules from this analysis. These can be succinctly stated. We describe both the Rydberg state and the continuum with which it is interacting in L-S notation:

1. If the Rydberg state and the continuum do not have the same values of L and S, then autoionization can only proceed through spin–orbit interaction, which usually will be weak and result in very sharp autoionization structure.

2. If autoionization is L-S allowed, and if the Rydberg electron of the Rydberg state and the free electron outgoing wave of the continuum state have the same value of l, autoionization will be rapid and result in very broad autoionization widths.

3. If autoionization is L-S allowed, but the orbital angular momentum quantum number of the Rydberg electron differs from that of the continuum electron (in our case, by ± 2), then the autoionization widths will be considerably narrower than for rule 2 but generally broader than for rule 1.

4. In the cases involving subvalence electron excitation, where the l quantum numbers of two electrons are involved, we shall generalize rules 2 and 3 so that the algebraic sum of $l_1 + l_2$ either changes by ± 2 or remains unchanged upon autoionization.

5. The profile index q is expected to be positive in most cases, but an exception can be anticipated when an s-like Rydberg electron goes into a d continuum wave.

We shall apply these rules, as well as another one discussed in Section IV.C dealing with core functions, to the experimental autoionization data available in the eight principal columns of the periodic chart in Section IV. Their utility should not be judged by the method of derivation but rather by their success or lack of success in systematizing a large body of data. They can equally well be regarded as having been arrived at by induction.

C. Integral Involving Core Functions

We consider two categories here:

1. A valence shell electron is excited, and concomitantly, the residual core is excited. The process of autoionization releases the excited electron and relaxes the core. This description is characteristic of valence p shell excitation in the noble gases, halogens, chalcogens, and Group IV elements. The integral involving the core functions will be of the form given in Eq. (13). The term

r^k couples this integral to the behavior of the Rydberg–continuum transition. Let us assume some simple form for these core wave functions, for example,

$$\phi_i^{(2)} \simeq \phi_c^{(2)} = N r_2^{\beta_e - r_2/a}. \tag{14}$$

Then,

$$R_2^k = \int_0^\infty r_2^k N^2 r_2^{2\beta_e - 2r_2/a} r_2^2 \, dr_2, \tag{15}$$

$$R_2^k = N^2 \frac{(k+2\beta+2)!}{(2/a)^{k+2\beta+3}}, \qquad k = 0, 2. \tag{16}$$

Now β is not likely to exceed 2 (for a d-like core orbital) and is typically 1 for the p-like cores we are considering in this section.

For the crude estimate to be made here, we take $a \approx a_0$. Then, the ratio of integrals for $k = 2$ (l-changing process) and $k = 0$ (l-unchanging process) becomes

$$\frac{R_2^k(k=2)}{R_2^k(k=0)} = a_0^2 \frac{6!}{4!} \left(\frac{1}{2}\right)^2 = 7.5 a_0^2, \tag{17}$$

or about one order of magnitude is regained for the $k = 2$ case ($1/r^3$ matrix element) compared to the $k = 0$ case ($1/r$ matrix element).

2. A subvalence shell electron is excited, for example, $s^2p^m \rightarrow sp^m np$. Now, the relaxation of the core itself (to s^2p^{m-1}) involves a change of one unit of angular momentum, and the Rydberg–continuum transition also involves $|\Delta l| = 1$. The net change of orbital angular momentum for the two-particle system will be $\Delta l = 0, \pm 2$. This is still consistent with the expansion of Eqs. (7) and (8).

However, the separation of Eq. (8) into two separate integrals [Eq. (11)] is more difficult to justify. Nonetheless, we shall see that the inference from the previous analysis ($\Delta l = 0$ corresponding to strong interaction, $\Delta l = \pm 2$ corresponding to weak interaction) is also exhibited by the experimental data in this case *if we identify* Δl *with the net change of orbital angular momentum of Rydberg and core electron upon autoionization.*

D. Improved Wave Functions

In Section III.C, we described the core wave functions ϕ_i and ϕ_c simply as $N r^{\beta_e - r/a}$. Here, we focus our attention on the variation of the core wave function with the atom.

The parameter a is a measure of the radius of the core. From Section

TABLE III Atomic Polarizabilities[a] ($\mathrm{Å}^3$)

N	O	F	Ne
1.10	0.802	0.557	0.3956
(0.542)[b]	(0.382)[b]		(0.196)[b]
P	S	Cl	Ar
3.63	2.90	2.18	1.6411
As	Se	Br	Kr
4.31	3.77	3.05	2.4844
Sb	Te	I	Xe
6.6	5.5	4.7–5.35	4.044

[a]From *Handbook of Chemistry and Physics*, 66th ed., compiled by T. M. Miller, CRC, Boca Raton, FL 1985.
[b]E. S. Chang and H. Sakai, *J. Phys. B*. **15**, L649 (1982).

III.C, we note that the relevant matrix element for configuration interaction involves a to some power, and therefore, it is important to make some semiquantitative assessment of this parameter. We describe two approaches:

1. The polarizability α of an atom is a rough measure of the volume encompassed by most of the charge cloud. In Table III, the polarizabilities of the atoms relevant to our discussion are listed. More relevant still would be the polarizabilities of the corresponding singly positive ions since it is this wave function we are trying to simulate. As an estimate of this trend, we have included entries corresponding to column V of the periodic table. Thus, N is isoelectronic with O^+, O is isoelectronic with F^+, and so on. In addition, we include in parentheses the polarizabilities of those atomic ions that are well established.

We can immediately note a striking variation between first-row atoms and heavier ones. There is a jump of a factor of 4–5 between first-row and second-row atoms, whereas the jump from second- to third-row atoms is 1.2–1.5 and that between third- and fourth-row atoms is smaller (and also less certain). We choose here to estimate the radius of the core a as $\alpha^{1/3}$. The radial matrix element involving the core wave functions goes at least as a^3 (when $\Delta l = 0$) and at least as a^5 (when $\Delta l = 2$). In other words, the configuration interaction matrix element V_2 goes as α^q ($q \geqslant 1$), and

$$\Gamma \propto \alpha^{2q}.$$

Thus, the difference in width between first- and second-row autoionization features that are otherwise allowed can be expected to be at least a factor of 20, whereas the corresponding factor between succeeding rows is more like ~ 1.5.

2. The preceding approach, although it helps to rationalize the disparity in autoionization widths between first-row and heavier atoms, begs the question, since it does not explain either the behavior of α or Γ but rather relates them. In this section, we seek the origin of the disparity.

It is well known that the polarizability of an atom can be expressed in the form (ref. 1, p. 59)

$$\alpha \propto \sum_{j\,\text{discrete}} \frac{f_j}{E_j^2} + \int_{\text{IP}}^{\infty} \frac{f(E)\,dE}{E^2}, \tag{18}$$

where f_j is the oscillator strength from the ground state to the state E_j. The two terms on the right side represent the contributions from the discrete spectrum and the continuum, respectively.

Rather complete information on the f_j and $f(E) = \sigma(E)$ exist for the noble gases (ref. 1, pp. 80, 84, 87, 90), and the complete expression is in good agreement with independently measured polarizabilities. Such a complete set is not yet available for the other atoms. The evidence from the noble gases is that significant contributions to α come from both the discrete and continuous spectrum. The ratio of the discrete component to the total is 0.244, 0.314, 0.368, and 0.446 for Ne, Ar, Kr, and Xe, respectively. Also, the very first transition(s), often called the resonance lines in these atoms, represent about 0.16–0.17 of the contributions to the total polarizability (discrete and continuum) in each instance. This resonance transition is $n\text{p} \rightarrow (n+1)\text{s}$. Since our analysis of the matrix elements describing the autoionization process has focused on *relative* changes rather than on absolute values, the resonance transition contribution to the polarizability correlates with resonance width as well as the total polarizability. However, a significantly better correlation can be noted if attention is focused on the partial contribution to polarizability from the lowest nd transition and, in particular, the component involving the $^2\text{P}_{1/2}$ cores ($\alpha\text{d}_{1/2}$). The partial contributions $\alpha\text{d}_{1/2}$ (in Å^3) for this transition are 0.00165 (Ne), 0.0587 (Ar), 0.0287 (Kr), and 0.1775 (Xe). Recalling the corresponding values of Γ_r from Section II.A and Table I (0.0143, 5.5, 2.5, and 4.5), we can see two improvements over the previous correlation:

1. The ratio of $\alpha\text{d}_{1/2}$ between Ar and Ne is larger than the corresponding α ratio and more nearly matches the Γ_r ratio.
2. There is a characteristic dip in Γ_r at Kr, which also occurs in $\alpha\text{d}_{1/2}$ but not in α.

The relatively large values of $\alpha\text{d}_{1/2}$ for Xe could be due to a breakdown in the coupling scheme and hence in the specific designation of this level. Of course, we should keep in mind that the proper correlation should be made with the polarizability properties of the ion core (Ne^+, Ar^+, etc.) and not the neutral atoms.

An alternative manner of characterizing the preceding discussion is to view the core functions ϕ_c and ϕ_i as being incompletely described by the simple configuration np^5 ($n = 2, \ldots, 5$). According to perturbation theory, the wave functions ϕ_c and ϕ_i can be expanded to include other configurations. The coefficients of these other contributing configurations are proportional to the matrix elements connecting initial and excited configurations and inversely proportional to the excitation energies. The matrix element in this case is the dipole matrix element. Carrying through this line of reasoning, we can show that V will be proportional to (matrix element/$\Delta E)^2$, which resembles the partial polarizability contribution we had focused on earlier.

One final point to be made in this section is that the polarizability in a given row (Table III) increases to the left, that is, as the valence shell becomes less occupied. If the correlation of resonance width and polarizability is appropriate, we might expect resonance widths that may be too narrow to measure in columns VII and VIII to become more easily measurable for column VI.

E. *Z*-Dependence of Resonance Width in Isoelectronic Sequence

We restrict our interest here to broad resonances that are known to narrow with increasing Z, since the sharp resonances will in most cases be too narrow for such comparative studies.

For such broad resonances, the analysis up to this point has focused on two interactions: one involving the Rydberg electron escaping into the continuum, with a hydrogenic matrix element going as Z/a_0, and therefore $\Gamma_r \sim Z^2$, and a core matrix element going roughly as $\alpha^{2q}, q \gtrsim 1$. If we were to confine ourselves to hydrogenlike wave functions for the core as well, then $\alpha \propto Z^{-4}$, and $\Gamma_r \sim Z^2(Z^{-4})^{2q} \sim Z^{2-8q} \sim Z^{-6}$, which is a very steep dependence indeed. However, a much more realistic description of the polarizability of the core is $\alpha \propto (Z-s)^{-4}$, where Z is the nuclear charge and s is a screening constant.[22]

With this approximation,

$$\Gamma_r \sim z^2(Z-s)^{-8q}. \tag{19}$$

The screening constant is a slowly varying function of Z in the NaI sequence investigated by Edlén,[22] but the appropriate value of s in a particular case (such as Xe, Cs^+, Ba^{2+}) is not generally available. One can estimate s from tabulated screening constants of the elements,[23] but these do not necessarily have the same meaning, as witness the difference between Edlén's[22] screening constants for the NaI sequence and Froese-Fischer's [23] values for Ne and Na.

IV. DISCUSSION

We have shown that the expansion of the operator e^2/r_{ij} for the class of problems cited previously is limited to two terms. We now wish to examine which of these terms is relevant to a particular series in each of the columns of the periodic chart.

A. Noble Gases

In an earlier analysis,[10] it was shown from symmetry arguments that Rydberg s electrons must go to continuum d waves, whereas Rydberg d electrons could go to either s or d continuum waves. Hence, the autoionization of d-like quasi-discrete states can be mediated by the $1/r$ operator, whereas the autoionization of s-like quasi-discrete states must be governed by the $1/r^3$ operator, which is seen to have an inherently smaller absolute magnitude and hence a narrower autoionization width. This is the characteristic behavior seen in Fig. 1 and quantified in Table I.

B. Halogen Atoms

The optically allowed ionization continuum between the onset of ionization and the first excited ionic state ($^1D_2$) consists of ^2P and ^2D. Rydberg s electrons bound to the ^1D core give composite ^2D states, which can readily autoionize to a ^2D continuum. However, this ^2D continuum involves d outgoing waves, and hence the matrix element governing this process is the weaker one, $1/r^3$. Rydberg d electrons bound to the ^1D core give composite ^2S, ^2P, and ^2D states. The ^2S states are "forbidden" to autoionize in *L-S* coupling and are enabled to do so only by the weak spin–orbit interaction. The ^2P and ^2D states are both permitted to autoionize by way of the stronger $1/r$ matrix element. The observations (Fig. 2 and Table I) match these conclusions.

In this section, we have neglected the possible autoionization structure that can occur between the fine-structure components $^3P_2$–$^3P_1$–$^3P_0$ near the ionization threshold. In Section IV.D, we describe the analogous $^3P_0$–$^3P_1$–$^3P_2$ region of the pnicogen atoms. As shown there, both cases can be expected to produce a complex autoionization structure, which will require detailed assignments for further analysis.

C. Chalcogen Atoms

The optically allowed ionization continuum between the ionization threshold and the first excited ionic state (^2D) consists of $^4S\varepsilon s\ ^3S$ and $^4S\varepsilon d\ ^3D$. Rydberg s electrons bound to the ^2D core give composite ^3D states, which are permitted to autoionize to a ^3D continuum but must change their orbital angular momentum by two units. Hence, the governing autoionization operator is

$1/r^3$. Rydberg d electrons bound to the ^{2}D core give composite, optically allowed ^{3}S, ^{3}P, and ^{3}D states. The ^{3}S states also require $\Delta l = |2|$ in the autoionization process and hence involve $1/r^3$. The ^{3}P states are forbidden to autoionize in *L-S* coupling, and hence the resonances should be very sharp. The ^{3}D states can autoionize with $\Delta l = 0$, the autoionization process involves the operator $1/r$, and the resonances should therefore be broad. The observations[8,9] summarized in Table I conform to these predictions.

The matrix elements evaluated using hydrogenic wave functions predict opposite signs for

$$\langle ns|1/r^3|\varepsilon d\rangle \quad \text{and} \quad \langle nd|1/r^3|\varepsilon s\rangle.$$

The definition of the resonance shape parameter, or profile index q, involves three matrix elements (see Section II.B). Two of them are transition matrix elements (in our case, optical dipole transitions) connecting the initial state with either the quasi-discrete state or the adjoining continuum. There is evidence from Hartree–Fock calculations[24] that these two dipole matrix elements will have the same sign as long as the energy of the transitions is not close to a Cooper minimum. This condition appears to be satisfied for the chalcogens.[24]

Under these circumstances, the sign of q would be determined by the sign of V, and hence $ns \rightarrow \varepsilon d$ and $nd \rightarrow \varepsilon s$ autoionizing resonance features may be expected to have opposite signs of q. In Section II.B, we have in fact presented experimental evidence for this behavior.

D. Group V (Pnicogen) Atoms

The ground state of the neutral atom is $^4S_{3/2}$. The ion core, with configuration s^2p^2, gives rise to the *L-S* states ^{3}P, ^{1}D, and ^{1}S, increasing in energy. The ^{3}P core has j components that have a fine-structure splitting[25] varying from 49.1 cm^{-1} ($^3P_1$–$^3P_0$) and 82.2 cm^{-1} ($^3P_2$–$^3P_1$) in NII to 3055.0 cm^{-1} ($^3P_1$–$^3P_0$) and 2604.0 cm^{-1} ($^3P_2$–$^3P_1$) in SbII and considerably more in BiII. The large gap between fine-structure components in the heavier atoms makes practical the observation of autoionization resonances between $^3P_0$ (ground state of the ion) and $^3P_2$, but such a study for the nitrogen atom requires rather high optical resolution. Hence, we consider first the case when the fine-structure splitting is negligible and subsequently the autoionization behavior between the fine-structure components.

1. *Autoionization Converging to* ^{1}D *and* ^{1}S

Rydberg s or d electrons bound to the excited core states ^{1}D or ^{1}S will give rise to doublet states, which are optically forbidden from the ground state in *L-S* coupling. The evidence in the case of nitrogen,[26] phosphorous,[27] and arsenic[28] is that no autoionization induced by optically allowed photoabsorption is observed corresponding to such transitions. Data are not yet available for antimony and bismuth in this energy region, but the HeI phetoelectron spectrum of bismuth[29] has a weak peak due to the process

$$\mathrm{Bi}(^4\mathrm{S}_{3/2}) + \mathrm{h}\nu \rightarrow \mathrm{Bi}^+(^1\mathrm{D}_2) + e,$$

whereas the lighter pnicogen atoms do not.[30] This is a clear indication that some photoionization to doublet states is occurring, in violation of *L-S* selection rules. It is therefore likely that photoabsorption to quasi-discrete doublet states and consequent autoionization will be observed for bismuth.

The optically allowed ionization continuum involving the ionic ground state consists of $(^3\mathrm{P})\varepsilon\mathrm{s}\,^4\mathrm{P}$ and $(^3\mathrm{P})\varepsilon\mathrm{d}\,^4\mathrm{P}$. The inner shell transition

$$\mathrm{s}^2\mathrm{p}^3 \rightarrow \mathrm{sp}^3 n\mathrm{p}\,^4\mathrm{P}$$

is both optically allowed and permitted to autoionize. It has, in fact, now been observed in nitrogen,[25] phosphorus,[26] and arsenic.[27] In the latter two systems, it manifests itself as a window resonance series.

The autoionization matrix element in this case would take the alternative forms

$$\langle \mathrm{sp}^3 n\mathrm{p} | e^2/r_{ij} | s^2 p^2 \varepsilon\mathrm{s} \rangle \quad \text{or} \quad \langle \mathrm{sp}^3 n\mathrm{p} | e^2/r_{ij} | \mathrm{s}^2\mathrm{p}^2 \varepsilon\mathrm{d} \rangle$$

In the first case, the Rydberg electron becomes an s-outgoing wave ($\Delta l = -1$), and a core p electron drops to a core s orbital ($\Delta l = -1$), and hence there is a net change of two quanta of angular momentum. In the second case, the Rydberg electron becomes a *d*-outgoing wave, and there is no net change of angular momentum. From the previous discussion, the second process should be stronger than the first, but it manifests itself in the nature of the outgoing wave rather than in the characteristics of differing Rydberg states. Also, the possibility exists for interference between the alternative exit channels, which might be observable in an angular distribution experiment involving the photoelectrons. The difficulty of such an experiment is exacerbated by the fact that these resonances tend to be windows.

2. *Autoionization between Fine-Structure Components* $^3\mathrm{P}_0$–$^3\mathrm{P}_1$–$^3\mathrm{P}_2$

This is the most complex of all the cases that fall within the scope of this chapter. The reasons are at least twofold:

a. There is convincing evidence[22,27,28] that the coupling departs from *L-S* and is more appropriately described by $[J_cK]$, or pair coupling, even for rather low values of n (J_c = total angular momentum of ion core, l = angular momentum of Rydberg electron, K = vector sum of J_c and l).

b. As a consequence, several optical dipole absorption processes and autoionizing transitions become permissible that would otherwise be forbidden by *L-S* selection rules.

To illustrate the nature of the problem, we consider the case of arsenic, for which the most detail exists at this time. Let us initially focus on the s-like transitions.

$$\ldots 4s^2 4p^3 (^4S_{3/2}) + h\nu \rightarrow \cdots \rightarrow 4s^2 4p^2 (^3P) ns \begin{cases} ^4P_{1/2,3/2,5/2} \\ ^2P_{1/2,3/2} \end{cases}.$$

Note that the fine-structure composition of the ^3P core is not designated, since then it would be overspecified. However, a centroid diagram (see ref. 22, p. 32) can be constructed (see Fig. 6), and it reveals a rather clean correlation: the $^4P_{1/2}$ series ([0] 1/2 in pair-coupling notation, where the

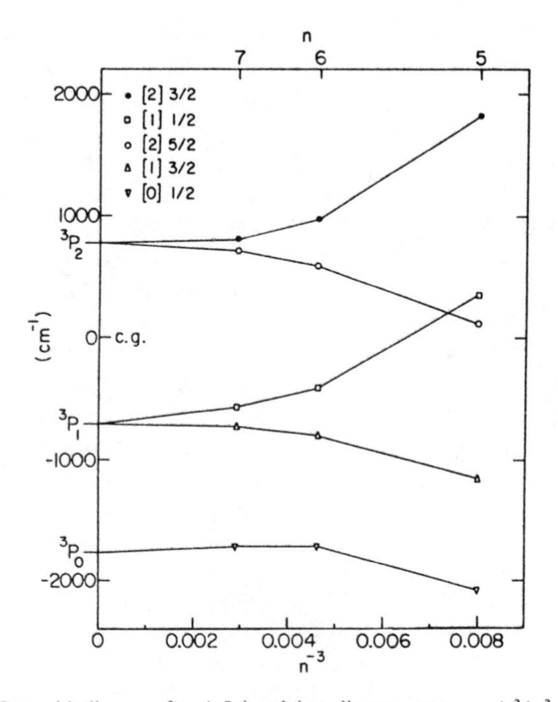

Figure 6. Centroid diagram for AsI involving discrete states $\cdots 4s^2 4p^2 ns$ for $n = 5, 6, 7$ displaying convergence to fine-structure components of ion (from data of ref. 32). Pair-coupling notation is used to label states.

bracketed number is K, the number that follows is total J) converges to $^3P_0$, the $^4P_{3/2}$ and $^2P_{1/2}$ to $^3P_1$ and the $^4P_{5/2}$ and $^2P_{3/2}$ to $^3P_2$. Thus far, the experimental data[28] have provided evidence for one s-like series converging to $^3P_1$ and one s-like series converging to $^3P_2$. Both are rather sharp autoionizing features, but they may each contain two components. At high values of n, pair coupling is appropriate, and the L-S optical dipole selection rule forbidding quartet–doublet transitions will not be rigorous.

Now let us consider the d-like transition

$$\ldots 4s^2 4p^3(^4S_{3/2}) + h\nu \rightarrow \cdots \rightarrow 4s^2 4p^2(^3P)nd \begin{cases} ^4F_{3/2,5/2,7/2,9/2} \\ ^4D_{1/2,3/2,5/2,7/2} \\ ^4P_{1/2,3/2,5/2} \\ ^2F_{5/2,7/2} \\ ^2D_{3/2,5/2} \\ ^2P_{1/2,3/2} \end{cases}$$

Of these 17 transitions, four (having $J = \frac{7}{2}$ or $\frac{9}{2}$) will be forbidden by the $\Delta J = \pm 1$ selection rule. The centroid diagram for the d-like transitions is shown in Fig. 7 in pair coupling notation, which is more appropriate for $n > 5$. One can readily appreciate the complexity that occurs with the d-like transitions. The photoionization experiment[28] is insensitive to the two series converging to $^3P_0$. There is evidence[28] for two of the four series converging to $^3P_1$. They may be blended. At high values of n, the centroid diagram appears to show bunching into two groups of two. The energy levels within each group may be too difficult to resolve. Of the seven optically allowed series converging to $^3P_2$, there is convincing evidence in the photoionization experiment for two and weaker evidence for four others. One of the two more definitive series has broad, asymmetric resonances.

A broad, asymmetric d-like resonance series converging to $^3P_2$ has also been observed by Joshi et al.[31] in a photoabsorption study of atomic antimony. Thus, the characteristic sharp s-like autoionization features and at least one broad, asymmetric d-like resonances series observed in the noble gases, halogens, and chalcogens in their heavier members (below the first row) appear to persist in the pnicogen atoms arsenic and antimony.

It is of interest to examine the possible autoionizing transitions within the pair-coupling approximation. On the left side of Table IV, the various s-like and d-like quasi-discrete states converging to $^3P_2$ are listed. They are described not only in pair-coupling notation but also with their approximate L-S designation, as given by Howard and Andrew.[32] The $J = \frac{7}{2}, \frac{9}{2}$ states are deleted, since they are rigorously forbidden in single-photon absorption. On the right side of Table IV, the various s-like and d-like continua having $^3P_0$ and $^3P_1$ cores are listed, subject to the same $\Delta J = \pm 1, 0$ restriction. Now we

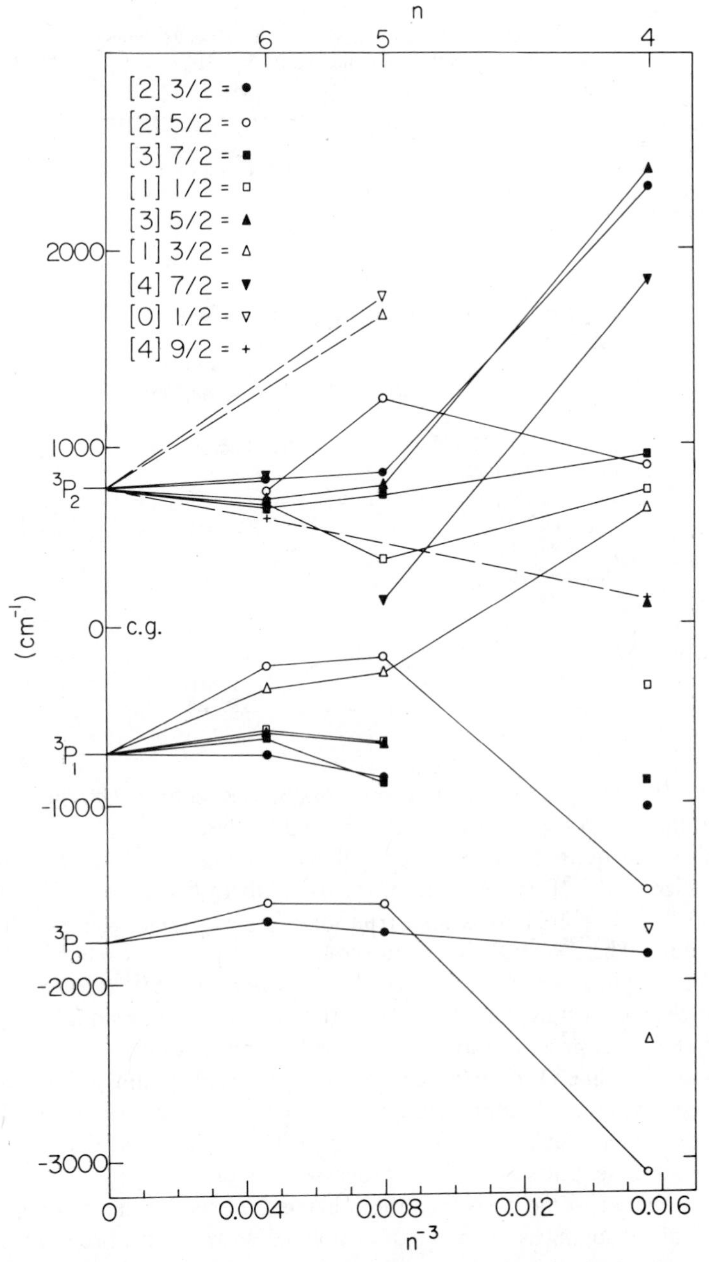

Figure 7. Centroid diagram for AsI involving discrete states $\cdots 4s^24p^2nd$ for $n = 4, 5, 6$ displaying initial complexity and subsequent convergence to fine-structure components of ion (from data of ref. 32). Pair-coupling notation is used to label states.

TABLE IV Relationship between Quasi-Discrete States Converging to $^3P_2$ and Continuum States with $^3P_0$ and $^3P_1$ Ion Cores for Arsenic

Quasi-Discrete States with $^3P_2$ Core	Continuum States
	(*a*) *With* $^3P_0$ *Core*
L-S J_cK	*L-S* J_cK
$^4P_{5/2} \cong ns2[2]5/2$	$^4P_{1/2} \cong \varepsilon s0[0]1/2$
$^2P_{3/2} \cong ns2[2]3/2$	
	$^4F_{3/2} \cong \varepsilon d0[2]3/2$
	$^4F_{5/2} \cong \varepsilon d0[2]5/2$
$^2P_{1/2} \cong nd2[1]1/2$	(*b*) *With* $^3P_1$ *Core*
$^2D_{5/2} \cong nd2[3]5/2$	
$^2D_{3/2} \cong nd2[2]3/2$	$^4P_{3/2} \cong \varepsilon s1[1]3/2$
$^4P_{5/2} \cong nd2[2]5/2$	$^2P_{1/2} \cong \varepsilon s1[1]1/2$
$^4P_{3/2} \cong nd2[1]3/2$	
$^4P_{1/2} \cong nd2[0]1/2$	$^2P_{3/2} \cong \varepsilon d1[2]3/2$
	$^4D_{1/2} \cong \varepsilon d1[1]1/2$
	$^2F_{5/2} \cong \varepsilon d1[3]5/2$
	$^4D_{3/2} \cong \varepsilon d1[1]3/2$
	$^4D_{5/2} \cong \varepsilon d1[2]5/2$

explore the various autoionization possibilities subject to the rigorous constraint that $\Delta J = 0$ between left and right sides.

The s-like quasi-discrete states (QDSs), having $J = \frac{3}{2}$ and $\frac{5}{2}$, can only autoionize to the $^3P_0$ continuum if the *ns* Rydberg becomes an εd outgoing wave (i.e. $\Delta l = +2$). This is also true for the $J = \frac{5}{2}$ QDS going to the $^3P_1$ continuum. The $J = \frac{3}{2}$ QDS can autoionize to $^3P_1$ with $\Delta l = 0$, but this is a doublet state in *L-S* notation. Since absorption to this state must occur from a quartet ground state, and decay from this state must proceed into a quartet continuum, this process may be weakened on both counts.

All of the d-like QDSs can decay into the $^3P_1$ continuum, and all but two can decay into the $^3P_0$ continuum, with $\Delta l = 0$. If we impose some weakening due to doublet–quartet transitions, the propensity-allowed decays become more restrictive, but some $\Delta l = 0$ processes remain.

Thus, there is some basis for rationalizing the broad d-like resonance and sharp s-like resonances, even in J_cK coupling. However, the broad resonance series in both As[28] and Sb[31] seem to display their breadth only after the members of those series exceed the $^3P_1$ threshold. This may be a clue about some additional restriction to autoionization.

E. Group IV Atoms

The neutral ground state is $s^2p^2\ {}^3P_{0,1,2}$. The ionic ground state is $s^2p\ {}^2P_{1/2,3/2}$. Since there are no higher states from this configuration, the only autoionization to be anticipated involving the valence p shell would occur between the ${}^2P_{1/2}$ and ${}^2P_{3/2}$ limits, similar to the noble gases but in reverse order. The other possibility is excitation from the s shell.

1. *Autoionization between Spin–Orpit Components* ${}^2P_{1/2}$–${}^2P_{3/2}$

For the sake of simplicity and clarity, we confine ourselves to the ${}^3P_0$ component of the initial state, which is the ground state. In *L-S* coupling, the optical dipole selection rules limit the excited states to ${}^3S_1$, ${}^3P_1$, and ${}^3D_1$.

The possible Rydberg states and continua built upon the fine-structure ion core states are

$$\begin{aligned}
\text{Ground state } {}^2P_{1/2} &+ ns(\varepsilon s) \rightarrow {}^3P_1, {}^1P_1 \\
&+ nd(\varepsilon d) \rightarrow {}^3P_1, {}^3D_1, {}^1P_1, {}^1D_1 \\
{}^2P_{3/2} &+ ns(\varepsilon s) \rightarrow {}^3P_1, {}^1P_1 \\
&+ nd(\varepsilon d) \rightarrow {}^3P_1, {}^3D_1, {}^1P_1, {}^1D_1
\end{aligned}$$

As in the case of the noble gases, the electric dipole selection rules rigorously require total angular momentum $J = 1$ for the quasi-discrete state. This restricts ϕ_s' to $j = \frac{1}{2}$ and ϕ_d' to $j = \frac{3}{2}, \frac{5}{2}$. Similarly, ϕ_s is restricted to $j = \frac{1}{2}$ and ϕ_d to $j = \frac{3}{2}$ [see Eqs. (5) and (6) for definitions of ϕ's].

The core wave functions ϕ_d and ϕ_i have $j = \frac{3}{2}$ and $j = \frac{1}{2}$, respectively. The operator $1/r_{12}$ can have only two terms in its expansion that will couple these core states, involving the Legendre functions Y_1 and Y_2. Since the parity of total initial and final states is the same, the matrix element must vanish for Y_1 (odd parity). The truncation to Y_2 also carries with it the selection rule

$$\phi_s'(j = \tfrac{1}{2}) \rightarrow \phi_d(j = \tfrac{3}{2})$$

and

$$\phi_d'(j = \tfrac{3}{2}) \rightarrow \phi_s(j = \tfrac{1}{2}), \qquad \phi_d'(j = \tfrac{5}{2}) \rightarrow \phi_d(j = \tfrac{3}{2}).$$

Two of these series involve $\Delta l = 2$ transitions and are expected to be narrow by our propensity criterion, while the remaining one does not require a change in angular momentum, and should be broad.

In Fig. 8*a*, the high-resolution photoabsorption spectrum of Brown et al.[33] for GeI is shown, in the relevant region, starting with ${}^3P_0$ and approaching the ${}^2P_{3/2}$ threshold. A magnified portion is shown[33] in Fig. 8*b*. Members of

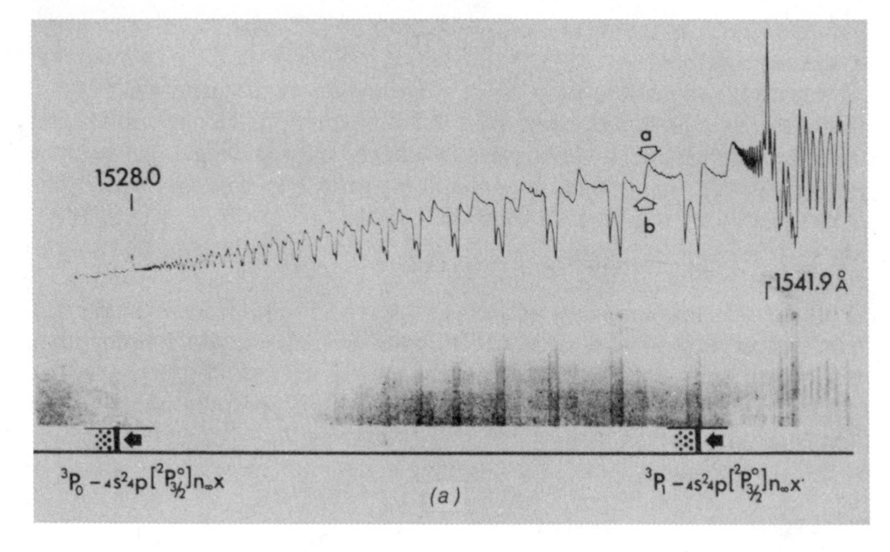

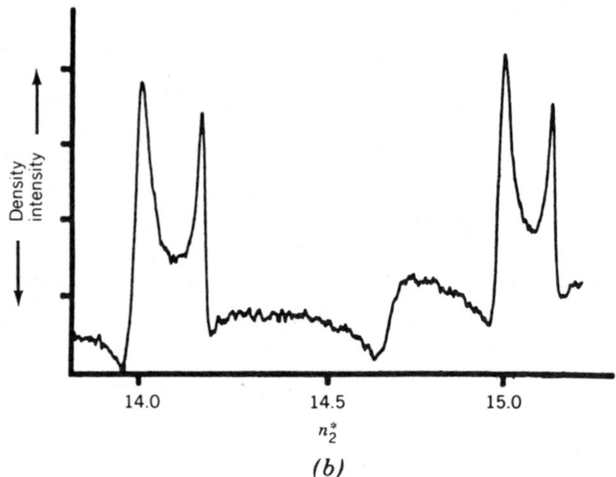

Figure 8. Densitometer traces of photoabsorption in atomic germanium. (*a*) Three Rydberg series corresponding to transitions $4s^24p^2\ ^3P_0 \rightarrow 4s^24p(^2P_{3/2})ns, nd$. (*b*) Enlarged section of region of (*a*) revealing shape of representative members of three series. Abscissa here is n_2^*, rather than wavelength, but this has negligible effect on the visual display of widths (Γ) and signs of q. (From ref. 33, with permission from the author and the publisher.)

three Rydberg series are apparent, two narrow and one broad. The broad series has $\Gamma_r \approx 4.1$ and can be identified as $\phi'_d\ (j = \frac{5}{2}) \rightarrow \phi_d\ (j = \frac{3}{2})$.

The narrow resonance series with members at $n^* = 14.147$ and $n^* = 15.148$ (Fig. 8*b*) has $\Gamma_r \leqslant 0.6$ and is identified with $\phi'_s\ (j = \frac{1}{2}) \rightarrow \phi_d\ (j = \frac{3}{2})$. The shape of this resonance series implies a negative value of q, consistent with an s → d transition. The narrow resonance series with members at 13.9928 and 14.9918 (Fig. 8*b*) has $\Gamma_r \cong 1.1$ and is identified with $\phi'_d\ (j=\frac{3}{2}) \rightarrow \phi_s\ (j=\frac{1}{2})$. This resonance series has a shape indicating a positive q, consistent with a d → s transition.

Absorption spectra of comparable quality have been obtained by this group for Si[34] and Sn,[35] but these publications do not include the densitometer traces necessary for carrying out such an analysis.

2. *Excitation from* s *Shell*

The inner shell transition $sp^2(^4P, ^2P)np$ is optically allowed to the states $^3S_1$, $^3P_1$, and $^3D_1$. Of these, the $^3S_1$ is forbidden to autoionize in an *L-S* framework, but $^3P_1$ and $^3D_1$ are both allowed. The matrix elements can be formally written

$$\langle sp^2np|e^2/r_{ij}|s^2p(^2P_{1/2,3/2})\varepsilon s, {}^3P\rangle$$

or

$$\langle sp^2np|e^2/r_{ij}|s^2p(^2P_{1/2,3/2})\varepsilon d, {}^3P \text{ and } {}^3D\rangle$$

In the first case, since the continuum is ^3P, the quasi-discrete state must also be ^3P. Also, there will be a net change in orbital angular momentum of 2, and consequently, this process is not favored.

In the second case, since the continuum can be ^3P or ^3D, the quasi-discrete state can also be ^3P or ^3D, and there is no net change of angular momentum. Hence, this process affords a relatively rapid mechanism for autoionization of both the ^3P and ^3D series. For atomic carbon, Carter and Kelly[36] have calculated by diagrammatic many-body perturbation theory that 68% of the continuum is ^3D; both their calculation and the experiment of Esteva et al.[37] indicate that the dominant autoionizing series is $(^4P)np\ ^3D$, although the series $(^4P)np\ ^3P$ is observable (see Fig. 9).

F. Group III Atoms

The neutral ground state is $s^2p\ ^2P_{1/2,3/2}$. The ionic ground state is $s^2\ ^1S_0$, and the corresponding optically allowed ionization continuum has $^2S_{1/2}$ and $^2D_{3/2,5/2}$ character. Obviously, no autoionization can arise until the inner s shell is accessed. We then have

$$s^2p \rightarrow spnp\ ^2S, ^2P, ^2D.$$

Of these quasi-discrete states the ^2P is forbidden to autoionize and should therefore be very sharp, whereas the ^2S and ^2D are allowed.

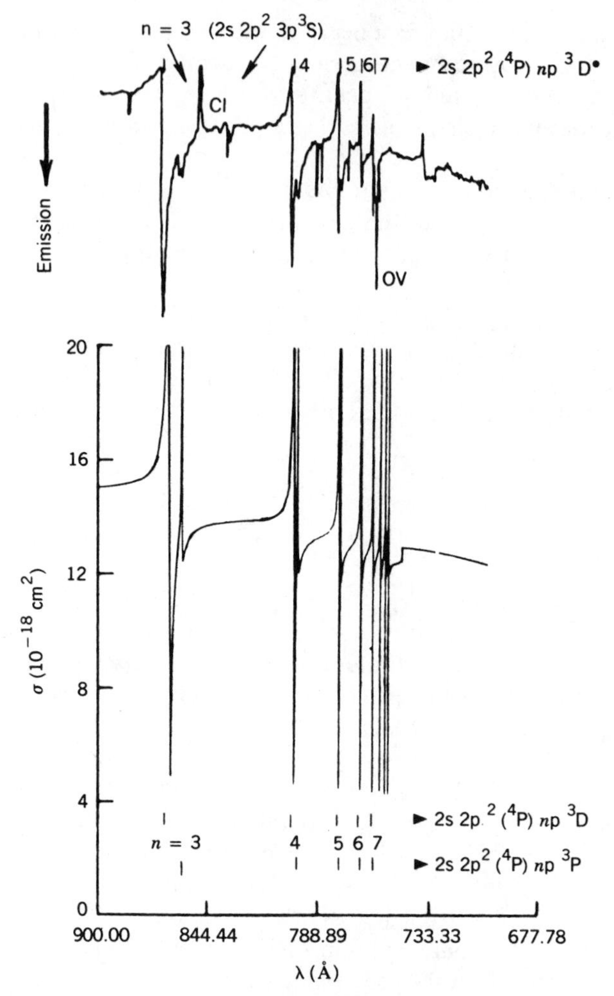

Figure 9. Photoabsorption of atomic carbon. Upper curve is densitometer trace from experiment by J. M. Esteva. Lower curve is calculation by S. L. Carter and H. P. Kelly. (From ref. 36, with permission from the author and the publisher.)

However, when the V matrix elements are written out, that is,

$$\langle \mathrm{sp}n\mathrm{p}|e^2/r_{ij}|\mathrm{s}^2(^1\mathrm{S}_0)\varepsilon\mathrm{s}, {}^2\mathrm{S}\rangle \quad \text{and} \quad \langle \mathrm{sp}n\mathrm{p}|e^2/r_{ij}|\mathrm{s}^2(^1\mathrm{S}_0)\varepsilon\mathrm{d}, {}^2\mathrm{D}\rangle,$$

it becomes apparent that the first case involves $\Delta l = -2$, while the second case involves $\Delta l = 0$. Therefore, the ^{2}S resonances should be narrower than the ^{2}D resonances.

Table V Reduced Autoionization Widths Γ_r(eV) for Series ^2S, ^2P, and ^2D in Group III Atoms

	^2S		^2P		^2D	
	Configuration	Γ_r	Configuration	Γ_r	Configuration	Γ_r
Al	$(3s3p^2)$	$0.025^{a,b}$	$(3s3p^2)$	0.00020^c	$(3s3p4p)$	$2.18^{b,d}$
				0.00005^c	$(3s3p5p)$	2.39^b
					$(3s3p6p)$	3.5^b
Ga	$(4s4p^2)$	0.083^e	$(4s4p^2)$	0.011^e	$(4s4p^2)$	2.83^e
In	$(5s5p^2)$	0.053^e	$(5s5p^2)$	0.085^e	$(5s5p^2)$	3.1^e
				0.033^e	$(5s5p6p)$	0.6^d
Tl	$(6s6p^2)$	0.04^f	$(6s6p^2)$	0.04^f	$(6s6p^2)$	1.33^g
					$(6s6p7p)$	$0.38^{d,g}$

[a] From J. L. Kohl and W. H. Parkinson, *Astrophys. J.* **184**, 641 (1973).
[b] From R. A. Roig, *J. Phys. B.* **8**, 2939 (1975).
[c] From G. G. Lombardi, B. L. Cardon, and R. L. Kurucz, *Astrophys. J.* **248**, 1202 (1981).
[d] From B. E. Krylov and M. G. Kozlov, *Opt. Spectr. (Engl.)* **47**, 579 (1979).
[e] From M. G. Kozlov and G. P. Startsev, *Opt. Spectr. (Engl.)* **24**, 3 (1968).
[f] From J. P. Connerade and M. A. Baig, *J. Phys. B.* **14**, 29 (1981); these authors calculate this state to be 60% $^2S_{1/2}$ and 35% $^2P_{1/2}$.
[g] M. G. Kozlov and G. P. Startsev, *Opt. Spectr. (Engl.)* **24**, 3 (1968).

Rather extensive experimental studies have been made on the Group III atoms in recent years, mostly by photoabsorption. The values of Γ_r deduced from those studies for Al, Ga, In, and Tl are summarized in Table V. Figure 10*a* is a photoabsorption spectrum of Al obtained by Esteva et al.[37] and compared with a calculation by Le Dourneuf et al.[38] Figure 10*b* is a composite of true photoionization (rather than photoabsorption) measurements on Ga, In, and Tl by Karamatskos et al.[39] For Al and Ga, the observations are in very good agreement with expectations. The values of Γ_r are 2–3 for the ^2D transitions, 0.025–0.083 for ^2S, and 0.0002–0.01 for ^2P. For in and Tl, the distinction between ^2S and ^2P becomes blurred, and the ^2D begins to diminish in width, though it is still by far the broadest. There is also some evidence that the first member of the ^2D series (the intrashell $nsnp^2\ ^2D$) has a larger value of Γ_r than the higher members for In and Tl. The tendency for the sharp peaks to become broader and the broad peaks to become narrower with increasing Z is suggestive of a progressive departure from *L-S* coupling, with some narrower and broader states mixing in *j-j* coupling.

A photoabsorption spectrum of boron has been obtained by Esteva et al.[37] Only the ^2D transitions are clearly visible. This series appears to be distinctly

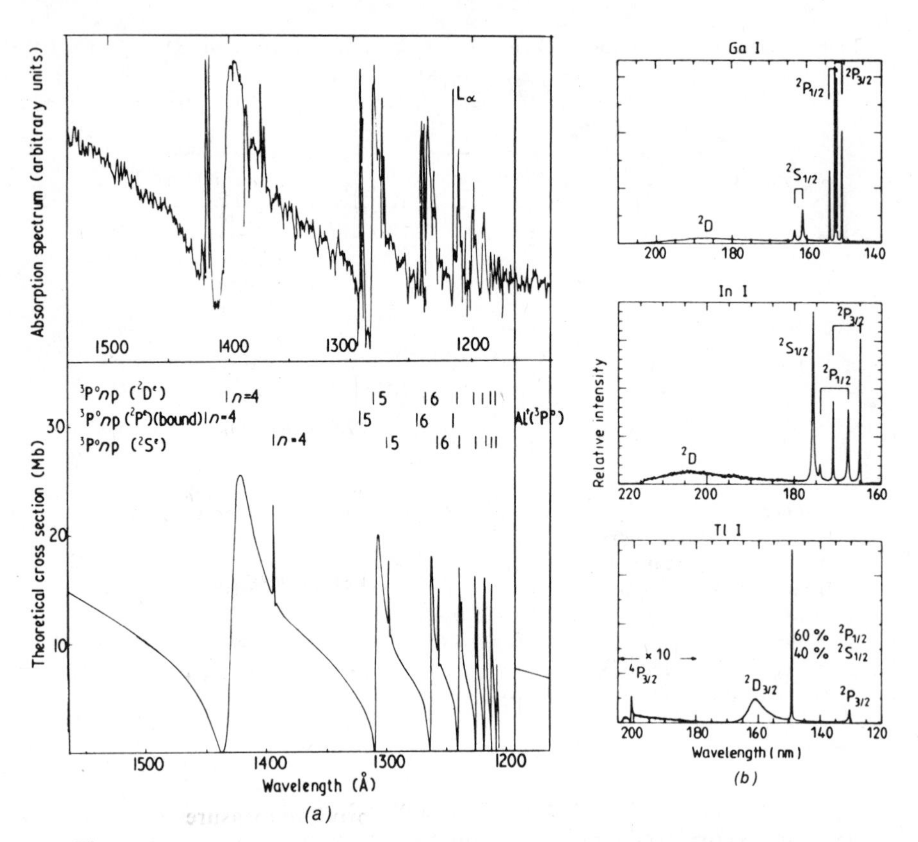

Figure 10. (*a*) Photoabsorption of atomic aluminum. Upper curve is densitometer trace from experiment by J. M. Esteva. Lower curve is calculation by M. LeDourneuf et al. (From ref. 38, with permission from the author and the publisher.) (*b*) Photoionization of atomic gallium, indium, and thallium. (From ref. 39, with permission from the author and the publisher.)

sharper than the corresponding one in Al and Ga, with an estimated value of $\Gamma_r \cong 0.6$–1.0. It is suggestive of the narrowing characteristic of first-row atoms but not as extreme as the change between the first-row elements O, F, and Ne and the second-row elements S, Cl, and Ar. However, the polarizability is also increasing to the left in the first row, and is 3.03 $(\text{Å})^3$ for boron, approximately 3.5 times that in oxygen.

G. Group II (Alkaline-Earth) Atoms

The neutral ground state is $s^2\ {}^1S_0$; the ground state of the singly charged ion is ${}^2S_{1/2}$. There are no higher valence ionic states, and hence the single-electron excitation leaves no mechanism for autoionization until inner shells are accessed. However, two-electron excitations and corresponding autoionization features are observed in this group, to which the helium atom can be added. Considerable attention has been directed to autoionization involving this group in the recent literature.

Here, we confine ourselves to two features in the autoionization of the alkaline-earth atoms that are relevant to the main points of this chapter.

1. In MgI, there is a sharp 3p*n*d series. In CaI and SrI, the corresponding series are broad. (See, e.g., ref. 40 and references therein.) There is also a change in the nature of the dominant series that occurs between MgI and CaI. This behavior can be rationalized by configuration interaction in the initial state, allowing some d^2 character to mix with the dominant s^2 configuration. With Ca, the proximity of the unfilled 3d orbital allows for significantly more d^2 character than with Mg.

2. Consider the two-electron excited states of Be, 2p*n*s and 2p*n*d. Both must autoionize to a continuum 2sεp, $^1P_1$. For the former case, the Rydberg *n*s electron becomes a p-outgoing wave ($\Delta l = +1$) and the 2p core electron relaxes to a 2s core state ($\Delta l = -1$), with a net $\Delta l = 0$. For the latter case, both Rydberg electron and core electron must lose a unit of orbital angular momentum in the process of autoionization, and thus, $\Delta l = -2$. As a consequence, we may expect the 2p*n*s series to be broad, and the 2p*n*d series to be sharp. Heretofore, we have encountered broad *n*d series and sharp *n*s series, which have been explained by the aforementioned analysis. In this case, the same analysis leads to the reverse behavior, and it conforms to experiment for both BeI and MgI. (See Fig. 11, from Esteva et al.[37])

H. Group I (Alkali) Atoms

The neutral ground state is $^2S_{1/2}$, the ionic ground state is $^1S_0$, and obviously no possibility exists for autoionization from the valence shell. Excitation from the subvalence p shell can be formally described as

$$p^6 s \rightarrow p^5 s ns, nd.$$

The matrix elements V for the ensuing autoionization may be written

$$\langle p^5 s ns(^2P) | e^2/r_{ij} | p^6(^1S_0)\varepsilon p, {}^2P \rangle$$

and

$$\langle p^5 s nd(^2P) | e^2/r_{ij} | p^6(^1S_0)\varepsilon p, {}^2P \rangle.$$

In the first case, both the Rydberg electron and the core s electron must increase their orbital angular momentum by one unit in the autoionization

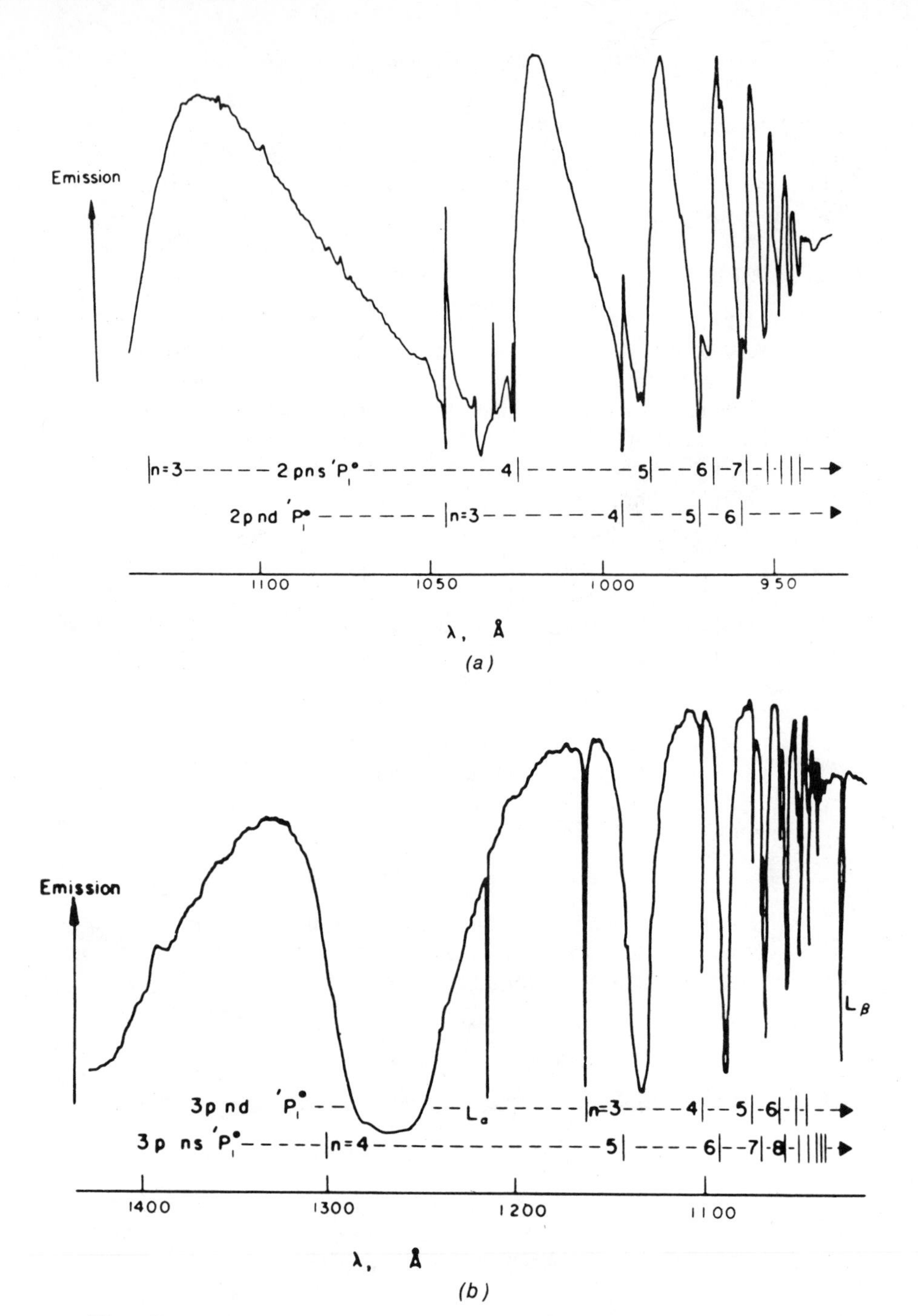

Figure 11. (*a*) Photoabsorption of atomic beryllium. (*b*) Photoabsorption of atomic magnesium. (From ref. 37, with permission from the author and the publisher.)

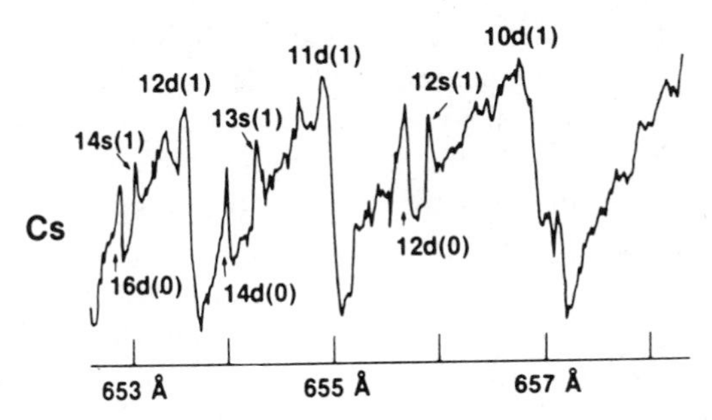

Figure 12. Photoabsorption of atomic cesium. (From ref. 41, with permission from the author and the publisher.)

process. Thus, the total change is $\Delta l = +2$, the $1/r^3$ operator is appropriate, and hence the s-like resonances should be narrow.

In the second case, the Rydberg electron loses one unit of angular momentum in the ionization process, compensating for the gain of angular momentum in the core. Hence, $\Delta l = 0$, and one of the d-like resonances should be broad. This parallels the behavior of the noble-gas resonances, as noted by the authors[41] of a recent experimental study of atomic cesium.

However, in the alkalis, the transition to $p^5 snd$ can give rise to ^2S optically allowed states that are forbidden to autoionize in *L-S* coupling due to the absence of a ^2S continuum. They can autoionize weakly due to spin–orbit interaction, and hence the alkalis should exhibit two narrow resonance series (one *n*s, one *n*d) and one broad *n*d resonance series. The photoabsorption spectrum of atomic cesium[41] (Fig. 12) shows just this behavior.

There is some evidence from photoabsorption studies that the corresponding region in Na has sharp *n*s and *n*d features[42] but that the *n*d resonance features in K become broad.[43] In this column of the periodic chart, the break between first-row elements and second-row elements should occur between Na and K, where the proximity of the unfilled 3d orbital becomes a factor.

V. SUMMARY

We have shown that a simple quantum mechanical description of the autoionization process in atoms, consistently applied, can explain the autoionization widths observed in every column of the periodic chart. The analysis, based on hydrogenic wave functions, leads to the conclusion that *l*-unchanging transitions should be distinctly more probable than *l*-changing autoionization processes. This often makes d-like Rydberg levels give rise to broad resonances and s-like Rydberg levels to sharp resonances, but in the alkaline earths, the reverse is predicted and occurs. The hydrogenic calculations give the correct order of magnitude for the reduced width Γ_r of the resonances. They further predict that the profile index q should have opposite signs for $n\text{s} \rightarrow \varepsilon\text{d}$ and $n\text{d} \rightarrow \varepsilon\text{s}$, $n\text{d} \rightarrow \varepsilon\text{d}$ and $n\text{s} \rightarrow \varepsilon\text{s}$, and the available experimental facts seem to support this observation.

While we do not claim that the mathematical considerations presented here constitute a proof, the available experimental data can be concisely rationalized by the principles derived from the approximate formulation and hence provide strong evidence for an underlying simplicity and generalization governing autoionization in atomic systems.

The domains of validity of several assumptions made in this chapter need to be tested with more accurate atomic wave functions and more sophisticated mathematical methods. These include the following:

a. The separability of Eq. (8) assumed in Eq. (11).

b. The testing of the signs of the matrix elements given in Table II. The values in Table II are based on hydrogenic wave functions and integration from zero to infinity. A more realistic case would involve a core and some modified coulomb functions. Column 4 of Table II ($\langle 3\text{s}|1/r^3|n\text{d}\rangle$) is the only one with negative values. Would this persist with more realistic wave functions?

c. The assumption in Section III.C, category 2, that the propensity rule deduced for a single-electron transition ($\Delta l = 0$ favored, $\Delta l = \pm 2$ unfavored) is applicable to the case where both core and Rydberg electrons change l, and Δl is to be interpreted as the net change in orbital angular momentum.

The arguments presented in this chapter place a great deal of emphasis on *L-S* coupling and selection rules derived therefrom. Multichannel quantum defect theory (MQDT) tries to ween us away from those concepts, and instead focusses attention on other parameters, the μ_i quantum defects and $U_{i\alpha}$ matrix. The emphasis on MQDT in the recent literature may be one reason why the observations in this chapter may not have been noted earlier.

This research was supported by the U.S. Department of Energy (Office of Basic Energy Sciences) under contract W-31-109-Eng-38.

References

1. J. Berkowitz, *Photoabsorption, Photoionization and Photoelectron Spectroscopy*, Academic, New York, 1979.
2. B. Ruščić, J. P. Greene, and Berkowitz, *J. Phys. B* **17**, L79 (1984).
3. B. Ruščić, and J. Berkowitz, *Phys. Rev. Lett.* **50**, 675 (1983).
4. R. Ruščić, J. P. Greene, and J. Berkowitz, *J. Phys. B* **17**, 1503 (1984).
5. J. Berkowitz, C. H. Batson, and G. L. Goodman, *Phys. Rev. A* **24**, 149 (1981).

6. P. M. Dehmer, J. Berkowitz, and W. A. Chupka, *J. Chem. Phys.* **59**, 5777 (1973).
7. P. M. Dehmer, W. L. Luken, and W. A. Chupka, *J. Chem. Phys.* **67**, 195 (1977).
8. S. T. Gibson, J. P. Greene, B. Ruščić, and J. Berkowitz, *J. Phys. B* **19**, 2825 (1986).
9. S. T. Gibson, J. P. Greene, B. Ruščić, and J. Berkowitz, *J. Phys. B* **19**, 2841 (1986).
10. K. Radler and J. Berkowitz, *J. Chem. Phys.* **70**, 216 (1979).
11. B. Ruščić, J. P. Greene, and J. Berkowitz, unpublished data.
12. J. Ganz, A. Siegel, W. Bussert, K. Harth, M.-W. Ruf, H. Hotop, J. Geiger, and M. Fink, *J. Phys. B* **16**, L569 (1983).
13. H. P. Palenius, R. E. Huffman, J. C. Larrabee, and Y. Tanaka, *J. Opt. Soc. Am.* **68**, 1564 (1978); also, R. E. Huffman, private communication.
14. H. G. Berry, L. P. Somerville, L. Young, and W. J. Ray, *J. Phys. B* **17**, 3857 (1984).
15. E. J. Knystautas, M. Brochu, and R. Drouin, *Can. J. Spectr.* **18**, 143 (1973).
16. U. Fano, *Phys. Rev.* **124**, 1866 (1961).
17. See, for example, E. U. Condon and G. H. Shortley, *The Theory of Atomic Spectra*, University Press, Cambridge, 1953, p. 174.
18. R. Smitt, Annual Report, Atomic Spectroscopy, University of Lund, Sweden, 1978, pp. 58–60.
19. M. J. Seaton and P. J. Storey, in *Atomic Processes and Applications*, P. G. Burke and B. L. Moisewitsch (eds.), North-Holland, Amsterdam, 1976, p. 163.
20. See, for example, U. Fano and J. W. Cooper, *Rev. Mod. Phys.* **40**, 441 (1968).
21. C. Froese-Fischer, private communication.
22. B. Edlén, *Handbuch der Physik*, Vol. 27, S. Flügge (ed.), Springer-Verlag, Berlin 1964, p. 125.
23. C. Froese-Fischer, *Atomic Data* **4**, 301 (1972); **12**, 87 (1973).
24. S. T. Manson, A. Msezane, A. F. Starace, and S. Shahabi, *Phys. Rev.* **A20**, 1005 (1979); S. T. Manson, private communication.
25. C. E. Moore, *Atomic Energy Levels*, Vols. I, III, NSRDS-NBS 35, U.S. GPO, 1971.
26. P. M. Dehmer, J. Berkowitz, and W. A. Chupka, *J. Chem. Phys.* **60**, 2676 (1974).
27. J. Berkowitz, J. P. Greene, H. Cho, and G. L. Goodman, *J. Phys. B* **20**, 2647 (1987).
28. J. Berkowitz, J. P. Greene, and H. Cho, *J. Phys. B* (in press).
29. S. Suzer, S. T. Lee, and D. A. Shirley, *J. Chem. Phys.* **65**, 412 (1976).
30. J. M. Dyke, S. Elbel, A. Morris, and J. C. H. Stevens, *J. Chem. Soc. Far. Trans. 2* **82**, 637 (1986).
31. Y. N. Joshi, V. N. Sarma, and Th. A. M. Van Kleef, *Physica* **125C**, 127 (1984).
32. L. E. Howard and K. L. Andrew, *J. Opt. Soc. Am. B* **2**, 1032 (1985).
33. C. M. Brown, S. G. Tilford, and M. L. Ginter, *J. Opt. Soc. Am.* **67**, 584 (1977).
34. C. M. Brown, S. G. Tilford, R. Tousey, and M. L. Ginter, *J. Opt. Soc. Am.* **64**, 1665 (1974).
35. C. M. Brown, S. G. Tilford, and M. L. Ginter, *J. Opt. Soc. Am.* **67**, 607 (1977).
36. S. L. Carter and H. P. Kelly, *Phys. Rev.* **A13**, 1388 (1976).
37. J. M. Esteva, G. Mehlman-Ballofet, and J. Romand, *J. Quant. Spectr. Rad. Trans.* **12**, 1291 (1972).
38. M. LeDourneuf, Vo Ky Lan, P. G. Burke, and K. T. Taylor, *J. Phys. B* **8**, 2640 (1975).
39. N. Karamatskos, M. Müller, M. Schmidt, and P. Zimmermann, *J. Phys. B* **17**, L341 (1984).
40. J. P. Connerade, M. A. Baig, W. R. S. Garton, and G. H. Newson, *Proc. Roy. Soc.* (*Lond.*) **A371**, 295 (1980).

41. V. Kaufman, J. Sugar, C. W. Clark, and W. T. Hill III, *Phys. Rev. A* **28**, 2876 (1983).
42. H. W. Wolff, K. Radler, B. Sonntag, and R. Haensel, *Z. Physik* **257**, 353 (1972).
43. M. W. D. Mansfield, *Proc. Roy. Soc.* (*Lond.*) **A346**, 539 (1975); more recent data have been obtained by K. Sommer, M. A. Baig, W. R. S. Garton, and J. Hormes, Abstracts of the 8th International Conference on Vacuum Ultraviolet Radiation Physics, Lund, Sweden, 1986, P. O. Nilsson (ed.), p. 44.

SYMMETRY AND ANGULAR MOMENTUM IN COLLISIONS WITH LASER-EXCITED POLARIZED ATOMS

ELEANOR E. B. CAMPBELL

Fakultaet Fuer Physik, Universitaet Freiburg, Freiburg, West Germany

HARTMUT SCHMIDT

Braun A. G., Wiesbaden, West Germany

and

INGOLF V. HERTEL

Fakultaet Fuer Physik, Universitaet Freiburg, Freiburg, West Germany

CONTENTS

I. Introduction
II. Kind of Questions One May Ask
  A. Optical Excitation Prior to Collision
  B. Inverse Collision
III. Well-Understood Case with Planar Symmetry
IV. Two Examples with Cylindrical Geometry
  A. Studies of Intersystem Crossings in Ca*–Rare Gas Collisions
  B. Associative Ionization
V. Spin–Orbit Interaction
  A. Nonadiabatic Transitions in $Na^+ + Na^*$ System
  B. Orientation Effects in $Na(3^2P_{3/2})$, $K(4^2P_{3/2})$ + Rare-Gas Systems
VI. Charge Exchange in $Na^+ + Na^*$
VII. Electronic-to-Rovibrational Energy Transfer
VIII. Conclusion
  References

I. INTRODUCTION

The systematic study of collisions involving electronically excited atoms and molecules started early in this century. When R. W. Wood investigated the fluorescence of alkali atoms in cells in 1911,[1] he remarked that "the intensity of the emitted light... was strongly reduced by the presence of foreign, non reactive gases. A quantitative investigation of these phenomena... appeared desirable." In 1979, nearly 70 years later, in a review on that subject, R. J. Donovan[2] stated that "the study of electronically excited states is in many respects still in its infancy and most of the chemistry that we know of at the present time relates to ground state atoms and molecules." Today we may describe this active field of research as just coming into its prime, and a wealth of interesting data is constantly revealed, due mainly to the broad variety of experimental tools that modern laser spectroscopy has made accessible. Whole books have been written on this subject,[3,4] so that we do not need to stress the importance of investigating these processes from both a fundamental point of view in understanding elementary steps in photophysics and photochemistry as well as for many areas of practical application such as plasma physics, laser development, reaction kinetics, or atmospheric physics. The use of lasers has also opened a completely new area of studies generally called laser-induced or laser-assisted collisions (the reader is referred to books on this subject,[5,6] which will not be treated in this chapter).

In this chapter we want to focus on a specific and—as we feel—particularly stimulating aspect of this field that has attracted increasing attention during recent years and leads to a new quality of insight into details of molecular dynamical processes: the study of collisions with laser-excited atoms in well-defined polarization states. By tuning polarized laser radiation to an atomic resonance transition, it is possible to excite a sizable fraction of atoms traveling in a beam into an excited state with a well-defined alignment of its magnetic substates (i.e., with an anisotropic charge distribution) if the laser light is linearly polarized, or even with a well-defined orientation (i.e., with a nonzero component of angular momentum) in the case of circular polarization. If one wants to exploit the specific polarization aspect for collision studies in its full beauty, one also needs to define an experimentally well-accessible coordinate frame with respect to which the atomic polarization may be adjusted and varied to probe certain dynamic aspects of the interaction. This is done in the most complete way in a differential crossed-beam scattering experiment in which the relative velocity (v_{rel}) before and after the collision (v'_{rel}) is determined, thus defining a geometry with planar reflection symmetry and well-defined momentum states in addition to the atomic substates. Such studies may be classified as *three-vector correlation* experiments (v_{rel}, v'_{rel}, P), where P is the initial polarization of the atom. They provide much more stringent tests for any scattering theory than rate constant, total cross section, or even differential cross section measurements. Even if the final velocity analysis (v'_{rel}) is omitted and one thus averages over all scattering angles and planes of orientation (*two-vector correlation*), the degree of detail that may be gleaned from such polarization studies is often still substantial even though the cylindrical symmetry of such a setup reduces the number of observables. It can be mentioned here only in passing that a completely analogous kind of polarization study is becoming more and more important in the investigation of molecular photodissociation, that is, in so-called half collisions. The interested reader is referred to the fundamental work of Greene and Zare[7,8] and to a most recent pioneering study by Houston.[9]

The feasibility of collision studies with laser-excited atoms in crossed beams was first demonstrated shortly after tunable continuous-wave (cw) dye lasers became available,[10,11] but it is only in recent years that these possibilities have been fully exploited to consequently improve our understanding of the dynamics of atom(ion)–atom,[12,13] atom–molecule,[14] and reactive scattering.[15–18] The latter will only be briefly touched on here. In this chapter we want to give a few illustrative examples of the kind of questions one may ask in this type of study, and we will develop and discuss some simple model concepts to help us understand the observed polarization phenomena. To keep the discussion focused, we will also not include the whole area of ionizing collisions (except for one example) and refer the reader to the pertinent literature describing the substantial progress made in recent years[19–28] in this field. But we will include some discussion on electronically elastic and on charge-exchanging collisions. So, we do not aim in any way at completeness, nor do we even try to give a representative survey. Rather, we concentrate on a few especially instructive cases where relatively complete experimental information is available and may be interpreted on the basis of both simple hand-waving arguments and rather rigorous calculations. Also, we want to place some emphasis on the most suitable choice of dynamic parameters that may be measured and from which intuitive insight into the collisional mechanisms may be derived. We finally want to indicate some prospects for future studies. Our choice of examples will of course be highly biased by the efforts in which our own laboratory has been involved. Specifically, we will discuss nonadiabatic processes in which one atom A containing an np electron is involved, for example, processes of the type

$$A(n\mathrm{p}) + B \rightarrow AB^* \rightarrow AB' \rightarrow A(n'l') + B' \pm \Delta \mathrm{KE}, \tag{1}$$

where the collision partner B, typically an atom, ion, or molecule, may also undergo a change of its internal state and the whole process may involve a change of the kinetic energy of the system. We have indicated in Eq. (1) that our emphasis will be on the molecular regime, that is, the situation that arises in low-energy collisions as long as the relative velocity of the interacting particles is small compared to the internal velocity of the atomic electrons involved in the process. Thus, by and large, the Born–Oppenheimer approximation is valid, and the processes may be viewed as evolving through a quasi-molecule AB in various states between which relatively localized transitions may occur.

As the title of this chapter indicates, our main emphasis will be on two polarization effects observed in these collisions, that is, we will study how the interaction process is influenced by the orbital alignment and the orientation of electronic angular momentum *prior* to the scattering event. It should be noted here in passing that this type of question is closely related to the broad field of atomic collision studies in which the atomic charge cloud *after* an interaction is probed by detecting the photon reemitted from a collisionally excited atom in coincidence with the scattered particle. By inverting Eq. (1), we may write this process as

$$A(n'l') + B' \rightarrow AB' \rightarrow AB^* \rightarrow A(np) + B \mp \Delta KE. \tag{2}$$

A wealth of experimental and theoretical data has been collected over the past decade in this field of *particle–photon coincidence* studies, which was pioneered by Macek and Jaecks[29] and Kleinpoppen and collaborators[30] (see also various reviews[31] and comments on this subject[32]). For more details the reader is referred to a series of comprehensive data compilations and surveys currently being prepared.[33]

Figure 1 and Eqs. (1) and (2) illustrate that the two approaches yield essentially identical information as far as the dynamics of inelastic processes are concerned, and in this chapter we will describe a given process occasionally in either forward or backward direction, depending on what appears more convenient. From the viewpoint of molecular dynamics and reactive processes (B being a molecule)—and that appears to be the most promising field of future research with these tools—the two methods are, however, quite different insofar as the inverse processes, Eq. (2), are normally not accessible to the experiment (B′ having undergone specific internal excitation, rearrangement, etc.). Thus, we restrict the present discussion for these processes to studies with laser-excited atoms [Eq. (1)].

This chapter is structured as follows: In Section II we first give a brief introduction to the terminology and ways of describing and preparing collisionally and laser-excited atoms in nonisotropic states, also indicating a few of the experimental ingredients of these studies. We will then outline the typical questions one may address and give a few experimental examples of the possibilities and problems in these polarization studies. In Section III, we will discuss collision-induced nonadiabatic transitions in the $Na^+ + Na^*$ system as a particularly clear and well-understood case with planar symmetry. Specifically, this will allow us to discuss the formation of the quasi-molecule

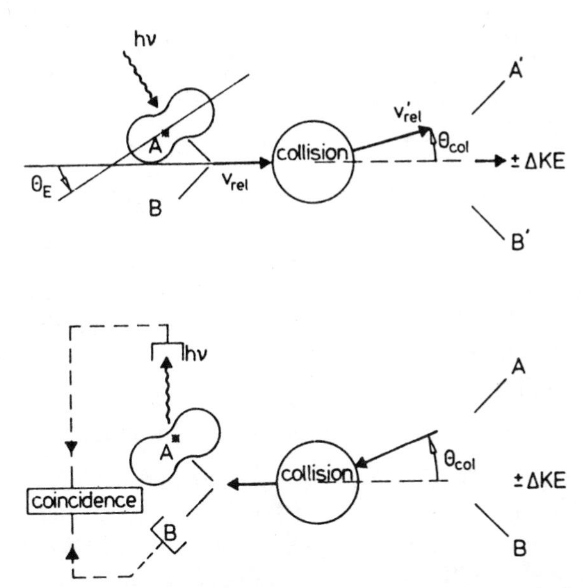

Figure 1. Schematic view of binary collision experiment with laser-excited, polarized atom A* and collision partner B (upper part) and its relation to inverse process studied in particle–photon coincidence experiment (lower part). Initial and final relative velocities of interactants (v_{rel} and v'_{rel}) define scattering plane. Kinetic energy may be gained or lost in the process ($\pm\Delta$KE).

Na_2^+ in different electronic states and the way that these molecular states evolve from the atomic p substates prepared at large internuclear distances—a theme that will follow us throughout the whole of the chapter. In Section IV, we will temporarily drop the planar symmetry and discuss studies of intersystem crossings in Ca*–rare-gas collisions and the associative ionization processes involving two laser-excited Na* atoms.

In Section V, a brief interlude on the importance of the spin-orbit interaction, which we otherwise neglect in the present discussion for simplicity, will broaden our understanding of potential complications in these studies.

In Section VI, resonant charge exchange will be discussed, again exemplified by the now familiar $Na^+ + Na^*$ system, a case that appears to offer itself for a fully quantitative understanding but has not yet reached this state of maturity.

Finally, in Section VII, we will elaborate on the very important and fundamental process of electronic to rotational–vibrational energy transfer where a great deal of understanding has been achieved during recent years by the combined efforts of molecular beam studies with laser-excited atoms and quantum chemistry. The probing of finer dynamic details by polarization studies is, however, still in its infancy in this case and nicely illustrates the problems one is just starting to overcome as well as the glimpses of insight that may be obtained in this way.

II. KIND OF QUESTIONS ONE MAY ASK

Before discussing any problems of the collision physics, we briefly outline the terminology we find most appropriate to describe excited atoms prepared in a p state: (a) either by optical excitation prior to the collision by tuning a cw laser to the appropriate resonance transition as in Eq. (1) or (b) in the inverse collision process given by Eq. (2).

A. Optical Excitation Prior to Collision

We recall from quantum mechanics textbooks that p-state atoms come in three varieties $|+1\rangle$, $|-1\rangle$, and $|0\rangle$ with magnetic quantum numbers $M = \pm 1, 0$, as illustrated in the upper row of Fig. 2. We call these the "physicists' basis states" (or the atomic basis), and we may create the $M = +1$ and $M = -1$ state by left- (LHC) or right-hand circularly (RHC) polarized light, respectively, propagating in the $+z$ direction while the $M = 0$ state is excited by linearly polarized light propagating in the x-y plane with its electric field vector parallel to the z axis. Remember that we assume atomic beam conditions in which these states are initially unperturbed by any collisions or external fields. Alternatively, we may find the dumbbell orbital description $|p_x\rangle$, $|p_y\rangle$, $|p_z\rangle$ for all three basis states, especially if we look in textbooks on chemical physics, as illustrated in the lower row of Fig. 2, and we call this alternative description the "chemists' basis states" (or molecular basis). They too may be created by linearly polarized light traveling in various directions with the electric vector parallel to the x, y, or z axis. Notice that both descriptions are equally valid; they may be transformed into each other by a simple linear superposition, and what is actually present in an experiment depends on the mode of preparation as outlined in the preceding. We do not go into the technicalities of the preparation process; we simply note that the situation described here is the ideal case, whereas for real atoms excited with real light we may often find less than 100% of any of these states and have to cope with a coherent or incoherent admixture of the other states as well. We will come to this point later when we discuss the experiments. We also may want to choose certain other directions of laser incidence or alignment of the E vector for deliberately probing polarization dependence of the process in Eq. (1). In all, one may thus describe the excited state by a 3×3 density matrix σ of which we will, however, only make use in passing.

We should note here that we have already labeled the axes in Fig. 2 in a

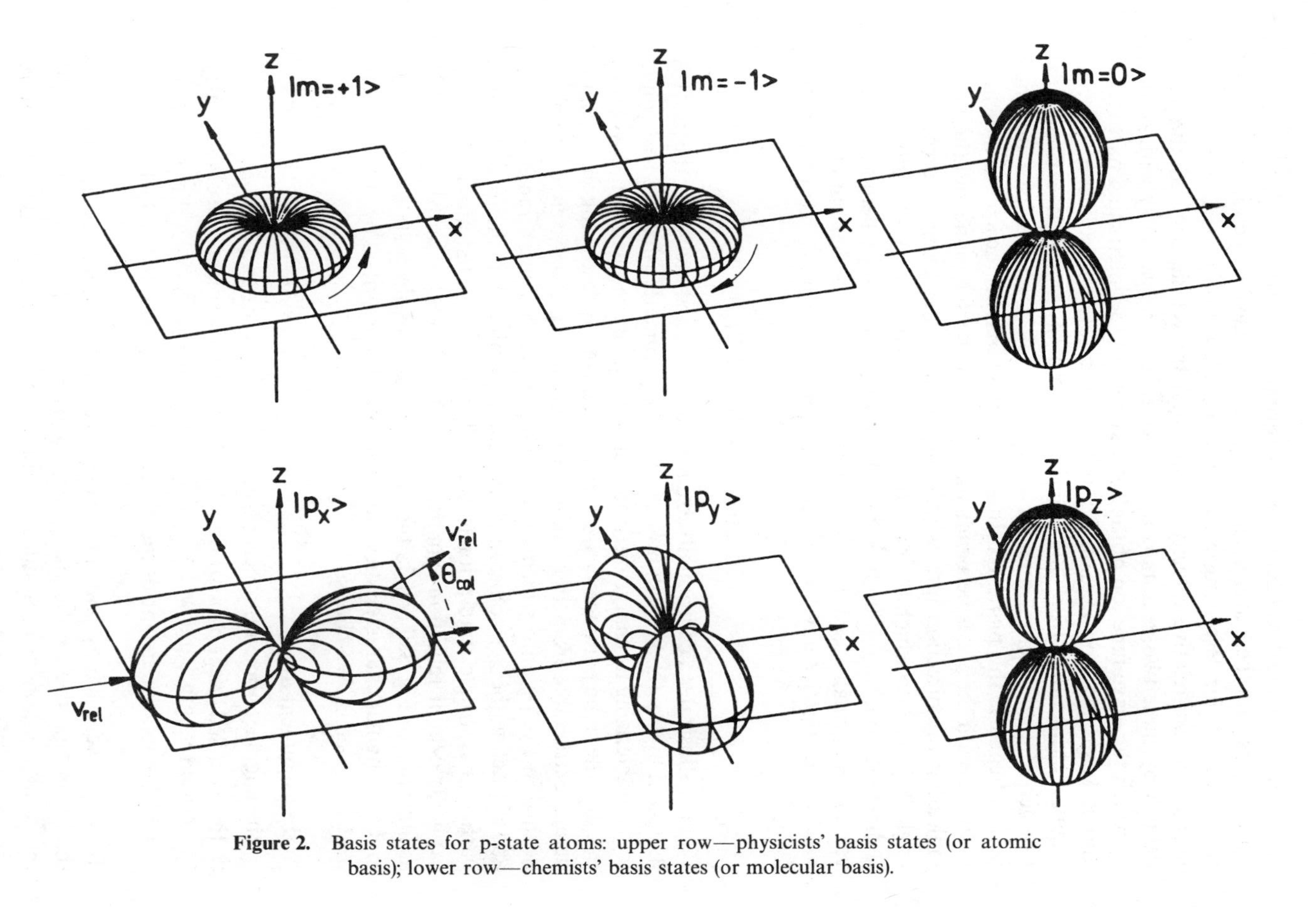

Figure 2. Basis states for p-state atoms: upper row—physicists' basis states (or atomic basis); lower row—chemists' basis states (or molecular basis).

special way. We must realize that we will eventually want to perform collision experiments with these atoms, so our description refers to a coordinate frame that is suitable for describing such collisions. In a collision experiment, where initial and final relative velocity (v_{rel} and v'_{rel}) are well determined, the plane defined by these two vectors is a plane of reflection symmetry with respect to which the wave function of the total system of the interacting particles A and B keeps its properties under reflection during the collision, whereas angular momentum from the relative motion of A and B can only be transferred into internal angular momentum of A and B in a direction perpendicular to this plane. It is thus convenient to label the axes as indicated in Fig. 2, with the z axis perpendicular to the scattering plane and the y axis indicating the direction into which the particles are scattered through an angle θ_{col} (i.e., into which their relative velocity is deflected). This convention is different from that often used in scattering calculations where the relative velocity is normally used as a z axis. For polarization studies our choice has, however, distinctive advantages and the term *natural frame* has been coined for it.[32]

B. Inverse Collision

Figure 3 illustrates a collisionally excited p atom immediately after the process indicated by Eq. (2). We will now discuss the parameters that are best suited for characterizing its charge distribution and angular momentum. It will turn out that the same parameters are very convenient for describing the dynamic details extractable from the studies with polarized, laser-excited atoms as defined in Eq. (1) onto which we will concentrate our attention. Again, v_{rel} and v'_{rel} define the x-y collision plane and, assuming that the system is in a particular isotropic state prior to the collision, the excited atom A(np) will explicitly exhibit reflection symmetry after the process described in Eq. (2), as indicated in Fig. 3. Then, in the most general case, the p-state distribution (only its angular characteristics are determined) is described by four independent parameters. We choose

- the alignment angle γ;
- the angular momentum perpendicular to the collision plane, $L_\perp^+$;
- the so-called linear polarization, that is, the relative difference of the charge cloud length l and width w as indicated in Fig. 3, $P_l^+ = (1 - w)/(l + w)$;
- and the relative height of the charge cloud, $\rho_{00} = h/(l + w + h)$.

These are very handy and intuitively acceptable parameters by which we shall later characterize the laser experiments in such a way that we describe the results obtained from studying the process in Eq. (1) as if they were obtained for the inverse process in Eq. (2). One nice feature of these parameters

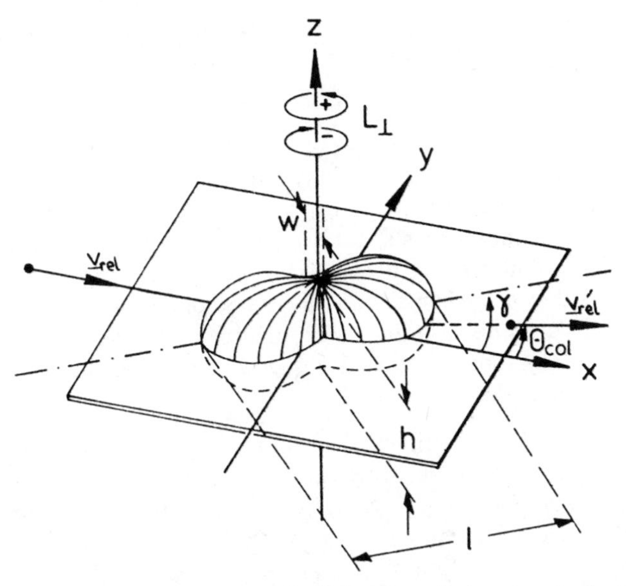

Figure 3. Charge cloud distribution of collisionally excited p atom in "natural" frame, giving definitions of alignment angle γ and angular momentum transfer $L_{\perp}^{+}$. Scattering plane is fixed by directions of incoming ($v_{\rm rel}$) and outgoing ($v'_{\rm rel}$) particles and l, w, and h refer to length, width, and height of charge distribution measured at intersections of main axes.

is that they may be easily related to the more formal description of the process needed when one wants to compare experimental results with theory: The collisionally excited atom is then described by a density matrix where the components $\rho_{MM'}$ are very simply related to these parameters; $L_{\perp}^{+} \propto \rho_{11} - \rho_{-1-1}$ and the off-diagonal elements are $\rho_{1-1} \propto P_l^{+} \exp(-2i\gamma)$. We will not elaborate on this here but rather discuss simple physical pictures illuminating the collision dynamics and resort to intuitively acceptable arguments as to how these parameters may be measured and interpreted. The reader interested in details is referred to the literature.[34-36] We only note here that the basis of evaluating experiments of the type laid down in Eq. (1) is to recognize[14,35] that the scattering intensity is given by

$$
\begin{aligned}
I(\theta_{\rm col}; \text{polarization}) &= \operatorname{Tr}(\sigma \cdot \rho) \\
&= \operatorname{Tr}(\rho \cdot \sigma)
\end{aligned}
\tag{3}
$$

where σ describes the optical preparation of the atoms prior to collision as in Eq. (1) while the inverse process, Eq. (2), is described by exactly the same relation for the detected intensity, except that σ now describes the detection system in the particle photon coincidence experiment.

Now let us discuss the physical significance of these measurable parameters from the viewpoint of molecular collisions. Here we will address three aspects of the molecular collision dynamics that may be studied.

1. The first potential of the laser polarization appears to be the possibility of creating molecular states of specific symmetry by choosing the appropriate direction of the electric vector of linear polarized laser light. The idea is thus to choose one of the chemists' basis states depicted in the lower row of Fig. 2 at will and in this way to switch on one of the molecular textbook Σ or Π states as they are schematically illustrated in Fig. 4, in connection with potentials typical in alkali–rare-gas or alkaline earth–rare-gas interactions: the p orbital may thus be aligned along (Σ) or perpendicular (Π) to the internuclear axis of particles A and B.

A very impressive experimental example exploiting this feature of the method has been demonstrated by the Pauly group.[37–41] They studied the differential elastic scattering of Na*(3p) and K*(4p) from Hg. One expects the typical rainbow oscillations of the scattering signal as a function of the scattering angle θ_{col}. Since both a Σ and Π potential will be involved, one expects two sets of rainbow oscillations. In this case, since the potentials of both excited states have rather deep well depths, one needs slightly supra-thermal energies (some electronvolts) for observing these rainbow structures. A typical set of experimental results is shown in Fig. 5: The expected two sets of rainbows are clearly seen, peaking at around 7° and 25°, respectively. Two sets of data are displayed with the laser electric vector parallel and perpendicular to the initial relative velocity of the colliding particles. According to the preceding argument, we expect that in the first case the

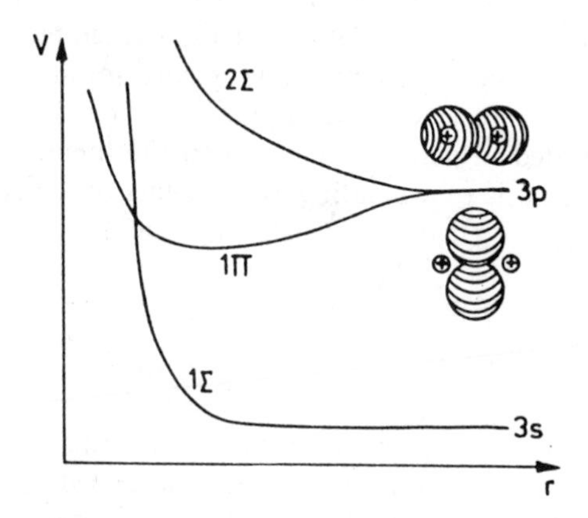

Figure 4. Typical potentials for alkali–rare-gas or alkaline earth–rare-gas interactions. The figure also shows how it is possible to "switch on" either the Σ or the Π molecular states.

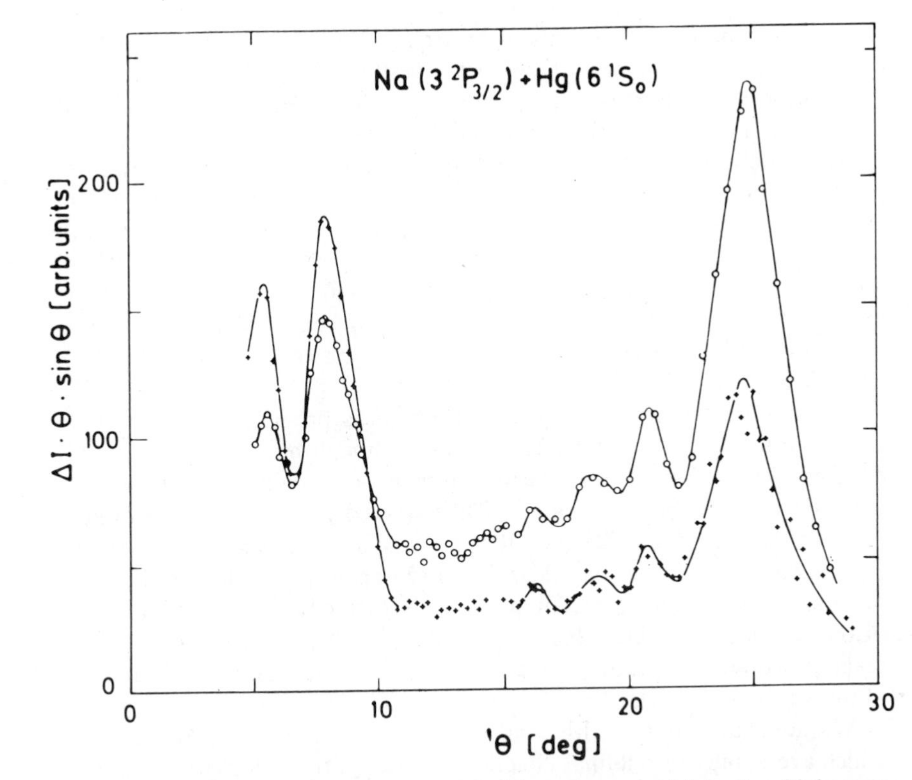

Figure 5. Difference ΔI (weighted with $\theta \sin \theta$) between scattered intensity with laser on and scattered intensity with laser off for Na* + Hg system as function of laboratory scattering angle θ for lab energy of 3.15 eV. Two plots show data collected with laser electric vector parallel (+) and perpendicular (○) to relative velocity of colliding particles. Reproduced from ref. 39.

NaHg(2Σ) state is populated predominantly whereas in the latter geometry the NaHg(1Π) state should be emphasized. The experimental results illustrate beautifully that this concept works, amplifying the rainbow at 7°(2Σ) and 25°(1Π), respectively. Since the Σ potential is somewhat flatter than the deep Π potential, this is precisely what one expects from semiclassical elastic scattering theory.

It seems that our intuitive concept works: We may switch the Σ and Π molecular states on and off by initially preparing the atom in the corresponding orbitals with appropriate laser polarization. Closer inspection of the data shows that things are not all that easy and that the hand-waving arguments given here are only a crude first approximation. The groups in Göttingen are continuing this type of work[12,42] and typically obtain good to excellent agreement between experimental and theoretical scattering signals obtained for specific experimental geometries and polarization of the exciting laser light. Similar studies have been performed by Morgenstern and collaborators.[43] Laser optical pumping methods have also been used successfully in the study of elastic collisions with excited rare-gas atoms.[44] Unfortunately, the powerful polarization studies of elastic alkali atom–rare-gas scattering have so far not yet fully eliminated the geometric factors of the experiment and derived all the measurable dynamically relevant parameters explicitly, as has been possible in other cases (see Sections III and VI). They have, however, led to a very quantitative determination of the potentials, and careful analysis of quantum mechanical close coupling calculations has revealed the essential mechanisms of the observed polarization effects (see Section V).

If one is daring and trusts the intuitive ideas presented here, one may venture into studies of reactive processes. Although not the subject of this chapter, we illuminate these perspectives by one pioneering experiment performed by Rettner and Zare.[15] They studied reactions of laser-excited $Ca(4s4p\,^1P_1)$ with HCl, Cl_2, and other gases. Without going into details, this experiment is briefly described as a crossed-beam experiment (with some remaining uncertainty in the initial relative velocities and intersection angles of the reactants) in which dispersed fluorescence from the products of the reaction is observed without, however, any further angular or velocity analysis of the scattered particles.

We thus have a setup with axial symmetry (as discussed in Section I) in which averaging over all final directions of the relative velocity alignment in the scattering plane and thus initial impact parameters occurs. One may expect that this reduces any polarization effect significantly, and before the experiment was done, one would even have been doubtful of observing any effect at all.

Figure 6 shows the experimental results for the reaction

$$Ca(4s4p\,^1P_1) + HCl \rightarrow CaCl + H. \quad (4)$$

The product CaCl may be found in the electronically excited $A^2\Pi_{3/2}$ or $B^2\Sigma_{1/2}$ states detected by observing their characteristic emission bands as displayed in the upper part of Fig. 6. We see that both the A and the B state are seen and may also be vibrationally excited. We do not elaborate on the details of these interesting results but emphasize the polarization effect of interest in our present discussion: The lower part of Fig. 6 reproduces the experimentally observed signal from the A and B states as one varies the direction of the electric E vector of the laser that excited the Ca prior to the collision. We focus on the two upper curves representing the stronger $\Delta v = 0$ bands. We clearly see the $\sin(\theta_E)$ or $\cos(\theta_E)$ oscillations as E is aligned parallel

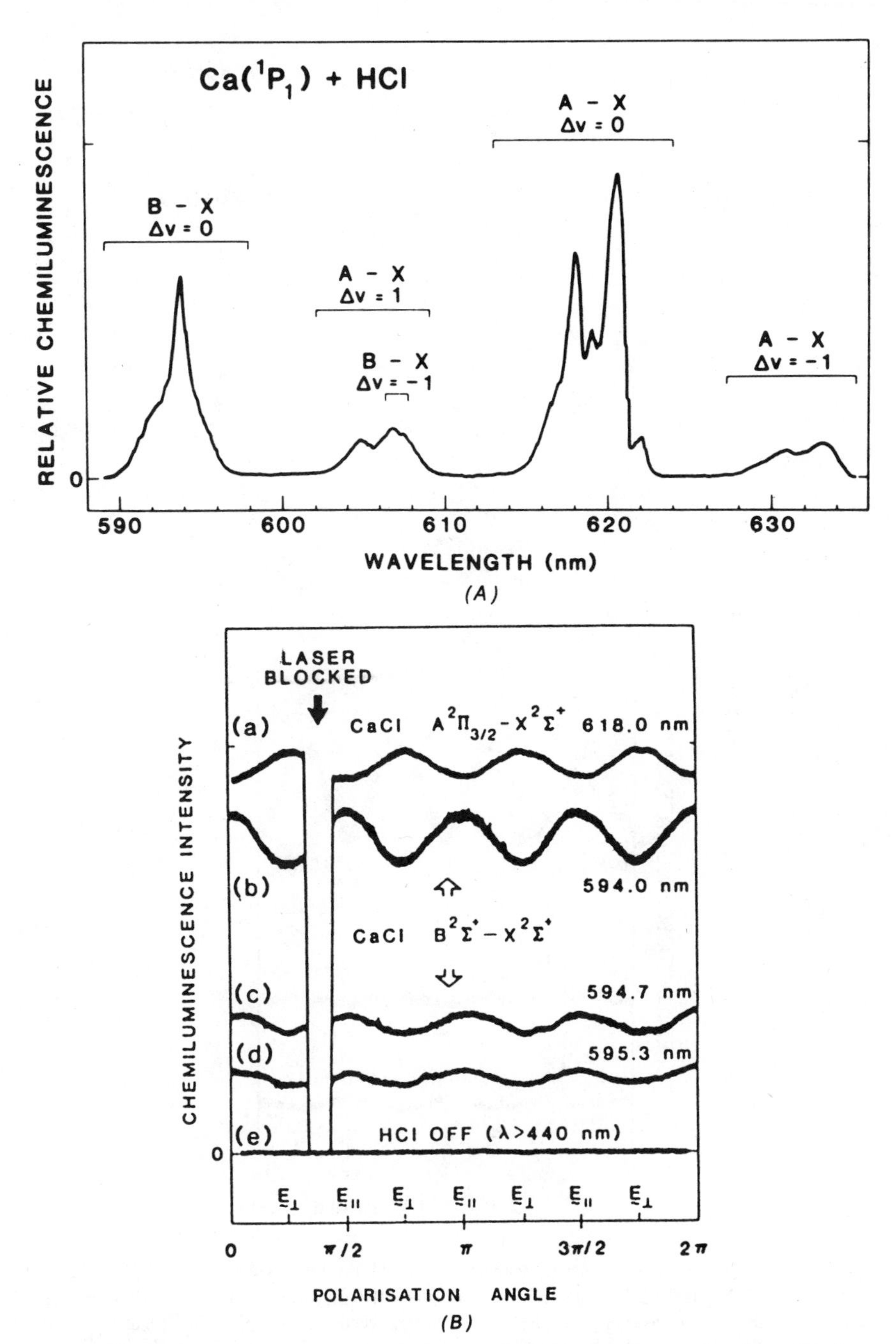

Figure 6. (*A*) Laser-induced chemiluminescence spectrum obtained by preparing Ca($^1P_1$) in beam impinging on HCl gas. (*B*) Variation of chemiluminescence intensity with direction of electric vector of exciting laser. Reproduced from ref. 15.

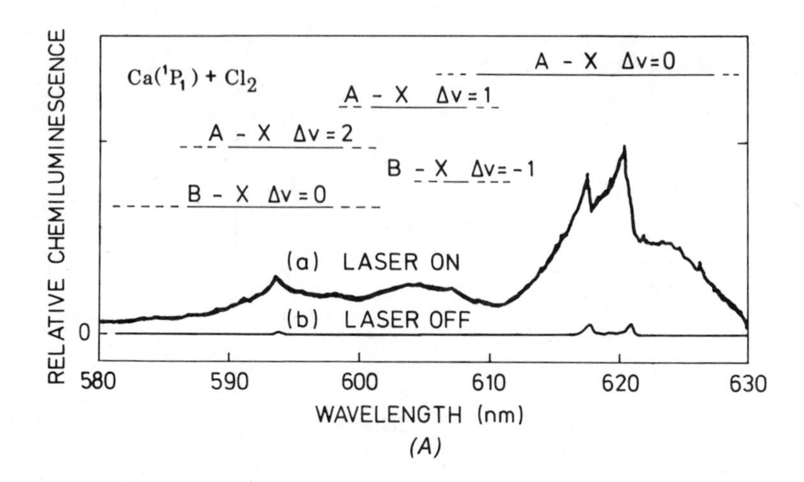

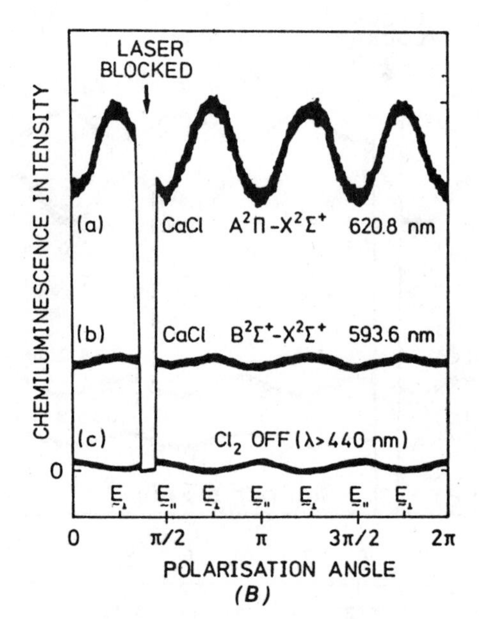

Figure 7. (*A*) Chemiluminescence spectra obtained (*a*) by preparing $Ca(^1P_1)$ in beam impinging on Cl_2 gas and (*b*) under same conditions but with laser blocked out. (*B*) Variation of chemiluminescence intensity with direction of electric vector of exciting laser. Trace (*c*) is due to scattered light. Reproduced from ref. 15.

or perpendicular to the initial relative velocity. It is a small, but pronounced, effect, and as one would have expected intuitively, the A state (Π) emits strongest when E is perpendicular, whereas the B state (Σ) gives a stronger signal for E parallel to the initial relative velocity. The only disconcerting feature seems to be that the effect is really very small (3.4% for the A state and 8% for the B state) even though the Ca(4s4p $^1P_1$) is prepared 100% in a pure dumbbell $|p_y\rangle$ or $|p_x\rangle$ state prior to the reaction.

However, it gets worse as we look at the reaction

$$\text{Ca}(4s4p\ ^1P_1) + \text{Cl}_2 \rightarrow \text{CaCl} + \text{Cl}. \tag{5}$$

Rettner and Zare's results for this case are reproduced in Fig. 7. Here the A state still behaves as expected, giving an even stronger polarization effect of 14%. However, the B (Σ) state in this case behaves counterintuitively, giving maximum fluorescence for perpendicular alignment of the Ca 4p dumbbell orbital as well, with a 6.5% anisotropy. This is somewhat disturbing.

Rettner and Zare go through a rather complex argument assuming plausible potential energy surface crossings of various molecular states based on a harpooning model to reconcile the intuitively attractive Π–Σ population via laser excitation with the experimental observation. We neither want to criticize these arguments here nor support them. We merely point out that the situation may be altogether much more complex than initially thought, and we will illustrate this by the examples discussed in the following sections. In fact, Rettner and Zare address one of the crucial questions of this kind of study quite clearly in their paper: the question of orbital following. So far we have assumed that the orbital preparation of the atom, which occurs at large internuclear distances a long time before the collision, maintains its symmetry as the quasi-molecule between the interacting particles is formed at close distance. What this implies is nicely illustrated in Fig. 8[15]: For small impact parameters b, a σ orbital aligned parallel to the center-of-mass system stays in this parallel alignment as the particles approach and thus inevitably forms a Σ molecular state (Fig. 8*a*). For large impact parameters, the same σ atomic orbital prepared at large internuclear distance may become either a Π or a Σ molecular orbital as the particles approach. Space-fixed motion (large rotational, i.e., Coriolis coupling) corresponds to nonadiabatic transitions and yields a molecular Π state (Fig. 8*b*). On the other hand, "orbital following," that is, rotation of the atomic charge cloud with the molecular axis, corresponds to the adiabatic case (large potential energy difference for the corresponding Σ and Π states) and gives rise to a Σ state in the quasi-molecule. What happens in a real collision will depend on the details of the adiabatic molecular potentials, the nonadiabatic coupling elements, the relative velocity of the interacting particles, the impact parameter, and if

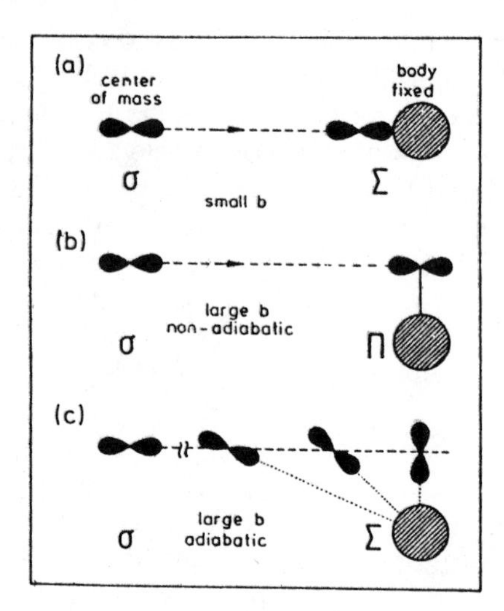

Figure 8. Relationship between center-of-mass (atomic) and body-fixed (molecular) frames. At small impact parameters (*a*), parallel center-of-mass alignment, σ, transforms to parallel body-fixed alignment, Σ. At large impact parameters, parallel center-of-mass alignment can give rise to (*b*) perpendicular Π body-fixed alignment for nonadiabatic behavior or (*c*) Σ alignment if orbital following occurs as in adiabatic behavior. Reproduced from ref. 15.

the collision partner B is a molecule (as in the above example), the rotational state of the molecule and its relative alignment with respect to the scattering plane. In any case, it is not trivial to decide a priori how the atomic orbital prepared by the laser polarization with respect to a space-fixed laboratory frame transforms under the influence of the collision into the molecular body-fixed frame. It will thus be a central issue of the following sections to address this question and to discuss simple, clean model cases for which complete experimental information may be obtained, and a more or less exact theoretical analysis seems feasible. One may then derive intuitive pictures for the dynamics of the electronic motion and see whether the "orbital steering" or "orbital stereospecificity,"[4,5] as one may call the concepts of selective molecular state population, is possible or not.

2. The second question we may ask in this context concerns the importance of reflection symmetry in collisions with polarized atomic species. It can be studied specifically in experiments with planar symmetry as introduced in Section III. To illustrate this, we have again depicted (in Fig. 9) the motion of a dumbbell p orbital. In this case, we have chosen a π (i.e., perpendicular) preparation at large internuclear distances and display the scattering plane explicitly. As indicated, we can have π^+ or π^- symmetry. Although, for both cases, the molecular potentials are identical, the dynamics may be vastly

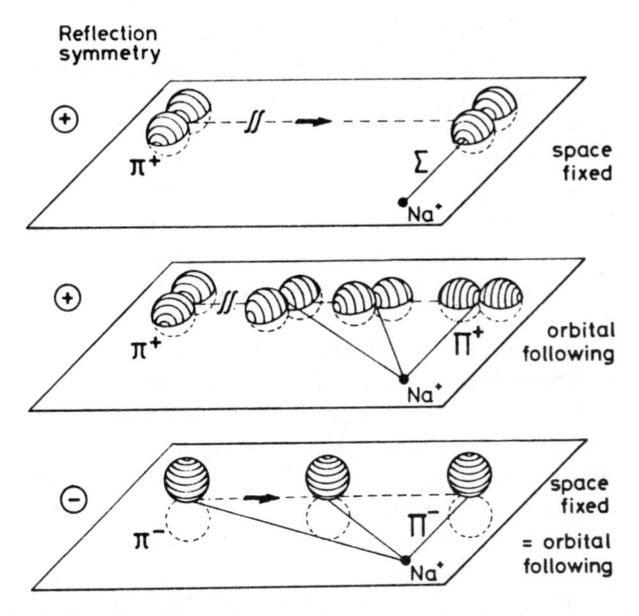

Figure 9. Motion of "dumbbell" p orbital for π (perpendicular) preparation at large internuclear distances. The π^+ orbital may become (*a*) Σ molecular state or (*b*) Π^+ molecular state at close internuclear distances depending on whether interaction is nonadiabatic or adiabatic, respectively. The π^- orbital (*c*) remains space fixed since reflection symmetry of wave function is conserved and molecular state must have Π^- character.

different, as illustrated in Fig. 9: The π^+ orbital may become a Σ or a Π^+ molecular state at close distances (Fig. 9*a*, *b*) depending on the details of the interaction, discussed in the preceding for the σ case. The π^- orbital, however, will stay space fixed since reflection symmetry of the wave function is conserved. From Fig. 9*c*, this implies, however, that the molecular state definitely has Π character, in contrast to the two possibilities occurring for the states with positive symmetry (Fig. 9*a*, *b*).[†] It is thus very challenging to probe pure Π dynamics by π^- orbital preparation and to add the complication of Σ–Π nonadiabatic couplings by π^+ and σ preparation of the atomic orbital. We should note here, however, that these concepts are strictly valid only for atom–atom(ion) collisions, whereas in the much more complex case of an atom–molecule interaction, the molecular symmetry enters into the dynamics as well and the relative motion of A and B no longer necessarily occurs in a

[†]The difference seen here is closely related to the well-known removal of the Λ degeneracy in molecular spectroscopy for large rotational quantum numbers (corresponding in our case to $b > 0$).

space-fixed scattering plane. We will discuss this situation in Section VII.

3. The third question, which arises in collisional polarization studies with laser-excited atoms, addresses circular polarization, that is, the possible asymmetries of the scattering signal that may arise for LHC or RHC polarized light exciting the atomic target. Again, a planar symmetry is the prerequisite for observing such a phenomenon, and scattering angles (different from zero degrees) have to be well defined in the experiment. In a simple-minded way, one may then expect, as depicted in Fig. 10, a different scattering signal for particles scattered to the right or to the left from a circular atomic state $|\mathrm{p}, M = \pm 1\rangle$. Equivalently, the signal at a given scattering angle θ_{col} may be different for LHC or RHC ($M = +1$ or $M = -1$) excitation of the atom. Although Fig. 10 intuitively suggests angular momentum considerations to determine the sign of such a left–right asymmetry, it turns out that for the molecular collision regime in which we are interested here, this kind of argument leads almost inevitably to the wrong conclusions.[46] Instead, one finds that the sign and magnitude of such an asymmetry critically probes the long-range interaction of the collision partners and also depends on the symmetry of the states as well as on the effectively attractive or repulsive nature of the interaction that essentially defines the reference frame for the experimental observation.[46] We will discuss this as well as give some examples in the following sections. However, for the high-energy regime where the relative velocity of the two particles is comparable to the electron velocity, such simple intuitive pictures based on angular momentum considerations are applicable within reasonable limits and may be put on a quantitative footing based on a stationary-phase argument when solving the coupled equations in the semiclassical trajectory model.[47–50] Such a picture may also be applicable at lower collision energies in the situation where A* is replaced by a molecule BC. Here it is the rotation of the molecule rather than that of the electron that should be considered. For the e–A* interaction, the model is useful when $1/\omega_{\mathrm{if}} \sim t_{\mathrm{col}}$, where ω_{if} is the atomic transition frequency and t_{col} the interaction time. For the molecular case, BC + D, we need $1/\omega_{\mathrm{rot}} \sim t_{\mathrm{col}}$, where $1/\omega_{\mathrm{rot}}$ is the time for one revolution of the molecule.

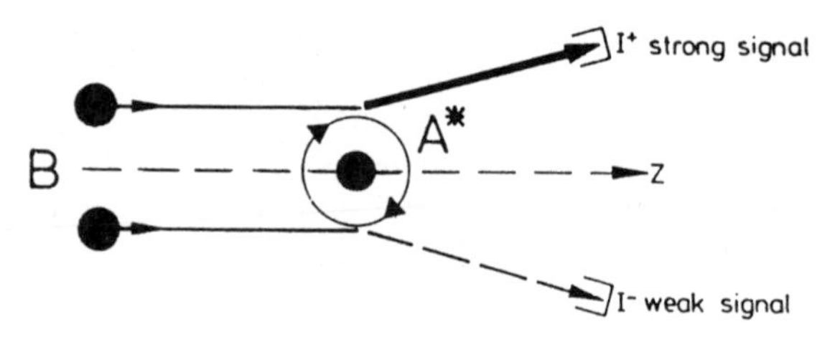

Figure 10. Different scattering signal is expected for particles B scattered to right to left from atom A* in circular atomic state.

The preceding discussion should have made it obvious that polarization studies may yield very detailed insights into the collision dynamics of molecular processes, which may, however, be much more complex than the typical educated intuitive first guess would predict. Thus, we will begin our discussion with a very simple and by now well-understood model case in which all three of the questions posed in the preceding may be interrogated thoroughly.

III. WELL-UNDERSTOOD CASE WITH PLANAR SYMMETRY

As an example of a system with planar symmetry, we focus on the now well-understood model case investigated in our laboratory in recent years[13,51–53]: the collision-induced nonadiabatic transitions in the Na^+–Na^* system. In particular, we will be concerned with the following processes:

$$\begin{aligned} Na^+ + Na^*(3p) &\rightarrow Na^+ + Na^*(3d), \\ Na^+ + Na^*(3p) &\rightarrow Na^+ + Na(3s). \end{aligned} \tag{6}$$

A schematic diagram of the apparatus used in the experiments is shown in Fig. 11. Briefly, sodium atoms in the scattering center are aligned or oriented with polarized laser light. The ion gun can be rotated about the atomic or

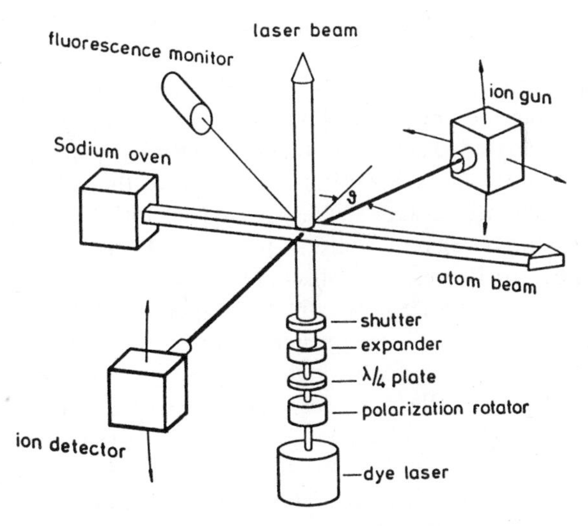

Figure 11. Schematic diagram of apparatus used to investigate collision-induced nonadiabatic transitions in $Na^+ + Na^*$ system.

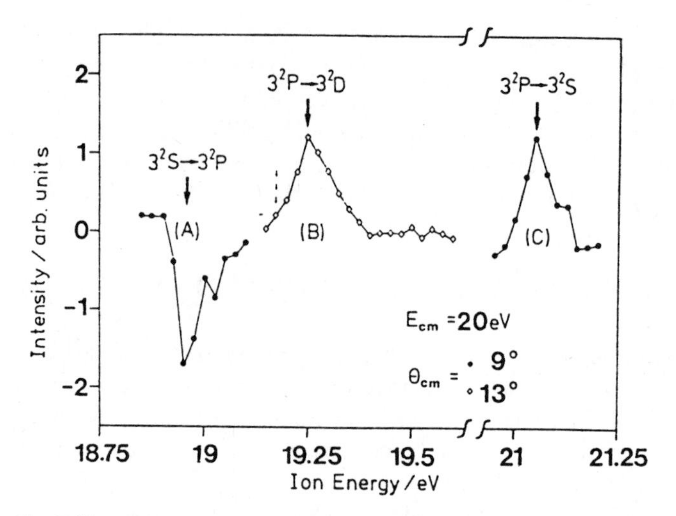

Figure 12. Difference energy loss spectra for scattering of Na^+ with kinetic energy $E_{cm} = 20$ eV from Na*(3p) at cm scattering angles of 9° (●) and 13° (◇). Difference spectra are obtained by subtracting Na^+ intensity without laser excitation from Na^+ intensity with laser excitation of Na target atoms. (*A*) and (*B*) denote collisional excitation processes; (*C*) deexcitation. Broken arrows show spectroscopic energy position of excitation process 3p→4p.

laser beam axes. Ions scattered into a particular angle (with resolution of 1°) are focused onto the entrance aperture of two 180° hemispherical analyzers in series. Those ions having the selected energy are detected by a particle multiplier. The ion detector can also be rotated about the axis of the atomic beam.

Figure 12 shows the energy loss spectra obtained from the scattering of sodium ions from sodium atoms at a collision energy $E_{cm} = 20$ eV. Measurements at each scattering angle are taken with and without laser excitation of the sodium atoms into the $3^2P_{3/2}$ state prior to the collision. The figure displays the difference between the respective Na^+ ion scattering signals. The spectra show three different collision processes. Processes (B) and (C) are indicated in Eq. (6). Process (A), collisional excitation of ground-state sodium atoms, is seen as a decrease in the difference spectrum for $\theta_{cm} = 9°$, which is a result of the decrease in the ground-state population due to the laser excitation. At the same scattering angle but increased final ion kinetic energy ($E'_{cm} = 21.05$ eV), the collisional deexcitation process (C) is observed. By increasing the scattering angle to $\theta_{cm} = 13°$, a distinct peak is seen at $E'_{cm} = 19.25$ eV, corresponding to the collisional excitation process. It should be noted that in the measurements shown the polarization and direction of the laser light exciting the atoms have been chosen for each process such that the cross section is at its maximum.

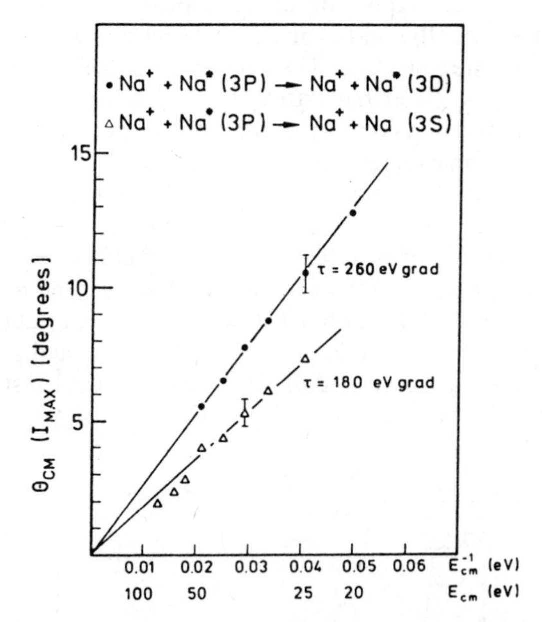

Figure 13. Center-of-mass angular position of maximum scattering intensity as function of $1/E_{cm}$ for collisional excitation (●), process B, and deexcitation (△), process C.

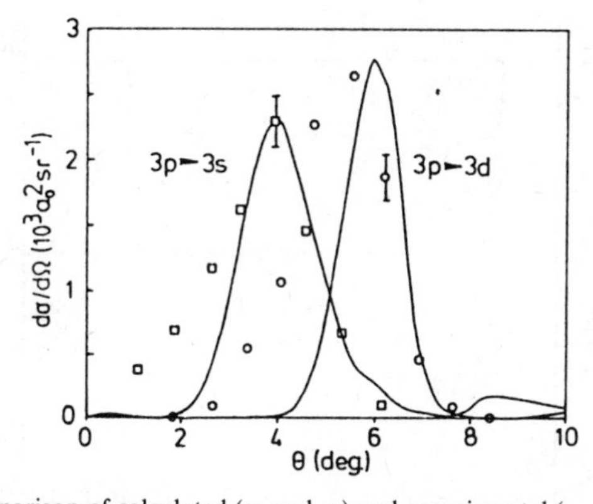

Figure 14. Comparison of calculated (□ and ○) and experimental (——) differential cross sections for superelastic (3p → 3s) and inelastic (3p → 3d) processes, respectively. Reproduced from ref. 54.

The measured angular position of the maxima in the differential cross sections for processes (B) and (C) at different collision energies is displayed in Fig. 13 as a function of $1/E_{cm}$. The data points show a linear dependence on $1/E_{cm}$ with slopes of 260 and 180 eV° for the excitation and deexcitation processes, respectively, illustrating that these processes always have the largest probability for the same impact parameter regardless of collision energy. The inelastic and superelastic differential cross sections are shown as a function of scattering angle θ_{cm} in Fig. 14 for a collision energy of 47.5 eV cm. As can be seen, they have a strongly peaked structure that can be explained with reference to the calculated potential energy curves for the molecular Na_2^+ ion[52] reproduced in Fig. 15. The structure results from rotational coupling at the crossings between the $1^2\Sigma_u$ ground and $1^2\Pi_u$ excited states (crossing C) for the deexcitation process and the $2^2\Sigma_u$ and $2^2\Pi_u$ excited states (crossing B) for the excitation process.

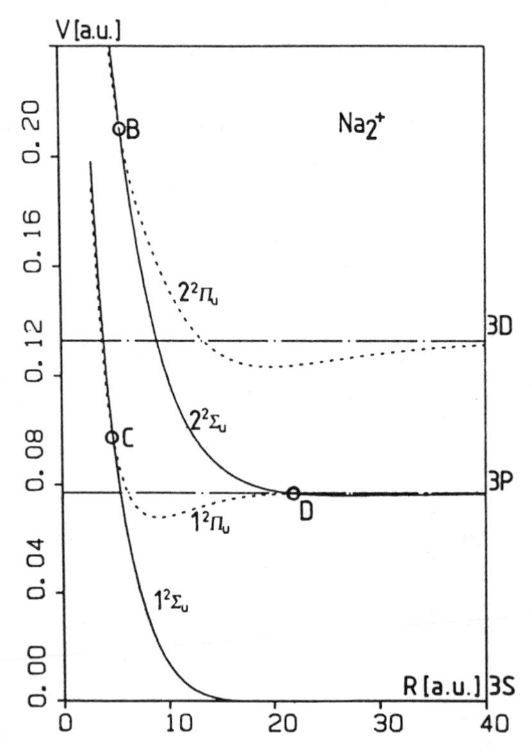

Figure 15. Plot of four potential curves contributing to p→s deexcitation through crossing C and p→d excitation through crossing B. Reproduced from ref. 52.

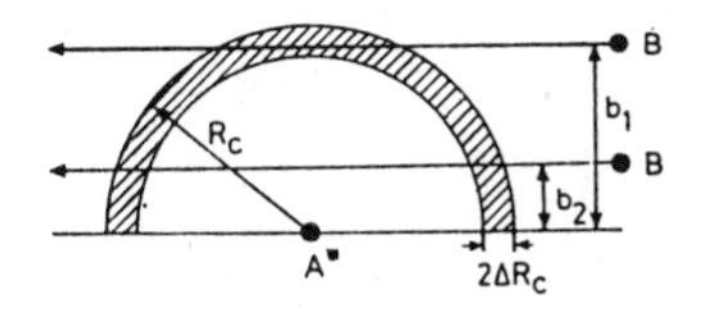

Figure 16. For rotational coupling, transition probability will be greatest for maximum impact parameter since overlap between impact parameter and crossing region is then at maximum. Cross section will therefore be most significant for range of impact parameters $b \pm \Delta b \approx R_c \pm \Delta R_c$.

For rotational coupling (dependent on vb/R^2), the transition probability will be greatest for the maximum impact parameter—in particular because there the overlap between the particle trajectory and crossing region is largest, as indicated in Fig. 16. Thus, we would expect the cross sections to be most significant for the largest impact parameters, $b \pm \Delta b \approx R_c \pm \Delta R_c$. As crossing B lies further up the repulsive wall than crossing C, we would also expect the maximum in the differential cross section for the collisional excitation process to appear at larger scattering angles than that for the deexcitation process as confirmed experimentally. From Fig. 14, we see that rotational coupling is the only important mechanism in this energy and angular range since there is no significant scattering signal for either the inelastic or superelastic processes at larger scattering angles. Semiclassical calculations[54] for the scattering processes based on the potentials shown in Fig. 15 give good agreement with the experimentally observed angular distributions shown in Fig. 14. We thus have a fairly clear picture of the "conventional" part of the scattering process but now have to ask ourselves how the alignment and orientation of the excited sodium atom affect the scattering signal and what additional insights into the scattering process this information gives us.

We will start by describing the experiments involving linear polarization. Figure 17 shows the scattering intensity for both processes as a function of linear polarization angle θ_E with the laser light propagating both perpendicular (upper part of Fig. 17) and parallel (lower part of Fig. 17) to the scattering plane. Let us first discuss the perpendicular case where the E vector lies *in* the collision plane. We clearly see that the asymptotic preparation of a π^+ state gives the strongest (3p → 3s) deexcitation signal while the preparation of a σ state is more important for the 3p → 3d excitation process. This is what we would expect if the molecular picture is valid: The π^+ orbital transforms into the $1^2\Pi_u^+$ molecular state,† enabling the crossing between the (3p) excited

†Fifty percent of the π^+ orbitals transform into the $1^2\Pi_g^+$ state. There is, however, no accessible relevant curve crossing in the g system.

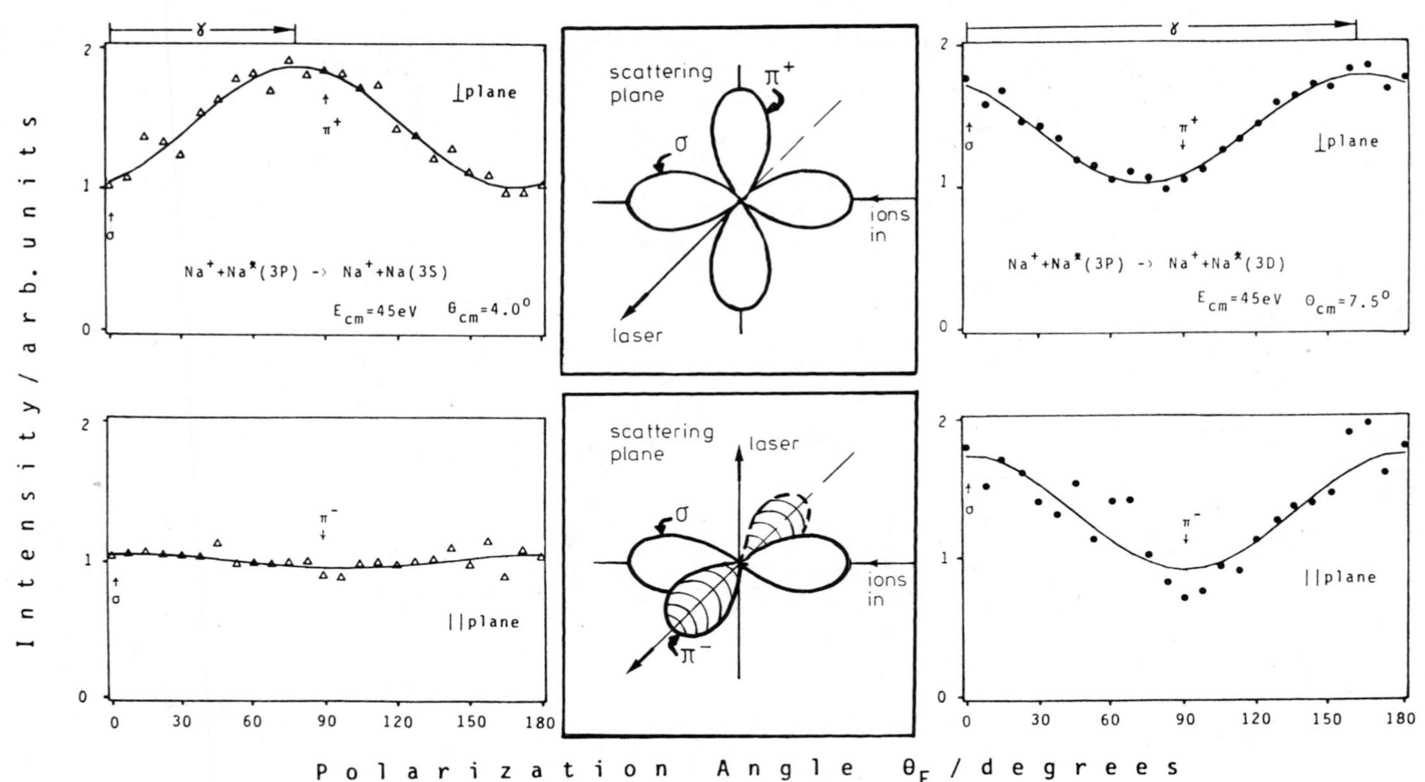

Figure 17. Scattered intensity as function of polarization angle of linearly polarized light, θ_E, propagating perpendicular (upper) and parallel (lower) to scattering plane.

and the (3s) ground state (crossing C in Fig. 15) to be reached. Similarly, a σ orbital populates the $2^2\Sigma_u$ molecular state enabling crossing B (Fig. 15) to be reached. For the laser light traveling parallel to the collision plane (and perpendicular to the incident ion beam), we are again able to prepare either a σ or a π orbital (lower part of Fig. 17). With the E vector parallel to the scattering plane and to the incident beam, the former is achieved ($\theta_E = 0$), and the scattering signal corresponds exactly to the equivalent situation in the upper parts of Fig. 17. In contrast, for $\theta_E = 90°$, a π^- orbital is prepared (rather than a π^+ if the E vector would lie in the scattering plane and be perpendicular to the incident beam), and we clearly see that for both the $3p \rightarrow 3s$ *and* the $3p \rightarrow 3d$ processes the scattering intensity is at a minimum. Reflection symmetry with respect to the scattering plane for the combined wave function of the Na atom and Na^+ ion (containing angular and spin momentum variables) is conserved during the collision.[55] Thus, we do indeed expect that the asymptotic preparation of a π^- orbital will not contribute to the $3p \rightarrow 3s$ deexcitation process since the final state has, by definition, positive reflection symmetry. Nor would we expect it to contribute to the $3p \rightarrow 3d$ excitation process since here it is the formation of a $^2\Sigma_u$ molecular state that is needed to reach the excited state, and whereas the preparation of a π^+ orbital can correlate with either a Π or a Σ molecular state, the preparation of a π^- orbital can give only a Π molecular state, as shown schematically in Fig. 9.

The reason that we observe any scattering intensity at all for the π^- preparation is that due to fine-structure and hyperfine-structure depolarization effects, we cannot prepare a pure "dumbbell."[56] Instead, for the system discussed here, we prepare a ratio of states of 2.5:1:1. This incomplete preparation is accounted for in the spirit of Fano and Macek[57] in the evaluation of the alignment parameters from the experimental data discussed in the following.

The differential cross section for scattering after excitation with linearly polarized light (incident perpendicular to the ion beam) can be written, in terms of our alignment parameters, as[35]

$$\begin{aligned} I(\theta_E, \theta_{cm}) = {} & \tfrac{1}{3} I_0 - \tfrac{1}{24} I_0 a_0(L) \\ & \times \{(1 - 3\rho_{00})[2\cos^2\theta_E - \sin^2\theta_E(1 + 3\cos 2\phi)] \\ & + 3P_l^+(1 - \rho_{00})[\cos 2\gamma(2\cos^2\theta_E - \sin^2\theta_E(1 - \cos 2\phi)) \\ & + 2\sin 2\gamma \sin 2\theta_E \sin\phi]\}, \end{aligned} \tag{7}$$

where $a_0(L)$ is the "alignment" parameter[56] (in this particular case it has the value $-\frac{2}{3}$) and ϕ is the azimuthal scattering angle, which for the experiment discussed here is 90° for the laser light propagating parallel to the scattering plane and 0° (180°) for the laser light propagating perpendicular to the scattering plane. Here, $I_0 = I_0(\theta_{\rm cm})$ gives the isotropic part of the differential cross section, that is, its value summed over the three possible orthogonal directions of the polarization vector

$$[(\theta_{\rm E}=0, \quad \phi=90°), \qquad (\theta_{\rm E}=90°, \quad \phi=90°), \qquad (\theta_{\rm E}=90°, \quad \phi=0°)],$$

and $\rho_{00}=\rho_{00}(\theta_{\rm cm})$, $\gamma=\gamma(\theta_{\rm cm})$, and $P_l^+ = P_l^+(\theta_{\rm cm})$ are the alignment parameters describing the spatial anisotropy as discussed here. We can also write this expression in terms of the relative differential cross sections for initial preparation of a σ, π^+, or π^- orbital ($\rho_{\sigma\sigma}$, $\rho_{\pi^+\pi^+}$, $\rho_{\pi^-\pi^-}$ with $\Sigma\rho_{ii}=1$):

$$\begin{aligned} I(\theta_{\rm E},\theta_{\rm cm}) = {} & \tfrac{1}{3}I_0 - \tfrac{1}{24}I_0 a_0(L)\{(1-3\rho_{\pi^-\pi^-}) \\ & \times[2\cos^2\theta_{\rm E} - \sin^2\theta_{\rm E}(1+3\cos 2\phi)] + 3(\rho_{\sigma\sigma}-\rho_{\pi^+\pi^+}) \\ & \times[2\cos^2\theta_{\rm E} - \sin^2\theta_{\rm E}(1-\cos 2\phi) + 2\tan 2\gamma \sin 2\theta_{\rm E}\sin\phi]\}. \end{aligned} \tag{8}$$

By fitting Eq. (7) to the experimental results (Fig. 17), we are able to obtain our three alignment parameters (γ, P_l^+, ρ_{00}) as displayed in Figs. 18 and 20.

A closer inspection of the upper parts of Fig. 17 shows that the optimal alignment angle γ is somewhat less than the 90° or 180° expected from such a simple model. This discrepancy can be explained by invoking[58] the concept of a "locking radius." A very similar concept is used to explain the interesting and closely related studies of collisional depolarization of atomic resonance radiation. The reader is referred to a recent review article by Burnett[59] for further details. The molecular picture implies that the charge cloud is attached to the internuclear axis. As the particles approach each other with impact parameters $b > 0$, the internuclear axis rotates. At short internuclear distances, the charge cloud follows this motion and rotates with the body-fixed molecular frame; that is, we have the situation depicted in Fig. 8c, "orbital following." However, at large internuclear distances where the interaction is weak (i.e., the Σ–Π potential energy difference is vanishing), the charge cloud has no reason to rotate with the internuclear axis and stays space fixed (Fig. 8*b*). We can obtain a reasonable idea of what happens by assuming that the two regions have a well-defined boundary; that is we assume strictly body-fixed motion for $R < R_{\rm L}$ and space-fixed motion for $R > R_{\rm L}$, where $R_{\rm L}$ is the locking radius. Of course, we have to bear in mind that, in reality, there will be no sharp boundary between the regions but rather a merging region where the charge cloud is neither strictly space fixed nor body fixed. The lower part of Fig. 18 illustrates how this concept can explain the experimental results.

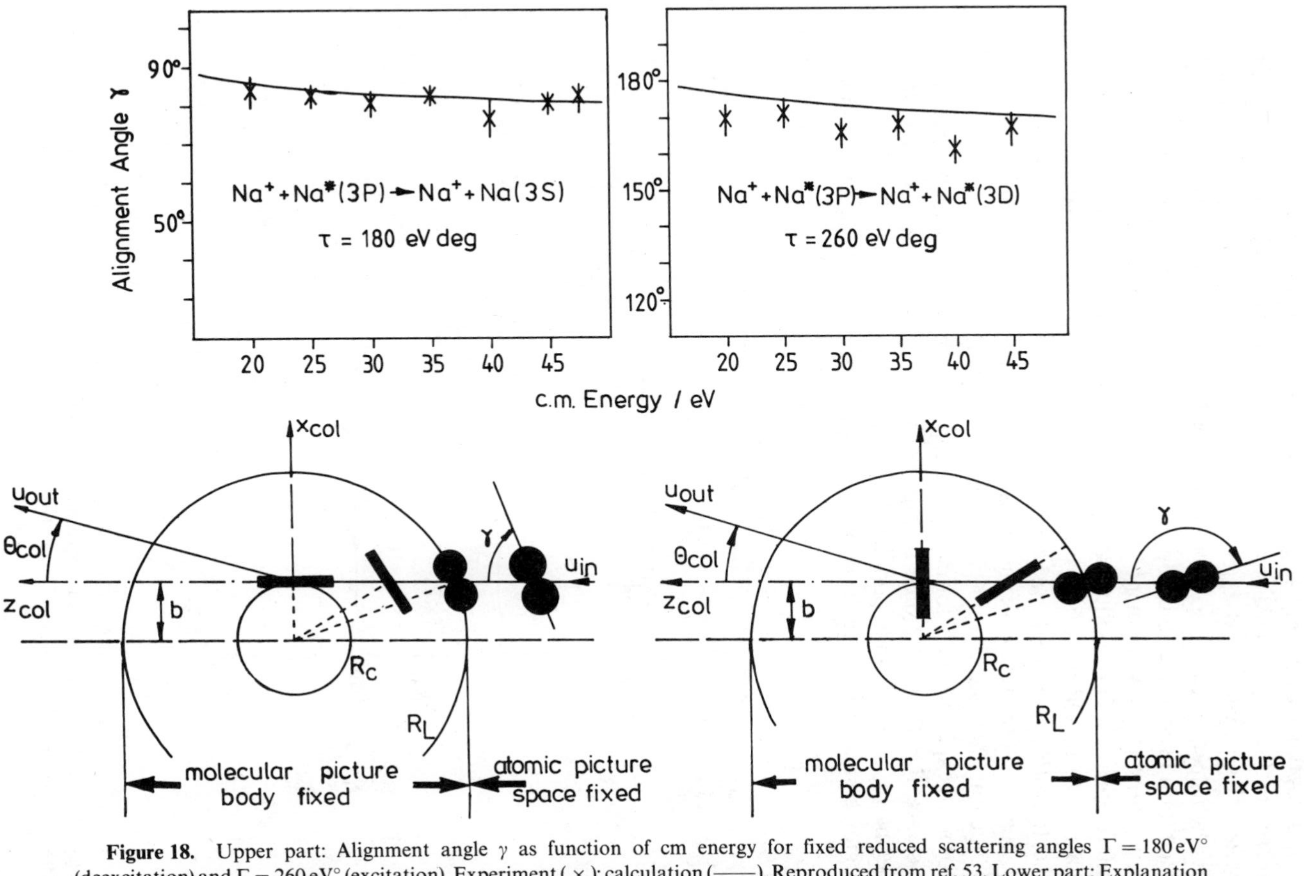

Figure 18. Upper part: Alignment angle γ as function of cm energy for fixed reduced scattering angles $\Gamma = 180\,\text{eV}^\circ$ (deexcitation) and $\Gamma = 260\,\text{eV}^\circ$ (excitation). Experiment (×); calculation (——). Reproduced from ref. 53. Lower part: Explanation of optimal alignment angle γ in terms of preparing Σ (right-hand side) or Π (left-hand side) molecular state at internuclear distance R_L, limiting region of body-fixed and space-fixed electronic motion.

The optimal alignment angle is shown as a function of incident kinetic energy in Fig. 18. Semiclassical calculations, described in detail in a recent review,[46] give very good agreement with the experimental values as shown in (the upper part of) Fig. 18. If we consider the time inverse of the $3p \rightarrow 3d$ excitation process (i.e., $3d \rightarrow 3p$), which was what was actually calculated, we can follow the development of the transition amplitudes as a function of internuclear distance as illustrated in the upper part of Fig. 19. The calculation starts with the 3d level fully populated. As the crossing is neared (5.6 a.u.), the transition amplitude $|a_{2\Sigma}|$ rises exponentially and then, after a short oscillation in the region of the crossing, stays fairly constant, confirming the deduction that the Σ molecular state is responsible for the transition as observed experimentally for the inverse ($3p \rightarrow 3d$) process. For an internuclear distance $R > 15$ a.u. (middle of Fig. 19), we see that the $3p\,1\Pi_u$ state also begins to be somewhat populated via $2\Sigma_u$–$1\Pi_u$ rotational coupling. This region coincides with the beginning of the decoupling of the system from the

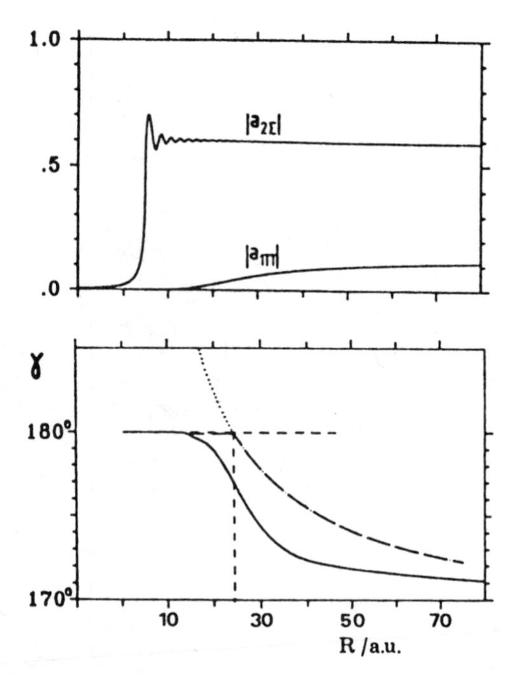

Figure 19. Upper part: Semiclassical calculations of development of transition amplitudes for $3d \rightarrow 3p$ deexcitation process as function of internuclear distance: $E_{cm} = 45$ eV, $b = 4.8$ a.u. Lower part: Semiclassical calculations of alignment angle γ as function of internuclear distance (——); alignment angle obtained by pure geometric projection of body-fixed orbital axis onto internuclear axis (–·–·).

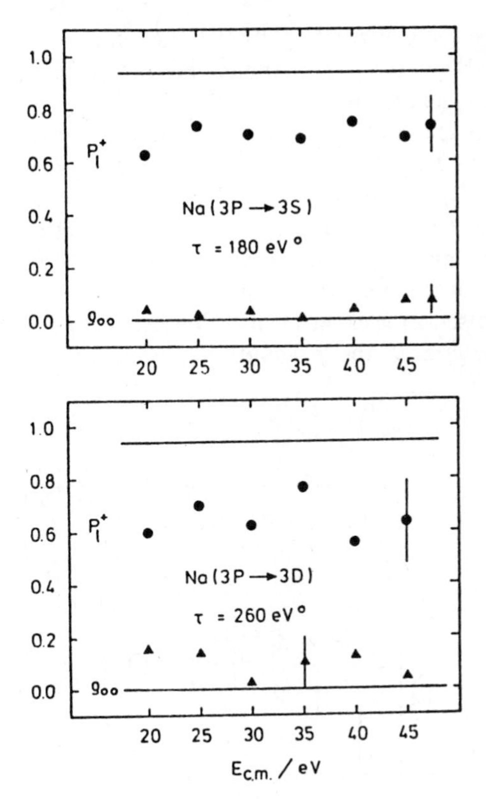

Figure 20. P_l^+(●) and ρ_{00}(▲) as function of cm scattering angle. Full lines are semiclassical calculations.

molecular (body-fixed) regime. The calculation of the alignment angle (lower part of Fig. 19) also shows the effects of the decoupling with a significant change from the original Σ ($\gamma = 180°$) position at $R = 15$ a.u.

The remaining alignment parameters, P_l^+ and ρ_{00}, are displayed in Fig. 20 for both processes as a function of collision energy. The results show no significant variation with collision energy in the range studied (or, indeed, with scattering angle at a particular collision energy).

As expected, ρ_{00}, which describes the height of the charge cloud along z (see Fig. 3), that is, the π^- contribution to the charge cloud, is close to zero within our limits of error. Remember that in evaluating the data, we have taken the incomplete "dumbbell" preparation into consideration and deduced parameters ρ_{00} and P_l^+ "as if" a pure dumbbell was prepared.

The semiclassical calculations of P_l^+ (Fig. 20), which assume full coherence, are consistently somewhat too large, indicating that there are several incoherent processes contributing to the transitions. The degree of incoherence observed[13] is in all probability due to the spin–orbit interaction, which we have neglected for simplicity but will discuss in Section V.

The P_l^+ results taken in conjunction with the results for γ (Fig. 18) tell us that for the 3p → 3s deexcitation process ($P_l^+ \approx 0.7$), it is not the pure dumbbell that is most effective for the transition. However, the preparation of Π states is six times more effective than the preparation of Σ states. Similarly, for the 3p → 3d excitation process ($P_l^+ \approx 0.8$), the preparation of Σ states is nine times more effective than the preparation of Π states.

Now we can discuss the experiments using circularly polarized light, which allows us to determine the orientation parameter $L_\perp^+$. This parameter,

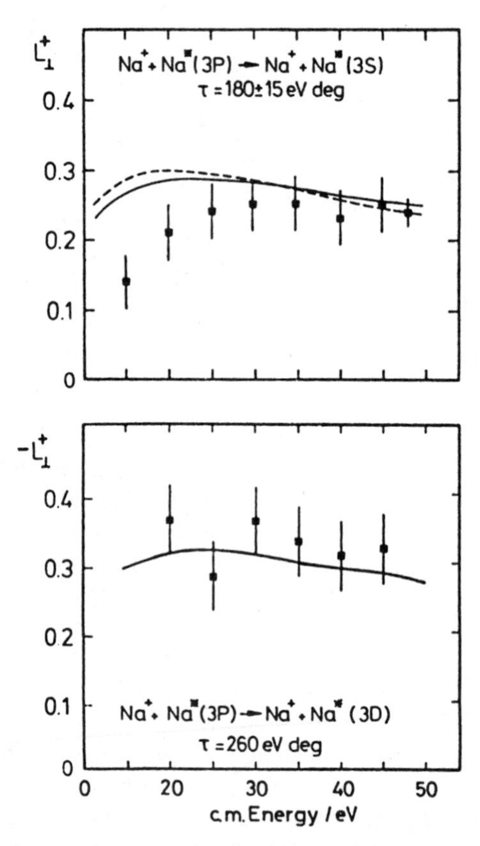

Figure 21. Angular momentum transfer $L_\perp^+$ as function of collision energy. Full line: straight line trajectory calculations. Dashed line: curved trajectory calculations.

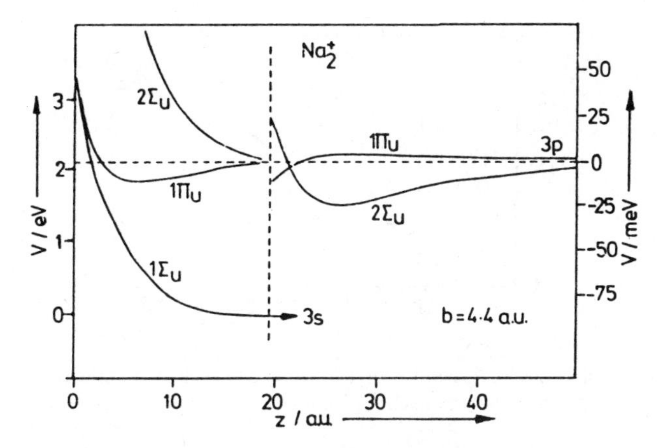

Figure 22. The lower-lying u-state potentials for Na_2^+ system with emphasis on long-range parts, plotted as function of $z = (R^2 + b^2)^{1/2}$. Note different energy scales.

which describes the transfer of angular momentum perpendicular to the scattering plane, is obtained from the difference in scattering intensity with LHC and RHC polarized light and is shown in Fig. 21. The main point to note here is the sign of the transferred angular momentum: positive for $3p \rightarrow 3s$, negative for $3p \rightarrow 3d$. The intuitive picture based on angular momentum considerations discussed in Section II is not valid in this case and in fact gives the opposite sign to that observed experimentally. We are well within an energy range low enough for the molecular picture to apply. If we look at the potential energy diagram again (Fig. 15), we see that it is only the population probability of the $1^2\Pi_u$ or $2^2\Sigma_u$ states that determines the cross section, and it is not a priori obvious how this should depend on the phase difference of $\pm 90°$ between the respective state population amplitudes for left and right circular polarization. The long-range potentials[52] (around and beyond point D in Fig. 15) are shown in Fig. 22 (using two energy scales). Here we can see that there is a crossing between the $2^2\Sigma_u$ and the $1^2\Pi_u$ curves at around $z = 22$ a.u. and a small but distinct splitting of the potentials for still larger internuclear distances. The semiclassical calculations[46] illustrated in Fig. 23 show that the angular momentum is mainly collected at large internuclear distances (i.e., in the space-fixed regime) and depends on the symmetry of the potentials, the effectively attractive or repulsive nature of the interaction, and the magnitude of the phase difference of the amplitudes accumulated on the Σ and Π potentials for internuclear distances greater than the locking radius R_L, which is approximately at the crossing between the two states ($z = 22$ a.u.).

In order to obtain a simple picture of how the angular momentum is

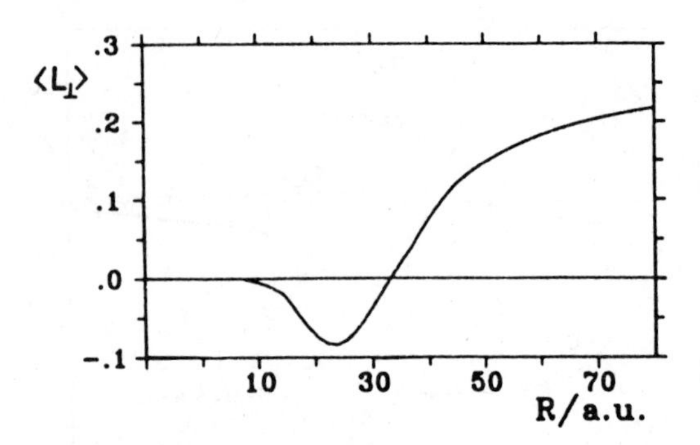

Figure 23. Semiclassical calculations of $L_{\perp}^{+}$ showing that angular momentum is mainly collected at large internuclear distances: $E_{cm} = 45\,\text{eV}$, $b = 4.8$ a.u.

created in the atomic electron charge cloud, we can consider the inverse process to that measured experimentally, $3s \rightarrow 3p$, which has been shown to give the same result.[46] Figure 24 illustrates the situation with the axis of the p orbital staying essentially space fixed for $R > R_L$. The body-fixed (x^b, z^b) coordinates for a time t_L (when R_L is reached) and for a later time t are displayed. If the charge cloud axis stays space fixed for $R > R_L$, we see that in terms of the molecular body-fixed frame this implies a $\Pi \rightarrow \Sigma$ transition

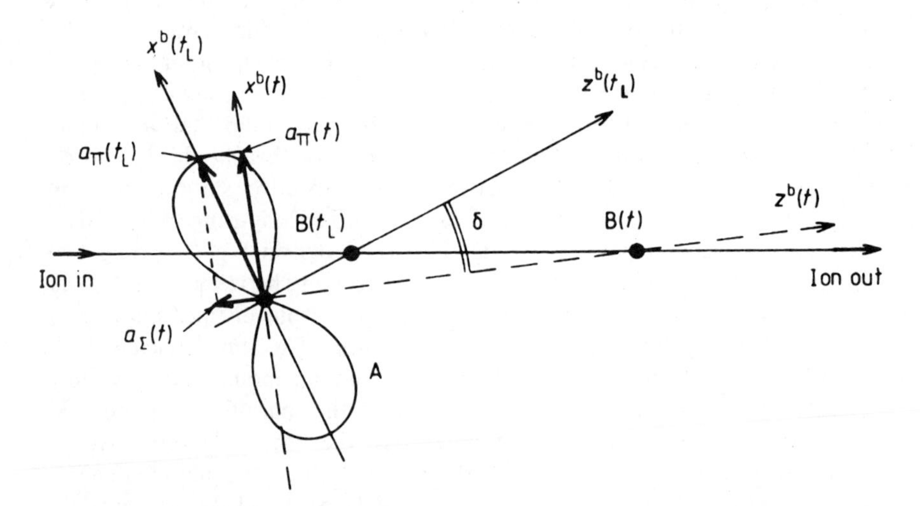

Figure 24. Model of p orbital space fixed for times $t > t_L$ (where t_L is time at which locking radius is reached). The Π and Σ components at t are indicated. They acquire different phases on long-range Π and Σ potentials. This has been drawn for $3s \rightarrow 3p$ excitation process, i.e., inverse of process measured experimentally. Reproduced from ref. 46.

with amplitudes $a_\Sigma(t)$ and $a_\Pi(t)$. The angular momentum $[= -2\,\mathrm{Im}(a_\Sigma a_\Pi^*)]$[46] then results from the different phase evolution on the Σ and Π curves. Both states have to be superposed at $t \to \infty$, and the angular momentum will be nonvanishing for a phase difference $\psi \neq 0, \pi, \ldots$.

Since $V_\Pi > V_\Sigma$ (see Fig. 22), the phase difference is positive and the interaction effectively repulsive, which for the 3s↔3p system gives us a positive value of $L_\perp^+$ as borne out experimentally (Fig. 21). As we start with a Σ state at R_L for the 3p↔3d system, the symmetry is reversed, and we find the opposite sign (consider Fig. 24) (i.e., a negative $L_\perp^+$) also confirmed by experiment.

The agreement between experiment and theory is extremely good, particularly for the 3p → 3d process. The remaining discrepancy between experimental and theoretical values for the deexcitation process, 3p → 3s, at low collision energies has been attributed to the effects of electron spin,[60] which will be discussed in Section V.

Thus, we see that a full set of measurements with both linearly and circularly polarized light combined with semiclassical calculations, used both as an aid and as confirmation of intuition, give us a fairly complete picture of the inelastic and superelastic scattering in this simple model case of $Na^+ + Na^*(3p)$.

IV. TWO EXAMPLES WITH CYLINDRICAL GEOMETRY

The kind of experiments described in the previous sections fall under the general category of three-vector correlations (v_{rel}, v'_{rel}, P), where in our case the polarization vector P stands for either alignment or orientation of the atomic target. We have seen how these differential cross-section experiments reveal a great amount of detail on the interaction dynamics of collision processes.

Unfortunately, such experiments are only possible for a very few examples, and in general, one has to be content with less detail. Often, one will have to be satisfied with measuring an angular integrated cross section. However, it is still possible to obtain useful information on the orbital stereospecificity of the process under investigation. In particular, it is possible in some cases to switch on the Σ or Π potentials, as we shall see in the following. The minimum requirement then is to have a well-defined axis of initial velocities with respect to which the polarization of the target atoms may be varied and thus determine a two-vector correlation (v_{rel}, P). In this case, only the linear polarization can have an influence on the outcome of the collision, since in such experiments with cylindrical symmetry all azimuthal angles are equally represented, and the effect of positive and negative helicity of the exciting light is canceled.

A. Studies of Intersystem Crossings in Ca*–Rare-Gas Collisions

Calcium is particularly well suited for alignment studies since it has no nuclear spin and, for the singlet states, one can in principle prepare the electron orbital in a 100% aligned state. This has in fact been demonstrated experimentally by observing the fluorescence of the laser-excited atoms.

In the experiments carried out in the Leone group,[61,62] the effect of the alignment of the initially excited Ca(4s5p $^1P_1$) state on the electronic energy transfer process

$$\mathrm{Ca}(4s5p\,^1P_1) + \mathrm{Rg} \rightarrow \mathrm{Ca}(4s5p\,^3P_j) + \mathrm{Rg} + \Delta\mathrm{KE} \tag{9}$$

is studied (where Rg stands for He, Ne, Kr, or Xe).

The process is considered to occur via crossings of potential curves correlating with the initial and final states. The transition between the singlet and triplet system is possible since neither the singlet nor the triplet states of calcium are purely coupled. Rather, each contains an admixture of the

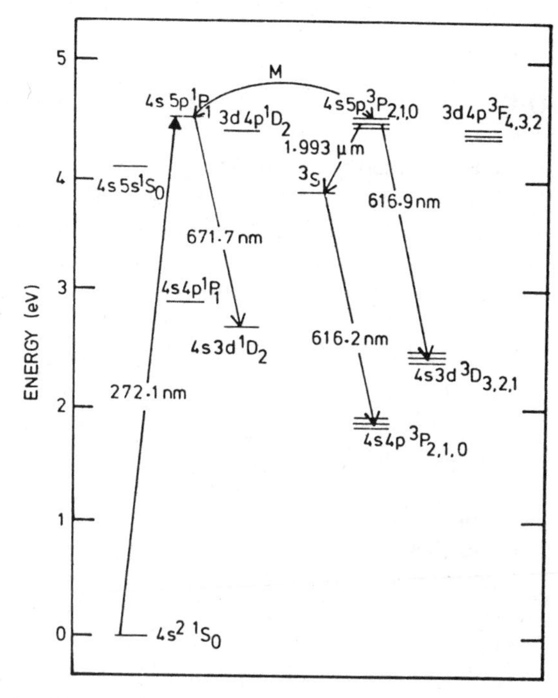

Figure 25. Partial energy level diagram for atomic calcium showing only transitions used in experiment. Reproduced from ref. 61.

other so that rotational coupling can mix Σ and Π states of both systems, just as in the previously discussed $Na^+ + Na^*$ system. A simplified energy diagram of the calcium atom is shown in Fig. 25. The pulsed laser is tuned to the $4s^2\,{}^1S_0$–$4s5p\,{}^1P_1$ transition at 272.1 nm. The product states, ${}^3P_j$, are observed on the $4s5p\,{}^3P_j$–$4s3d\,{}^3D_j$ transitions at 616.9 nm and on the cascade transition $4s5p\,{}^3P_j$–$4s4p\,{}^3P_j$ at 616.2 nm.

The experiment consists of crossed beams of calcium and the rare gas, a pulsed laser for exciting and aligning the $Ca^*(4s5p\,{}^1P_1)$ state, and a photomultiplier tube to monitor the fluorescence from the $Ca^*(4s5p\,{}^3P_j)$

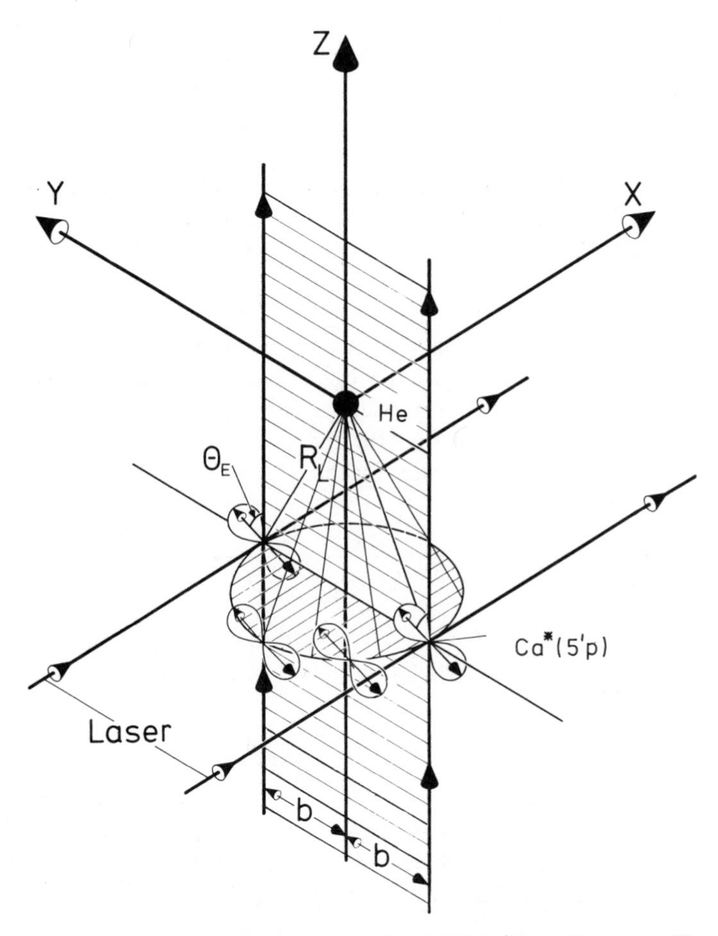

Figure 26. Experimental collision geometry for $Ca^*(4s5p\,{}^1P_1) + He$ process. The z axis is parallel to incoming He beam. The Ca* orbital is aligned with angle θ_E relative to z axis. Locking radii are displayed for same impact parameter but different azimuthal angles.

state. It measures an average over impact parameter, azimuthal angle, and relative velocity. Figure 26 depicts in a schematic view the collision geometry for such experiments with axial symmetry. We see that the orbital prepared with the laser at infinite internuclear separations may look more like a Π or more like a Σ orbital in the molecular frame, depending on the azimuthal angle ϕ.

Therefore, to interpret experimental results quantitatively in terms of the language presented in the previous sections, we have to average the evaluation formula [Eq. (7)] over all azimuthal angles. This gives the azimuthally averaged differential cross section, still as a function of the polarization angle θ_E of the linearly polarized light exciting the atoms:

$$I(\theta_{cm}, \theta_E) = \tfrac{1}{3} I_0 - \tfrac{1}{24} I_0 a_0(L) \times [1 - 3\rho_{00} + 3P_l^+(1 - \rho_{00})\cos 2\gamma](3\cos^2\theta_E - 1). \tag{10}$$

We see that the differential cross section has its maximum or minimum at θ_E of zero or 90° (depending on γ) as we expect for symmetry reasons and that $I(\theta_{cm}, \theta_E) = \frac{1}{3} I_0$ for the magical angle $\theta_E = 54.7°$, independent of the alignment parameters.

In addition, the actual experiment sums $I(\theta_{cm}, \theta_E)$ over all scattering angles, giving $I(\theta_E)$, and all that we can determine experimentally is an averaged "polarization" (or sensitivity parameter as called by Zare and collaborators) S, which is usually defined as the relative difference of the scattering signal for θ_E of zero and 90°. From Eq. (10) and with $a_0(L) = -2$ in this case of optimal alignment, which is possible for the $^1P_1$ case discussed here, we derive

$$S = \frac{I(0°) - I(90°)}{I(0°) + I(90°)} = \frac{1 - 3\langle\rho_{00}\rangle + 3\langle(1 - \rho_{00})P_l^+ \cos 2\gamma\rangle}{3 - \langle\rho_{00}\rangle + \langle(1 - \rho_{00})P_l^+ \cos 2\gamma\rangle}. \tag{11}$$

The quantities in angle brackets are averaged over all scattering angles and weighted with the respective differential cross section $I_0(\theta_{cm})$:

$$\langle x \rangle = \int x I_0(\theta_{cm}) \sin\theta_{cm}\, d\theta_{cm} \Big/ \int I_0(\theta_{cm}) \sin\theta_{cm}\, d\theta_{cm}. \tag{12}$$

One may, alternatively, express the measurable sensitivity parameter (11) in terms of the relative integrated cross sections for preparation of a σ, π^+, or π^- orbital, the diagonal density matrix elements $\rho_{\sigma\sigma}$, $\rho_{\pi^+\pi^+}$, and $\rho_{\pi^-\pi^-}$:

$$S = \frac{1 - 3(\langle\rho_{\pi^-\pi^-}\rangle + \langle\rho_{\pi^+\pi^+}\rangle - \langle\rho_{\sigma\sigma}\rangle)}{3 - (\langle\rho_{\pi^-\pi^-}\rangle + \langle\rho_{\pi^+\pi^+}\rangle - \langle\rho_{\sigma\sigma}\rangle)}. \tag{13}$$

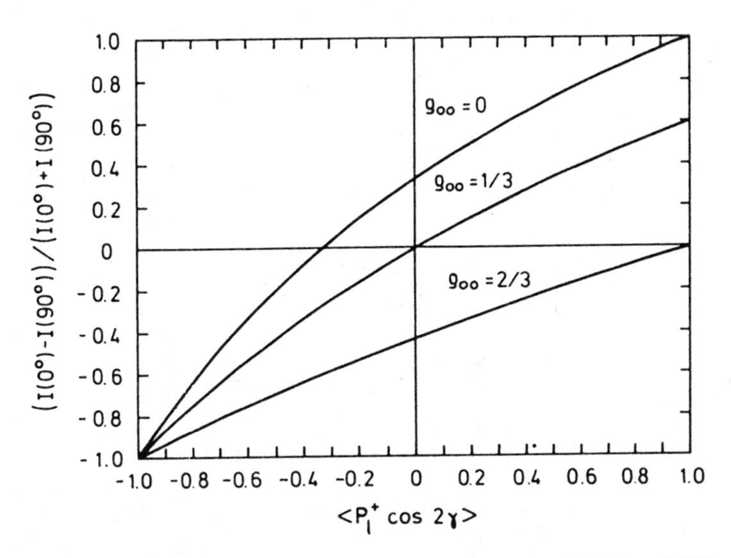

Figure 27. Averaged polarization of Ca atom, $S = [I(0°) - I(90°)]/[I(0°) + I(90°)]$, plotted as function of averaged cosine of alignment angle, $\langle P_l^+ \cos 2\gamma \rangle$, for different values of ρ_{00}.

To obtain some feeling for the significance of the sensitivity parameter S, we plot it in Fig. 27 as a function of the averaged value $\langle P_l^+ \cos 2\gamma \rangle$, assuming ρ_{00} to be a constant parameter. The clearest picture is given for $\rho_{00} = 0$ (no spin–orbit effects and full conservation of reflection symmetry in the atom) and $P_l^+ = 1$, describing a pure dumbbell. As we can see, only the two extreme values of the averaged alignment angle give full anisotropy $\langle \cos 2\gamma \rangle = 1$ and $\langle \cos 2\gamma \rangle = -1$, corresponding to pure σ or pure π^+ scattering, respectively. The sensitivity parameter S vanishes for $\langle \cos 2\gamma \rangle = -\frac{1}{3}$ corresponding to an alignment angle $\gamma = 54.7°$, again the magic angle. As ρ_{00} is increased, the σ–π^+ contribution to the scattering is decreased; that is, the averaged polarization becomes more negative for a particular $\langle \cos 2\gamma \rangle$. Of course, when $\rho_{00} = 1$, we have pure π^- scattering, and $\langle S \rangle = -1$ for all averaged alignment angles.

The experimental results[62a] are shown in Fig. 28, which gives the relative probability for the energy transfer [Eq. (9)] as a function of the laser alignment angle θ_E for the different rare gases studied. For helium and, especially, neon there is a clear maximum at an angle of approximately 90° corresponding to a perpendicular alignment of the Ca* p orbital with respect to the initial relative velocity. Thus, for Ca* + He, Ne, the asymptotic preparation of a π orbital is preferred for the $^1P_1$–^3P_j transition. With Xe we have a maximum at 0°, indicating the dominance of the asymptotically prepared σ orbital. No dependence on alignment angle is observed for Kr.

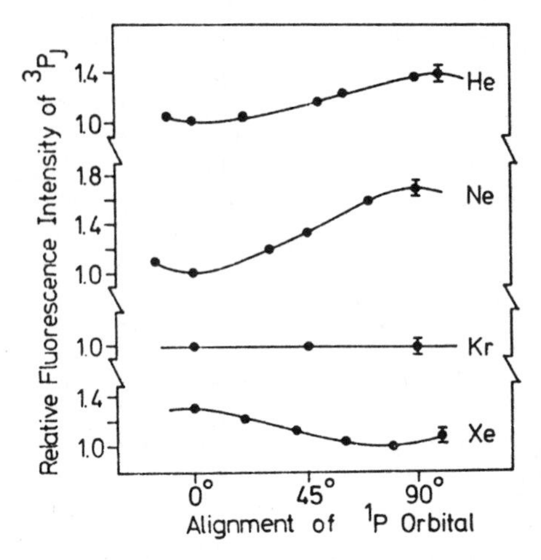

Figure 28. Relative fluorescence intensity as function of alignment angle for $Ca^*(4s5p\,^1P_1) + Rg \rightarrow Ca^*(4s5p\,^3P_j) + Rg$ process. Reproduced from ref. 62a.

The polarization dependence for Ca* + He can be understood by considering the potential energy curves of the Ca*–He quasi-molecule formed in the collision. Figure 29 shows two sets of model potentials adopted from Alexander and Pouilly,[63] both indicating the essential mechanism of the process to proceed via a $^1\Pi$–$^3\Sigma$ crossing at around 8.5 a.u. while the long-range Coriolis coupling between the $^1\Sigma$ and $^1\Pi$ curves will be responsible for the locking into the molecular frame. Thus, we have the situation encountered in Section III with a π orbital prepared at large internuclear distances scattering as a Π molecular state, that is, orbital following. A theoretical treatment using the Landau–Zener curve crossing model and invoking the locking radius concept[64] shows the correct trends but predicts a magnitude for the alignment effect which is considerably larger than that observed experimentally. It would, however, be interesting to use the "Alexander potentials"[63] in such a treatment. In the strict locking picture, which is certainly too oversimplified in the present case, one may use Fig. 27 to attribute an effective locking radius to the experimental observation of $S = -0.17$ $[I(90°)/I(0°) = 1.4]$. We read there that $\langle P_l^+ \cos 2\gamma \rangle = -0.48$, assuming that we have a pure dumbbell ($P_l^+ = 1$) and no spin–orbit effects ($\rho_{00} = 0$). If we further assume that only impact parameters close to the crossing radius R_c contribute to the process, as expected for rotational coupling, we can estimate the ratio of locking radius to crossing radius, $R_L/R_c \sim 1.95$ (making use of the relation $\cos \gamma = b/R_L$). If we take the crossing radius R_c to be 8.5 a.u. (Fig. 29), this gives a locking radius $R_L \sim 17$ a.u.

Alexander and Pouilly[63] performed quantum mechanical calculations

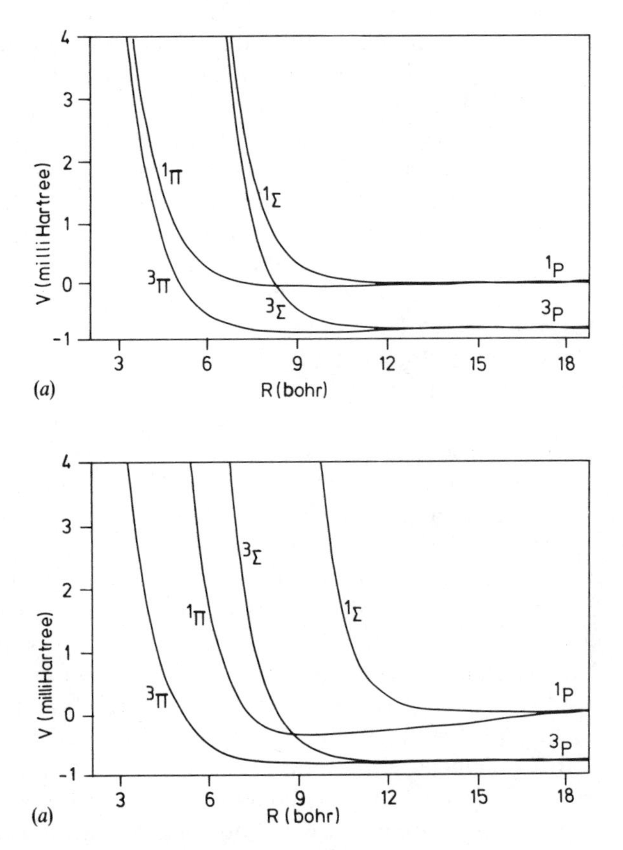

Figure 29. Calculated molecular potentials[63] for Ca* + He. Upper part: Potentials that give best agreement with experimental data. Lower part: Potentials with increased Σ–Π splitting for ^1P state leading to enhanced polarization effects.

using reasonable assumptions for the interaction potentials, since at the present time there is no hope for ab initio or even pseudopotential calculations in such a rather complex system for the high-lying states involved. All potentials were chosen such that the total cross section is well reproduced by the computation (i.e., the inner $^1\Pi$–$^3\Sigma$ crossing is essentially kept at the same internuclear distance), whereas the Σ–Π splitting was modified so that the polarization effect could be simulated.

Under these conditions, the calculations were able to reproduce the experimental results with the potentials shown in Fig. 29*a* giving an average polarization $S = -0.2$ (experimental value, -0.17 ± 0.06). In contrast, the potential curves depicted in Fig. 29*b* give $S = -0.47$, which seems to illustrate that the observed polarization effects are a direct probe of the dependence of the Σ–Π splitting for the $Ca^*(^1P_1)$ state on the internuclear distance. Although the crossing radius remains essentially unchanged at around 8.5 a.u., the different Σ–Π splittings assumed lead to a greatly increased polarization effect. This may be the reason for the larger ratio observed with Ne. Collisions with the heavier rare gases may have contributions from additional curve crossings or the Π and Σ potential curves may have a qualitatively different appearance for the heavier, more polarizable collision partners. One effect of this may be that the most effective impact parameter becomes substantially larger so that the ratio of b/R_L can no longer be assumed small compared to 1. Then a situation may arise that resembles the nonlocking condition illustrated in Fig. 9. It may also be possible that the spin–orbit interaction starts to play a more dynamic role for these heavy rare gases.

Thus, it seems that the quantum mechanical results support our intuitive picture of the locking concept. Unfortunately, for the time being, this cannot be gleaned directly from the details of the calculations. The coupling between the atomic electron angular momentum and the angular momentum of the relative motion of the interacting particles prevents at closer inspection a one-to-one translation of quantum mechanical formulation and semiclassical intuitive model. Also, the computed coupling elements do not give a clarifying view. It is nevertheless quite remarkable that the quantum mechanical calculation leads to the same results as one would expect from the intuitive concepts and also predicts the same trends as one increases the splitting between Σ and Π potentials asymptotically. Thus, we may well suspect that the quantum mechanical calculations contain the essential elements of the semiclassical picture, even though it is not obvious. A conclusive discussion would be possible if one could compare the evolution of the atomic charge cloud (as we have seen it in the semiclassical calculations for the $Na^* + Na^+$ case; Fig. 19) as a function of internuclear distance calculated in both semiclassical and quantum mechanical manner. The latter would involve a projection of the $j - l$ coupling scheme onto the basis of atomic orbital states using a density matrix formalism. It would be very interesting to watch the two pictures merging as one increases the energy from thermal, where the quantum approach may be essential, up to some 100 eV, where the semiclassical calculations should give correct results.

Finally, we briefly mention some very recent results of Bussert and Leone,[62b] who were able to study the inverse process, exciting the $^3\Sigma$ state and probing the fluorescence of the $^1\Pi$ state after collisional energy transfer induced by He. Naively, one would expect that in the locking picture the σ orbital preparation at large internuclear distances might give the maximum cross section since a look at Fig. 29 shows that we need to populate the $^3\Sigma$ potential in order to reach the relevant curve crossing that leads us to the

$^1\Pi$ potential. A somewhat closer inspection of the situation shows, however, that the optical preparation of a $^3\Pi$ state with linearly polarized light parallel to a given z axis prepares this state to 100% in the Π states; that is, for cylindrical symmetry, one obtains an average 50% in each of the π^+ and π^- states, whereas the perpendicular polarization leads to 50% σ and 50% π state population (i.e., 25% in each of the π^+ and π^-). This is a consequence of the corresponding Clebsch–Gordan coefficients describing these states in the *L-S* coupling scheme† (for a general and formal description, see, e.g., Appendix A of Andersen, Gallagher and Hertel[33]). Thus, the correct prediction of the locking model is also to expect that the perpendicular polarization gives the maximum cross section. This is indeed observed experimentally.[62b]

Again, the close coupling calculations confirm this value using the potentials of Fig. 29*a*.[65] Thus, we see that the locking model gives a reasonable, although not yet quantitative, prediction that agrees with that of the quantum mechanical results. Although the locking concept is certainly oversimplified and cannot be expected to give exact predictions at low energies and especially if spin–orbit coupling is involved, one finds a surprisingly good correlation with the intuitive expectations and reality. As will be seen in the section on spin–orbit interaction, these semiclassical models do not seem to hold at low energies for the description of the angular momentum transfer. It appears, however, that the charge cloud alignment is less sensitive to such complications. This would be a rather comforting aspect from the viewpoint of orbital specific stereochemistry.

It thus seems very worthwhile and important to devote some more effort to clarifying these issues.

B. Associative Ionization

Collisional ionization processes have been extensively investigated since their discovery in the 1920s.[66] Here we only want to discuss the polarization effects observed in associative ionization for one particular collision system but refer the reader to a wealth of work on both Penning and associative ionization involving rare-gas atoms (see, e.g., refs. 19–28). The associative ionization process discussed here is one of the simplest examples of a system involving two electrons excited with polarized laser light,

$$\mathrm{Na^*(3p) + Na^*(3p) \rightarrow Na_2^+ + e}, \tag{14}$$

and has become the standard case of recent, rather elaborate studies.[67–79]

† $|^3P_1, M_j = 0\rangle = \frac{1}{2}(|S=1, M_s=1, L=1, M_L=-1\rangle - |S=1, M_S=-1, L=1, M_L=1\rangle)$.

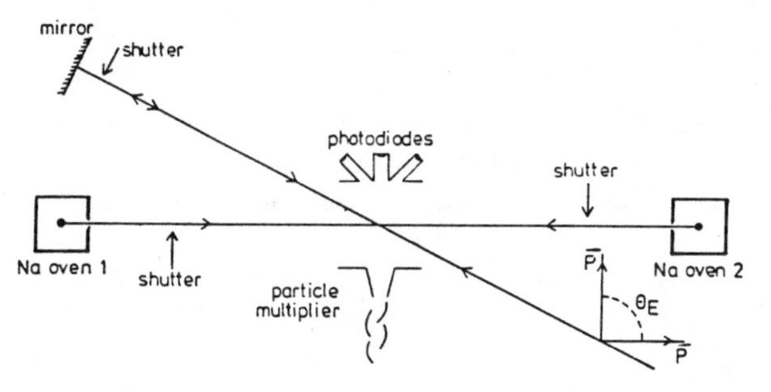

Figure 30. Schematic diagram of Morgenstern experiment to study associative ionization with laser-excited, polarized Na*(3p) atoms. Reproduced from ref. 79.

It has been shown[68] that the observed ion signal takes the simple form

$$I(\theta_{\mathrm{E}}) = R_0 + R_1 \cos(2\theta_{\mathrm{E}}) + R_2 \cos(4\theta_{\mathrm{E}}), \tag{15}$$

where θ_{E} is the angle between the electric vector of the polarized light and the atom beam and the R_i are functions of the products of the ionization amplitudes. Earlier experiments [67,70–72] used a single beam of sodium atoms. As a result of the velocity distribution of the atoms within the beam, the excited Na atoms could collide with each other. Results obtained by two different groups with different beam velocity distributions[72] indicated that the polarization dependence of process (14) changed when the collision velocity was varied from subthermal to thermal velocities. The most recent results from the Morgenstern and Weiner groups[75,79] using two counter-running thermal beams substantiate this by showing an increasing and very interesting polarization dependence at increasing collision velocities. The Morgenstern experimental setup is shown in Fig. 30. Two thermal beams of Na atoms effusing from ovens in opposite directions intersect one another. The beams are crossed perpendicularly by linearly polarized laser light. All ions created in the interaction region are extracted by a small electric field and are detected by a particle multiplier. The fluorescence light is monitored by photodiodes. The Doppler effect is used to excite atoms with a well-defined velocity, and the ion production rate for different collision velocities is measured as a function of the angle θ_{E} between the linear polarization vector of the laser light and the direction of the atomic beams.

It should be noted that in this case the results are discussed in terms of the "collision" reference frame, that is, with the z axis aligned in the direction of the atomic beams. Thus, exciting the $M_{\mathrm{L}} = 0$ substate aligns the charge cloud in the z direction and produces a σ orbital (see Fig. 2). Results are shown in Fig. 31 for four different collision velocities.[79] The ion yield shows

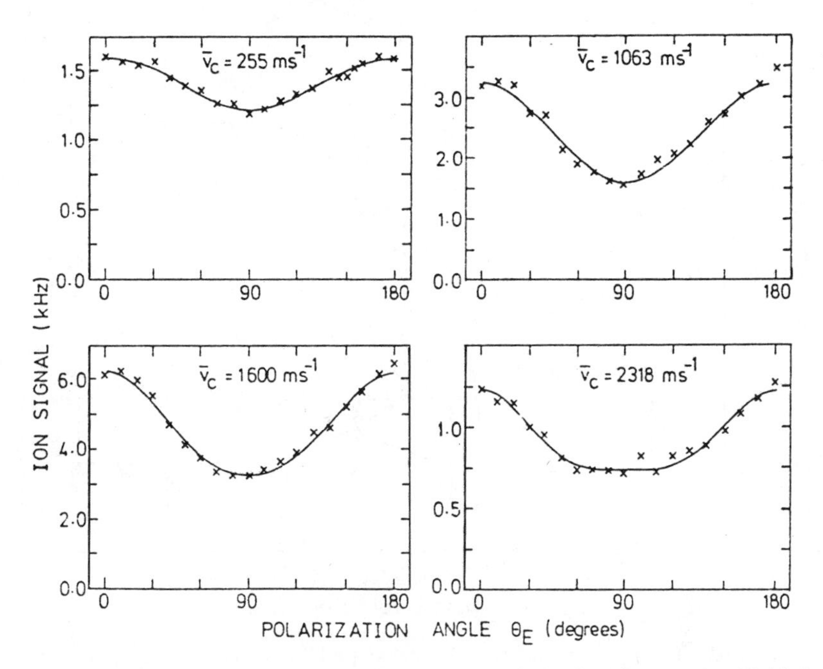

Figure 31. Polarization dependence of associative ionization for process Na*(3p) + Na*(3p) → Na_2^+ + e^- at four different collision velocities. Reproduced from ref. 79.

a strong dependence on polarization angle, which becomes more pronounced as the collision velocity increases. The maxima appear at 0° and 180°, indicating that asymptotic preparation of σ orbitals is most effective for associative ionization. What is particularly interesting is that at the highest collision velocity investigated (2318 m/s) the minimum in the curve is flattened out, and we see the increasing importance of the $\cos(4\theta_E)$ term in Eq. (15), which arises from the fact that we have two excited p electrons, not just one, as in the previous examples.

From fitting the experimental data to the theory,[68] one obtains the cross sections for associative ionization as a function of collision velocity for different initial preparations of the atomic states.[79] These are illustrated in Fig. 32 and show some interesting effects. For this system, it is not yet entirely clear whether a description in the $|lm_l, lm_l'\rangle$ basis or one in the $|jm_j, jm_j'\rangle$ basis (including the effects of the electron spin) is the more appropriate. For the sake of convenience and simplicity, we will consider the former, referring only to the spatial part of the atomic states (i.e., to the well-known p dumbbells).

At the lowest collison velocity investigated (255 m/s), where the ionization

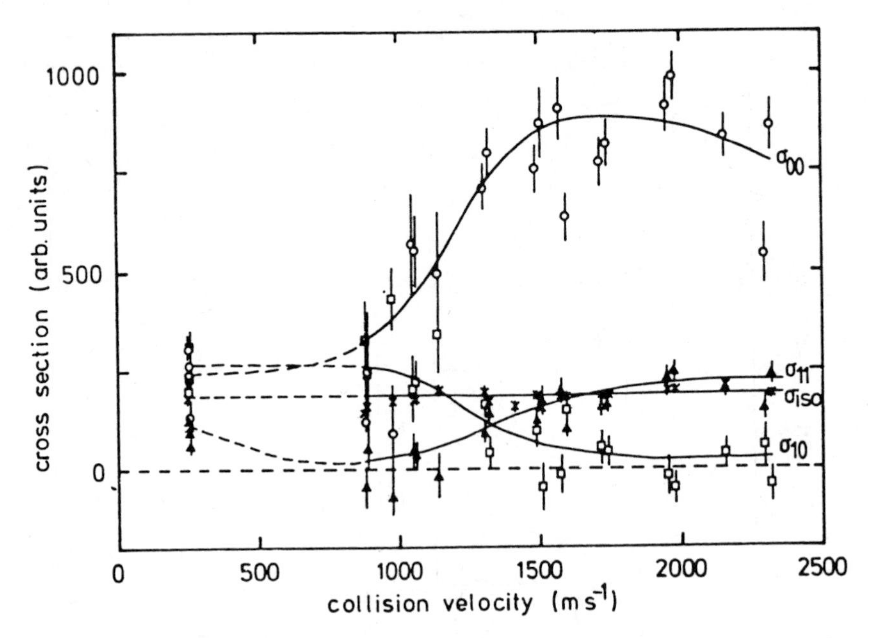

Figure 32. Cross sections for $|M_L| = 0, 0$ (○) $|M_L| = 1, 0$ (□), and $|M_L| = 1, 1$ (▲) as function of collision velocity. Reproduced from ref. 79.

is due to collisions between atoms in the same beam, the cross sections obtained for both atoms initially prepared in a pσ state ($|M_L| = 0, 0$) and those with one atom in a pπ state and one in a pσ state ($|M_L| = 1, 0$) are dominant. However, there is only about a factor of 2 between the value of these cross sections and that obtained for both atoms initially in a pπ state ($|M_L| = 1, 1$).

As the collision velocity is increased, the results become much more interesting. At velocities $\bar{v}_c > 1000$ m/s, having both atoms in the $|\mathrm{p}\sigma\rangle$ state is clearly the most effective preparation for associative ionization. The strong increase of this cross section at about $\bar{v}_c = 1000$ m/s indicates that an endothermic process is involved and that a potential barrier of about 40 meV has to be overcome before the ionization continuum can be reached. The fact that the cross section does not vanish below this velocity indicates that ionization can also occur via another potential curve crossing for which the threshold is lower or the process is exothermic.

The $\sigma\pi$ and $\pi\pi$ orbital configurations also lead to associative ionization, although to a lesser extent. However, the cross sections for these processes show very different behavior.

The cross section obtained from the initially prepared $\sigma\pi$ configuration decreases significantly for velocities greater than 1000 m/s. This indicates that ionization takes place at small internuclear distances via an attractive potential curve, as is well known for a variety of reaction processes (the absorbing sphere model[80]). At low collision velocities, the attraction will lead to small distances of closest approach even for fairly large impact parameters, whereas at higher collision velocities, the attraction will be less effective, and a crossing with the ionization continuum at small internuclear separations will no longer be reached.

Finally, the cross section obtained by preparing both atoms in $|p\pi\rangle$ states shows a more complicated behavior. It was explained[79] by assuming that at least two potential curves contribute: one slightly attractive with a crossing at very small internuclear distances and the other repulsive with a barrier of approximately 60 meV.

The Weiner group[71,74] has recently extended the experimental measurements to study the effect of the Na(3p) polarization on the rotational angular momentum alignment of the product Na_2^+ ions. These researchers do this by using a Nd^{3+}:YAG-pumped dye laser tuned to 570 nm to photodissociate the dimer ions. The ratio of the parent ion depletion signal for parallel and perpendicular polarizations of *this* laser gives a measure of the angular momentum alignment of the Na_2^+ as described in detail elsewhere.[71] Jones and Dahler[78] used earlier experimental results obtained from collisions within a single sodium beam[67] (and thus with relatively low collision velocity) to try to determine which molecular states of the Na_2 molecule contributed to the associative ionization process. They concluded that a ${}^1\Sigma_g^+$ state with the dominant molecular orbital configuration σ_g^2 contributes most with smaller contributions from at least one other state. With this in mind, the Weiner group developed a semiclassical model[74] based on the idea of a "locking" radius since here, as in the previous examples, asymptotic preparation of σ orbitals appears to correlate with Σ molecular states, and we have the orbital following situation again. By varying the locking radius R_L and the relative rate coefficients of axial, σ^2, transverse, π^2, and mixed, $\sigma\pi$, orbital coefficients, they were able to fit their experimental data for both the ratio of associative ionization rates for different alignment angles of the linearly polarized laser light and the spatial anisotropy observed in the distribution of angular momentum vectors of the product Na_2^+. Their analysis showed that the σ^2 configuration (and thus the Σ molecular states) is six times more efficient for associative ionization than the π^2 configuration with the contribution from mixed, $\sigma\pi$, configurations being negligibly small and gave a lower limit for the locking radius of 25 Å.

Thus, we see that even this simplest of examples with two active electrons has proved to be extremely complicated, with many potential curves contributing and those contributions varying significantly with collision velocity. It does, however, seem that an orbital following picture is also valid here and that Σ molecular states dominate the scattering. There are some calculated potential curves available for this system,[81] but much more theoretical work is required before the participating potentials can be identified.

It will be very interesting and exciting to follow the future developments in the understanding of this complicated and fascinating system.

V. SPIN–ORBIT INTERACTION

We now return to experiments with planar symmetry and discuss the effect of the spin–orbit interaction, which we have neglected so far. This has been and is continuing to be investigated thoroughly for electron scattering on laser-excited atoms using both spin-polarized[82–85] and spin-unpolarized electrons (e.g., refs. 86–89) and for the scattering of excited alkalis or alkaline earths from rare gases.[12,42,90–92] For a recent review on spin–orbit effects in gas-phase chemical reactions, see ref. 93. Here, only two well-understood cases will be discussed, with particular relevance to the polarization effects of concern.

A. Nonadiabatic Transitions in $Na^+ + Na^*$ System

In Section III, we attributed the slight discrepancy between theory and experiment to the effect of the electron spin–orbit coupling. In particular, the small but finite value observed for ρ_{00} (see Fig. 20) indicates that reflection symmetry of the orbital atomic wave function is not wholly conserved, which could be caused by spin flip processes. At the energies investigated experimentally, the spin precession time due to the spin–orbit interaction can be comparable to the total interaction time of the Na^+ with Na. To determine the effect of this process on the collisions, semiclassical calculations were carried out using the coupled-channel method in the impact parameter approximation and including the dynamics of the electron spin coupling to the heavy-particle motion.[60] The long-range potentials, leading asymptotically to $Na^+ + Na^*(3^2P_{3/2,1/2})$, used in the calculations are shown in Fig. 33.

In the region where the difference in the potentials ($|V_\Sigma - V_\Pi|$) is less than the spin–orbit splitting of an isolated Na atom in the 3^2P state (i.e., at large internuclear distances), Hund's coupling case (c)[94] is valid and Ω, the projection of the total angular momentum J onto the internuclear axis, is a good quantum number.

In order to solve the coupled equations derived from the relevant Schrödinger equation, the radial and rotational (Coriolis) coupling elements have to be calculated. It was shown in Section III that the main structure observed was due to rotational coupling at crossings B and C in Fig. 15. The

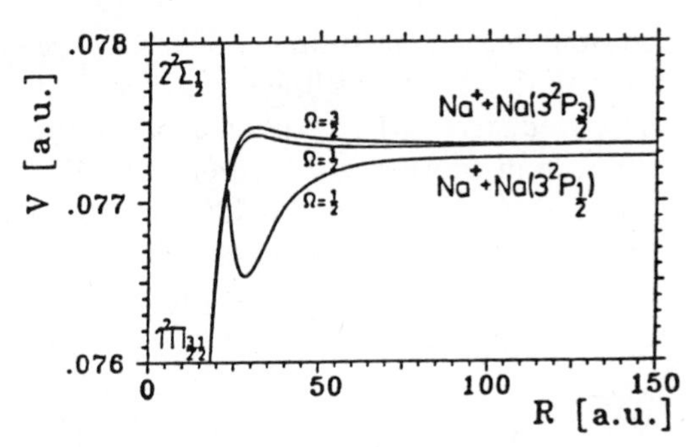

Figure 33. Plot of long-range Na_2^+ potentials leading asymptotically to $Na^+ + Na^*(3^2P_{3/2,1/2})$. For internuclear distances R, where $V_{el}(R) < V_{so}$, Hund's coupling case (c) applies, and projection Ω of total electronic angular momentum J is good quantum number. Reproduced from ref. 60.

polarization effects were also shown to be dominated by rotational coupling, mixing the states at larger internuclear distances around and beyond crossing D as the wave function developes on the 1Π and 2Σ potentials. If one leaves the pure orbital picture to include spin–orbit coupling, one has to rediagonalize the Hamiltonian. The calculations show that the relevant coupling at large internuclear distances, where Hund's case (c) is valid, are now expressed as a radial coupling term between the $\Omega = \frac{1}{2}$ states, as depicted in Fig. 33. The dependence of this calculated radial coupling on the internuclear distance R is displayed in Fig. 34. The strong "spike" at $R \approx 23$ a.u. originates from the avoided crossing between the two $\Omega = \frac{1}{2}$ states (see Fig. 33). The broad peak occurring at $R \approx 60$ a.u. occurs in a region where the Σ–Π splitting is of the same order of magnitude as the spin–orbit splitting.

The angular momentum transfer provides the most stringent test of the theoretical calculations since it is sensitive to the long-range interaction; that

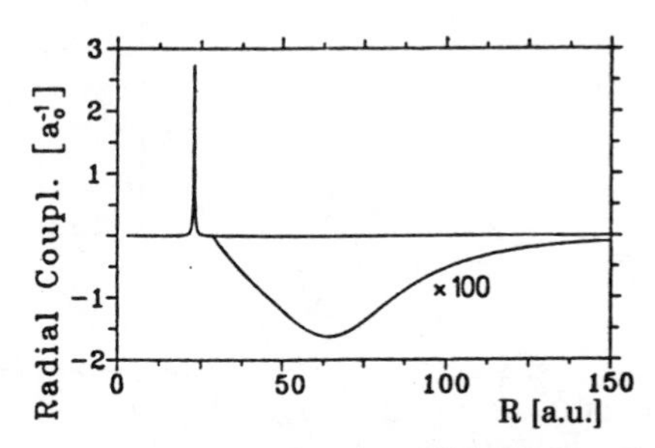

Figure 34. Radial coupling element as function of internuclear distance displaying strong peak at $R = 23$ a.u. and smaller peak in region where Σ–Π splitting is of same order of magnitude as spin–orbit splitting. Reproduced from ref. 60.

is, it probes the collision for the longest time, which may become comparable to the spin precession time. Also, it has been measured with higher accuracy than the alignment parameters.[52] The experimental and theoretical results for the angular momentum transfer, $L_\perp^+$, for both the collisional excitation and deexcitation processes are shown in Fig. 35. The calculations including the influence of the electron spin are in excellent agreement with the experimental results. The oscillations predicted by the calculations for the excitation process $3^2P \rightarrow 3^2D$ have been tentatively assigned to the effect of

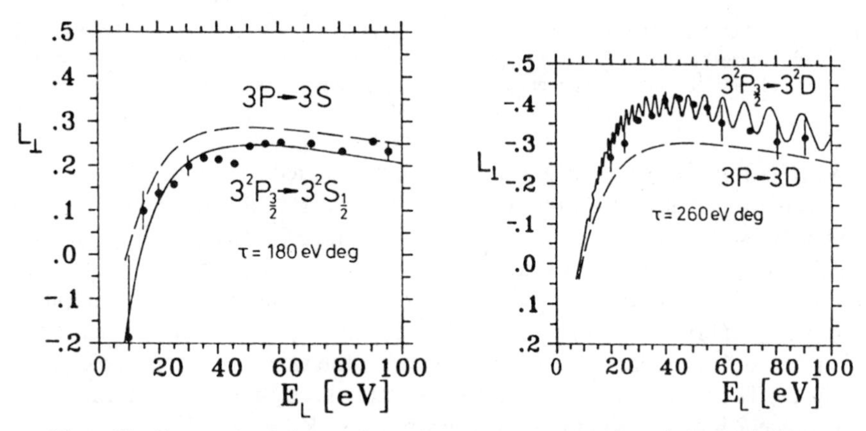

Figure 35. Energy dependence of angular momentum transfer $L_\perp^+$ for $Na^*(3^2P_{3/2}) \rightarrow Na(3^2S)$ collisional deexcitation (*a*) and $Na^*(3^2P_{3/2}) \rightarrow Na^*(3^2D)$ collisional excitation (*b*) processes. Full circles are experimental data. Dashed line: Calculations in L picture. Full line: Calculations in the J picture.

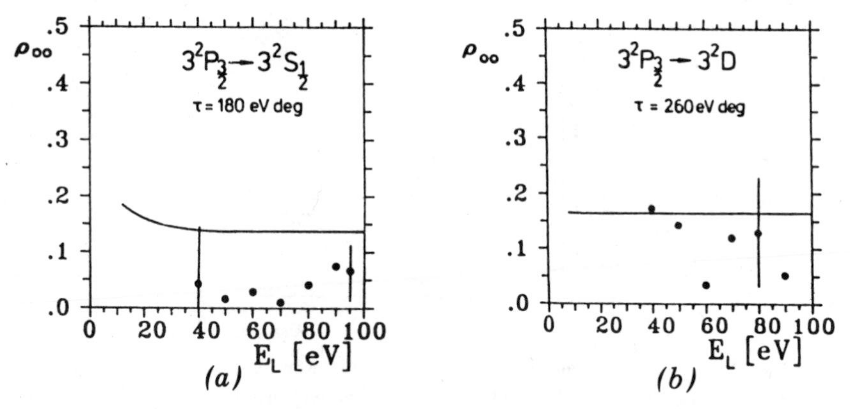

Figure 36. Alignment parameter ρ_{00}. Full lines are calculations.

the precession of the electronic orbital angular momentum.[60] Unfortunately, the experimental resolution is not sufficient to confirm or deny the presence of oscillations in the data.

In Fig. 36, the calculations of the alignment parameter ρ_{00} are compared with the experimental results. The agreement is good, and although the computed results for the deexcitation process are too large, they do show the correct trend being smaller in magnitude than the calculations for the excitation process.

Thus, we see that the spin–orbit interaction can play a significant role, particularly at low collision energies (for this system, $E_{cm} < 30\,eV$) and has to be included for a complete picture of the collision process.

B. Orientation Effects in $Na(3^2P_{3/2})$, $K(4^2P_{3/2})$ + Rare-Gas Systems

Düren and co-workers[12,42] have systematically investigated the left–right asymmetry (related to the angular momentum transfer $L_\perp^+$) in the differential scattering of circularly polarized $Na^*(3^2P_{3/2})$ and $K^*(4^2P_{3/2})$ atoms from rare-gas targets at and around thermal energies. The scattering signal observed adds elastic scattering and fine-structure changing collisions. They use the familiar crossed-beam technique with a cw dye laser to excite the alkali atoms. The scattered alkali atoms are detected with a hot-wire detector that can be moved around the scattering region. The detector can be placed on both sides of the alkali beam, allowing a useful cross check of the experimental results since the asymmetry between measurements with LHC and RHC polarized light as measured at a given angle should be the same as the asymmetry between measurements with the same polarization but on opposite sides of the beam.

Figure 37 shows the difference signals ($I^{ex} - I^{ground}$) for the collision pair K–Ar with the detector on both sides of the beam. The coarse structure of the signal is dominated by the rainbow oscillations caused by scattering on the excited $A^2\Pi$ potential. Since the ground-state scattering has no structure for angles larger than 5° and only contributes a nearly constant background, all the features in the difference signal can be attributed to the scattering in the excited state. For angles less than about 5°, the difference signal becomes negative because in this region the ground-state rainbow maximum leads to a larger scattering signal than from the excited state.

The asymmetry is seen to be strongly oscillating and appears as an additional oscillatory structure in one of the two signals compared with the other one. As expected, the two excitations exchange their roles if the detector is placed on the opposite side of the beam.

Asymmetry was also observed for the other systems investigated (Na^*, K^* + Ne, Ar, Kr, and Xe), although the particular details of the asymmetry vary considerably for the different combinations. Some interesting systematic

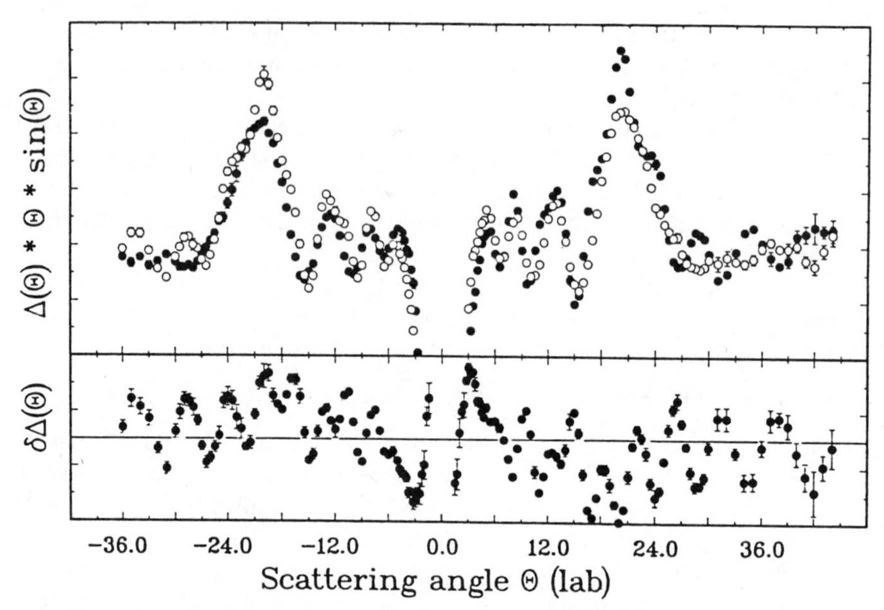

Figure 37. Difference signal ($I^{ex} - I^{ground}$) as function of laboratory scattering angle θ for K* + Ar for excitation with RHC (○) and LHC (●) polarized light directed perpendicularly to scattering plane (upper panel). Position of detector with respect to primary beam indicated by sign of θ. Differences between these scattering signals (lower panel) show oscillatory structure of RHC–LHC asymmetry. Statistical error represented by error bars if larger than symbol used for graph: $E_{col} = 125$ meV. Reproduced from ref. 42.

features were observed: The Na–Ar and K–Ar results are very similar; for both Na and K + Xe, the supernumerary rainbow oscillations in the individual fine-structure cross sections are shifted with respect to each other; and for the Na–Kr and Na–Xe systems, the cross section for excitation with LHC polarized light is a factor of 2 greater than that for RHC polarized light, a phenomenon not observed in the corresponding potassium systems.

Using the accurate potentials available for the K–Ar system[95,96] and a close coupling calculation, Düren et al. were able to fit the experimental data, as shown in Fig. 38. The model calculations have been angle averaged with a Gaussian profile to simulate the experimental resolution. By this averaging procedure, the "rapid osillations" in the theoretical cross sections are smeared out. The total cross sections calculated for both RHC and LHC polarized light result from scattering in three channels: the completely elastic channel ($j' = j, m_j' = m_j$), the fine-structure inelastic channel ($j' = \frac{1}{2}$), and the m_j inelastic channel ($j' = \frac{3}{2}$, $|m_j'| = \frac{1}{2}$). The different contributions are shown in Fig. 39. In the large-scattering-angle region, the elastic and fine-structure transition

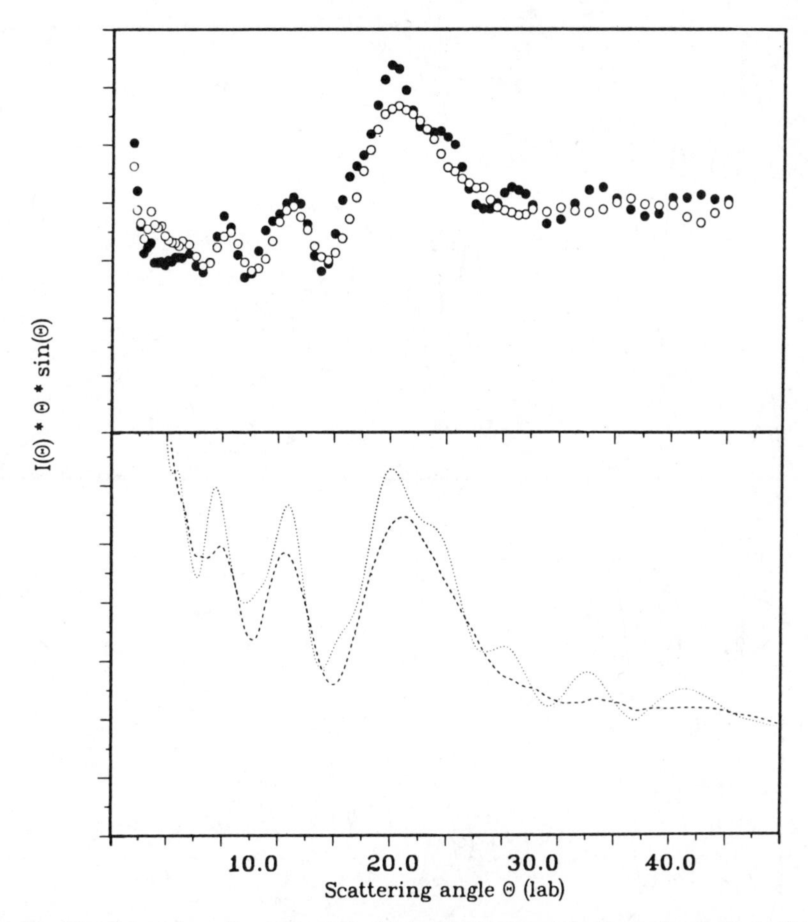

Fig. 38. Comparison of experimental and calculated cross sections for K–Ar using potentials from ref. 96 (RHC: ○ and – –; LHC: ● and ···). Reproduced from ref. 42.

cross sections dominate. Trajectories involving the $A^2\Pi$ and $B^2\Sigma$ potentials interfere and result in Stueckelberg oscillations in the angular dependence of the cross sections, with the elastic and the inelastic one being opposite in phase. A closer look at the large-angle oscillations shows that the additional oscillations observed for LHC polarized excitation (Fig. 37) have the same period as the pattern in the individual cross sections. Comparing the pattern for the two excitation schemes (RHC and LHC), we see that the oscillations for the two main partial cross sections are shifted by about 5° when changing from RHC to LHC excitation. Also, the oscillations are slightly shifted with respect to each other; for the RHC case, they are opposite in phase, but for

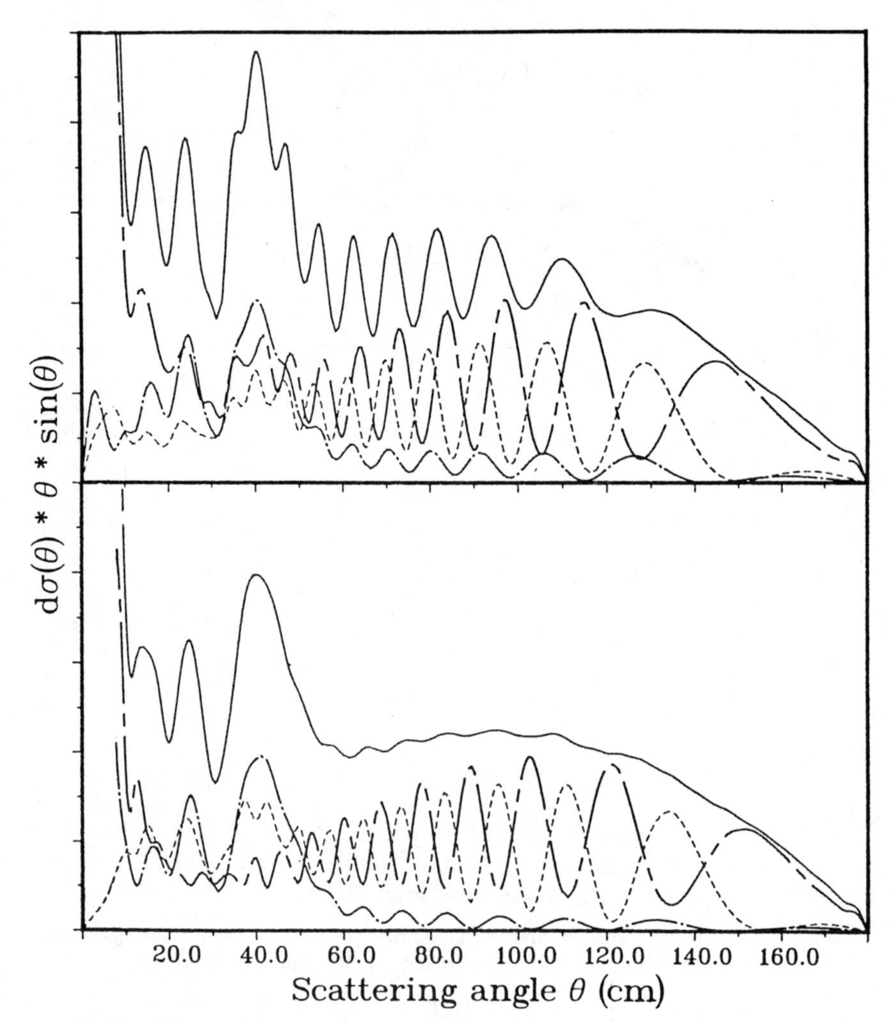

Figure 39. Calculated differential cross section for $K(4^2P_{3/2})$–Ar as function of center-of-mass scattering angle ($E_{col} = 125$ meV). Total and partial cross sections shown for LHC (upper) and RHC (lower): ——, total cross section; – – – –, j and m_j elastic part; - - -, fine-structure inelastic cross section ($j' = \frac{1}{2}$); —·—, j elastic and m_j inelastic part. Reproduced from ref. 42.

the LHC case, they are shifted by one-eighth of a period. This phase shift is induced by the incomplete spatial symmetry of the entrance channel. The appearance of the total cross section depends of course on the phase relation of the partial cross sections. For the RHC case, the oscillations cancel completely, giving the rather smooth, structureless shape of the total cross section for large scattering angles. However, for LHC polarization, the phase relation is such that some residual oscillation appears in the total cross section.

The observed asymmetry varies strongly with the different alkali–rare-gas systems investigated. However, Düren and Hasselbrink[42] were able to show that a model taking into account the different ratios of the alkali fine-structure splittings and the depth of the minimum in the anisotropic part of the potential that determines the coupling between the molecular states is able to reproduce most of the experimental results.

Thus, we have a situation here where the consideration of the spin–orbit interaction is essential for the understanding of the observed effects with circularly polarized excitation in these alkali–rare-gas systems.

Düren has devoted a great deal of ingenuity and computer power to unravel the different contributions to the left–right asymmetry and to reveal the physical origin of the underlying angular momentum transfer. In doing so, the "natural coordinate frame" (see Fig. 3) used throughout this chapter again proved to be very useful. Düren has shown that by choosing the z axis perpendicular to the collision plane one may break the partial wave expansion of the total scattering wave function into two subsets, which do not couple, each of which belongs to a set of atomic states with well-defined, opposite reflection symmetry: the $|j=\frac{3}{2}, m=\frac{3}{2}\rangle|j=\frac{3}{2}, m=-\frac{1}{2}\rangle$ and the $|j=\frac{1}{2}, m=+\frac{1}{2}\rangle$ state belonging to one set and the remaining three states to the other. Both sets are subject to slightly, but distinctively, different interaction potentials, notably, among others, different centrifugal potentials. Düren has shown that this difference in the centrifugal potential is a necessary condition for the left–right asymmetry to occur, whereas the other ingredients, such as spin–orbit coupling, determine its shape and magnitude.

In the spirit of our previous discussion, we could interpret these different centrifugal potentials as connected with the effectively repulsive or attractive behavior of a trajectory if a semiclassical picture were to be used. And we could again say, in the spirit of Komohto and Fano,[97] that the attractive or repulsive nature of the interaction is one determinant for the sign of the orientation—although, of course, it does not allow us to predict this sign, which in fact changes with the scattering angle.

In total, these spin–orbit effects seem to be clouding the nice clear picture of polarization effects outlined in the previous sections. One should, however, bear in mind that the left–right asymmetry (i.e., the angular momentum transfer) is a quantity that is highly sensitive to small phase changes. Such phase changes can occur even at very large internuclear distances, where the main structure of the atomic charge cloud remains undisturbed. Thus, from a more chemical viewpoint, where one is perhaps more interested in this shape (i.e., the σ or $\pi^{\pm}$ character of the orbitals involved), there is hope that such effects are less significant, and one may still use the more intuitive pictures, such as the locking concept, even though they may be oversimplified. At present, the situation is not finally clarified, and further experimental and theoretical studies analyzing linear polarization effects are needed.

VI. CHARGE EXCHANGE IN $Na^+ + Na^*$

We now return to our simple model sodium system, this time to investigate the polarization effects in resonant charge exchange between laser-excited sodium atoms and singly charged positive sodium ions.

A schematic drawing of the apparatus is shown in Fig. 40. An energy-selected Na^+ beam crosses a sodium atom beam at right angles. The sodium atoms in the scattering center are excited to the $Na^*(3^2P_{3/2}, F = 3, M_F)$ states by either circularly or linearly polarized light. A position-sensitive microchannel plate detector is used to detect the charge-exchanged atoms, any residual ions being removed from the neutral beam by a deflecting electric field.

The scattering geometry is shown in Fig. 41 for both linearly and circularly polarized light. The figure depicts the charge distributions of excited atoms projected onto the detection surface of the channel plates. With such a detector, it is possible to measure both the polar (θ_{col}) and azimuthal (ϕ)

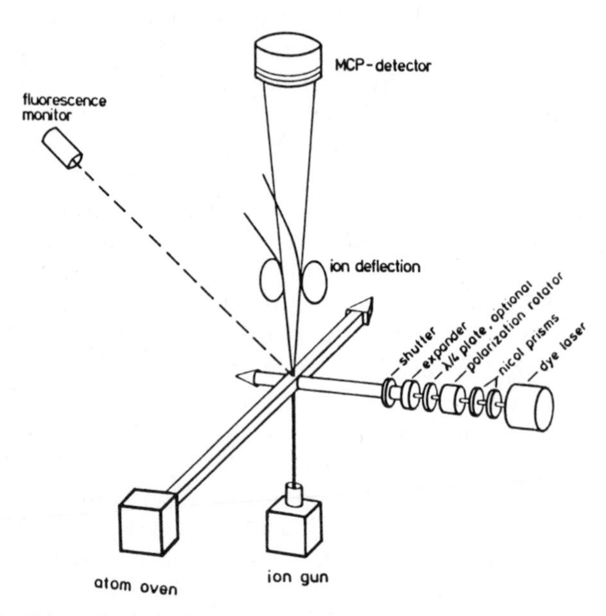

Figure 40. Schematic diagram of apparatus used to investigate resonant charge transfer in $Na^+ + Na^*(3p)$ system.

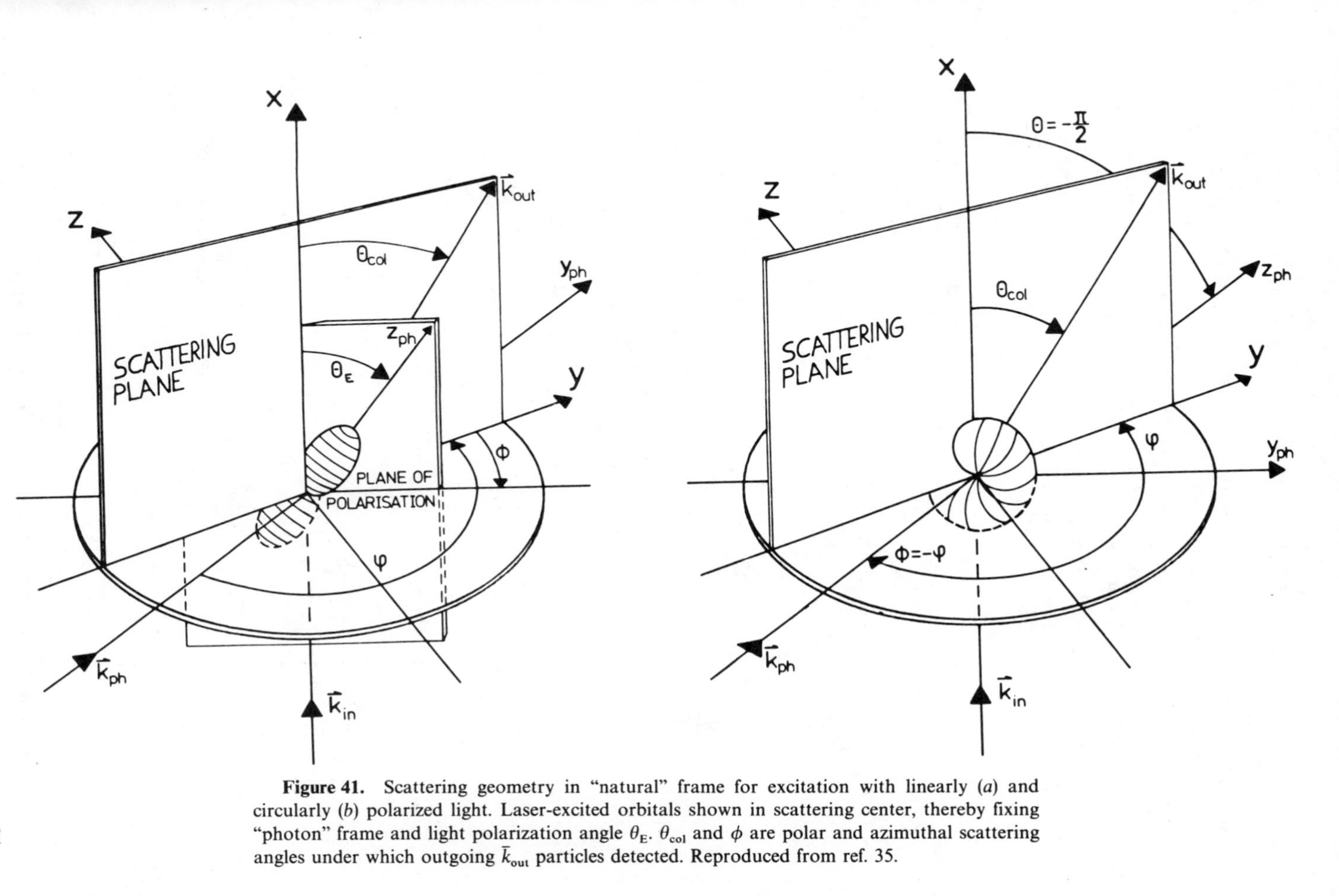

Figure 41. Scattering geometry in "natural" frame for excitation with linearly (*a*) and circularly (*b*) polarized light. Laser-excited orbitals shown in scattering center, thereby fixing "photon" frame and light polarization angle θ_E. θ_{col} and ϕ are polar and azimuthal scattering angles under which outgoing $\bar{k}_{out}$ particles detected. Reproduced from ref. 35.

angular distributions simultaneously. We obtain the four alignment and orientation parameters by fitting Eq. (7) and the corresponding equation for excitation with circularly polarized light to the experimental data.[35]

The charge transfer differential cross sections for both ground- and excited-state sodium for $E_{cm} = 100\,\mathrm{eV}$ are shown in Fig. 42. To obtain these plots, we have integrated over the azimuthal angle ϕ and averaged over the polarization state of the atom. The differential cross section for excited sodium

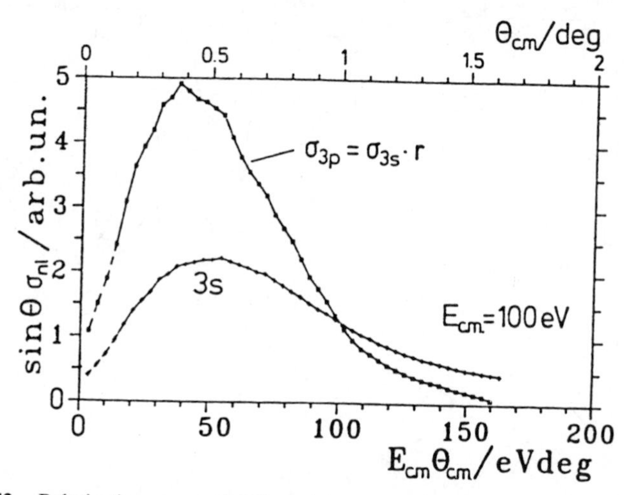

Figure 42. Polarization-averaged differential cross sections σ_{gr} and σ_{ex} for charge transfer from ground- and excited-state sodium. Reproduced from ref. 35.

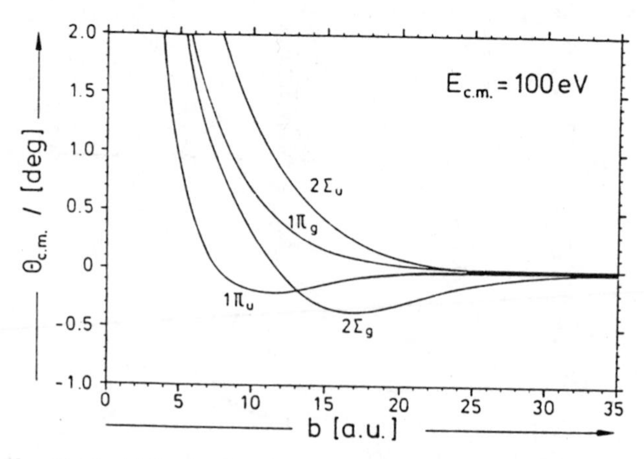

Figure 43. Classical deflection functions for elastic scattering of Na^+ from $Na^*(3p)$.

shows a broad maximum at about 35 eV°. If we consider the classical deflection functions appropriate to this process (Fig. 43) we see that scattering from two of the potentials, $1\Pi_u$ and $2\Sigma_g$ should give rise to rainbow maxima at approximately 20 and 40 eV°, respectively. The broad peak will be seen to be a combination of these two rainbow contributions.

The alignment parameters (γ, P_l^+, ρ_{00}) determined by linear polarization studies are plotted in Fig. 44 for $E_{cm} = 100$ eV. At small scattering angles, $\tau = E_{cm}\theta_{cm} > 20$ eV°, the alignment angle γ is very small, indicating that

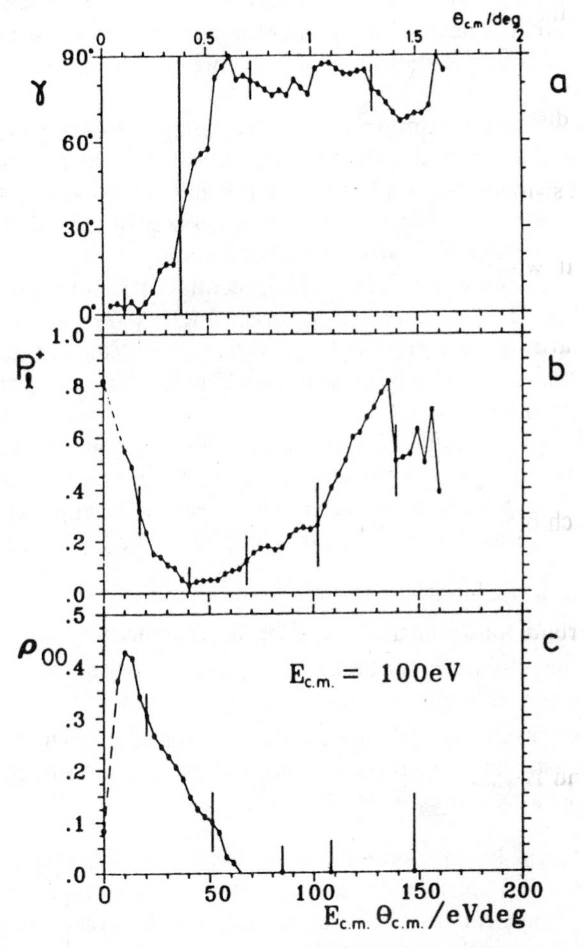

Figure 44. Alignment parameters γ, P_l^+, and ρ_{00} for $E_{cm} = 100$ eV. Upper and lower abscissae show scattering angle $\theta_{cm} = \frac{1}{2}\theta_{lab}$ and reduced scattering angle $\tau = E_{cm}\theta_{cm}$, respectively. Reproduced from ref. 35.

a pσ atomic orbital prepared at a large internuclear separation of the interacting particles is most efficient for the charge transfer process. This is a substantial effect, as shown in Fig. 44*b*, with $P_l^+ \approx 0.4$ at $\tau = 15\,\text{eV}^\circ$, which implies that p$\sigma$ is more than twice as effective for charge transfer than pπ^+. As the scattering angle increases, both orbitals become more and more equivalent until at around $\tau = 40\,\text{eV}^\circ$, P_l^+ almost vanishes, and γ is no longer well defined, as indicated by the large error bar computed in this range of scattering angles. The anisotropy increases again as the scattering angle is increased further. However, this time, γ is nearly orthogonal to the situation at small scattering angles. For $\tau > 60\,\text{eV}^\circ$, we find $\gamma = 80^\circ$ and $P_l^+ \approx 0.8$, which means that a nearly pπ^+ preparation of the atomic orbital at large internuclear distances is about nine times more effective for the process than the pσ orbital.

The rapid switching from pσ at small scattering angles to pπ^+ dominance at larger scattering angles occurs at around the same reduced scattering angle, $\tau \approx 40\,\text{eV}^\circ$, at which the maximum of the differential cross section was observed (see Fig. 42). However, in the discussion of the differential cross section, we attributed the pronounced maximum to a combination of a rainbow caused by the molecular $1\Pi_u$ potential at small scattering angles ($20\,\text{eV}^\circ$) and by the $2\Sigma_g$ potential at larger angles ($40\,\text{eV}^\circ$). Thus, it would appear that a pσ preparation at large internuclear distances populates the Π states while a pπ^+ orbital transforms into Σ molecular states as the particles approach each other.

This is a different situation from that observed in the inelastic and superelastic transitions with the same collision partners discussed in Section III. There, orbital following appeared to be a reasonable approximation. The charge transfer process is rather more complex in the following ways:

both g and u potentials participate,

both Σ and Π states may be found in the exit channel,

several different impact parameters contribute to one scattering angle and may be very different in value (see Fig. 43), and

the impact parameters of importance are typically much larger than in the inelastic and superelastic cases and are comparable in magnitude to the locking radius R_L (≈ 20 a.u.).

The concept of a locking radius can no longer be meaningful for impact parameters $b \approx R_L$. Consider, for example, that we have prepared a pσ orbital at a large internuclear distance and are interested in very small scattering angles, $\approx 0.2^\circ$. From the deflection function (Fig. 43), we see that on the $2\Sigma_u$ curve this implies an impact parameter $b \approx 20$ a.u., which is about the magnitude of the locking radius. Thus, the orbital will move space fixed until it reaches the distance of closest approach, $R = b$, and the system will find itself in a Π_u state so that we are left with the situation depicted in Fig. 9*a*. In this way, the observed pσ dominance at small scattering angles in combination with the theoretically expected $1\Pi_u$ rainbow can be qualitatively understood.

The angular dependence of ρ_{00}, the relative probability of the π^- contribution to the charge transfer, shown in Fig. 44*c* confirms this. Since reflection symmetry is conserved, transitions to Σ states are possible for this initial preparation of the atomic orbital, as illustrated in Fig. 9*c*, and we thus have an image of pure Π scattering. This is largest for the small scattering angles and reaches almost zero for $\tau > 60\,\text{eV}^\circ$ as we go beyond the maximum of the differential cross section. This clearly shows that the predominance of

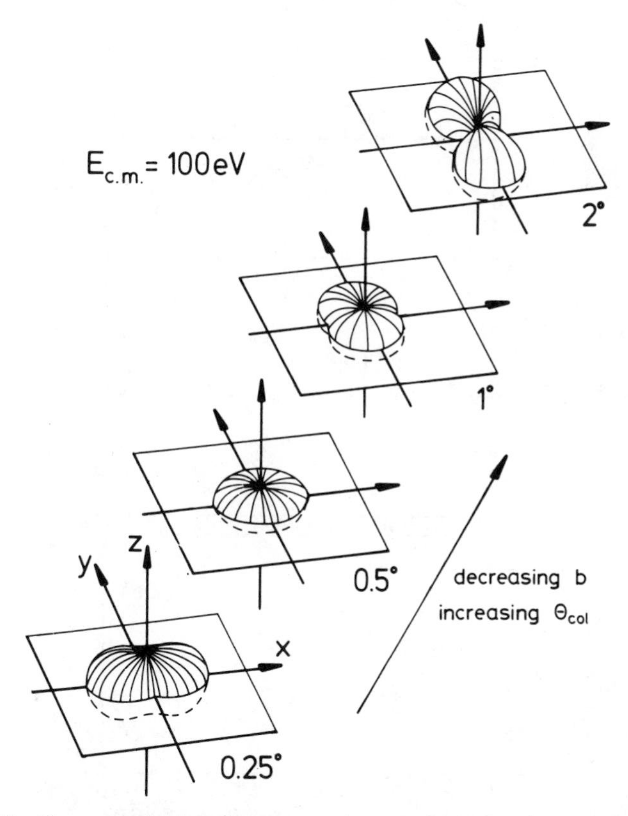

Figure 45. Charge distribution after resonant charge exchange from isotropically populated Na*(3p) state as derived from alignment parameters (Fig. 44) for different scattering angles with $P_l^+ = 2|\rho_{1-1}|/(1-\rho_{00}) = (1-w)/(1+w)$ and $\rho_{00} = h/(1+w+h)$. Reproduced from ref. 35.

the pπ^+ orbital preparation at larger scattering angles has to be seen as a combined effect of the Σ potentials and rotational coupling.

Figure 45 summarizes the experimental results derived from the studies with linearly polarized light showing the shape of the charge cloud at different values of $\theta_{c.m.}$.

The angular momentum transfer $L_\perp^+$ obtained by studies with circularly polarized light is plotted in Fig. 46 for three different collision energies. The effect is fairly small but significant and shows an interesting structure. At small scattering angles, $L_\perp^+$ is positive, reaching a maximum of 0.04–0.12$\hbar$ at around $\theta_{cm} \approx 1°$. It then rapidly changes to become negative, crossing the axis at $\approx 1.1°$, and appears to reach a plateau with a value of about $-0.08\hbar$.

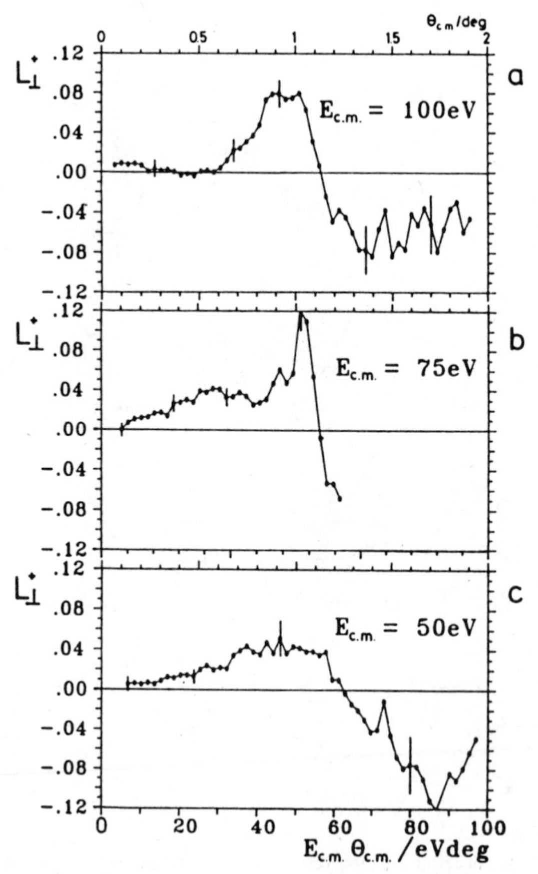

Figure 46. Angular momentum transfer $L_\perp^+$ for collision energies E_{cm} of 100, 75, and 50 eV. Reproduced from ref. 35.

We saw in Section III that relatively large values of $L_\perp^+$, up to ≈ 0.4, were observed in the inelastic (superelastic) studies and that for a Π final state the sign was positive, but for the same conditions and a Σ final state the sign of $L_\perp^+$ was negative. In the charge transfer process, there are contributions from both Π and Σ final states, and we could expect the two contributions to cancel, explaining the small values of $L_\perp^+$ observed. However, we do see a small positive $L_\perp^+$ in the low-angle scattering regime ($\leqslant 1°$), where the scattering is dominated by the Π potentials, and a small negative value at larger angles ($> 1.1°$), where the Σ potentials dominate. Further studies are in progress.

VII. ELECTRONIC-TO-ROVIBRATIONAL ENERGY TRANSFER

In this section we want to discuss, in some detail, the effects of the alignment and orientation of the atomic charge cloud in a thermal nonadiabatic molecular collision process:

$$A^* + BC \rightarrow (ABC)^* \rightarrow (ABC)^\# \rightarrow A + BC^\#, \tag{16}$$

where $^\#$ indicates rotational–vibrational excitation, that is, the electronic-to-rovibrational (E–VRT) energy transfer from an electronically excited atom A* to a diatomic molecule BC. This general process is a very important model case in molecular dynamics, and substantial progress has been made in its understanding in recent years (see, e.g., refs. 80 and 98–101). Time-resolved studies and the use of lasers[102,103] have allowed the product state distribution to be identified, whereas crossed-molecular-beam[104] experiments have given the angular distributions of the scattered products[105] as well as angular and energy-resolved cross sections for the quenching of Na* by various diatomic[106–109] and triatomic and polyatomic[110,111] molecules.

Alignment and orientation studies have been carried out in our laboratory for the Na* + N_2 system.[14] The results of these experiments provide a rigorous test of the available theory as well as give some insight into the rather complicated collision process.

A schematic view of the apparatus is shown in Fig. 47. It consists of crossed beams of sodium atoms and nitrogen molecules. The scattered sodium atoms are detected by a hot-wire detector after velocity analysis with a mechanical selector. The molecular beam is supersonic with an angular divergence less than $\pm 1°$. For the Na* + N_2 system, the initial center-of-mass kinetic energy is $E_{cm} = 0.16$ eV. The estimated angular resolution of the experiment is $\pm 3°$, whereas the energy resolution is better than one vibrational energy quantum of N_2 (~ 0.3 eV). The laser beam used to excite the Na atoms intersects the two beams at right angles propagating either *perpendicularly* to the scattering

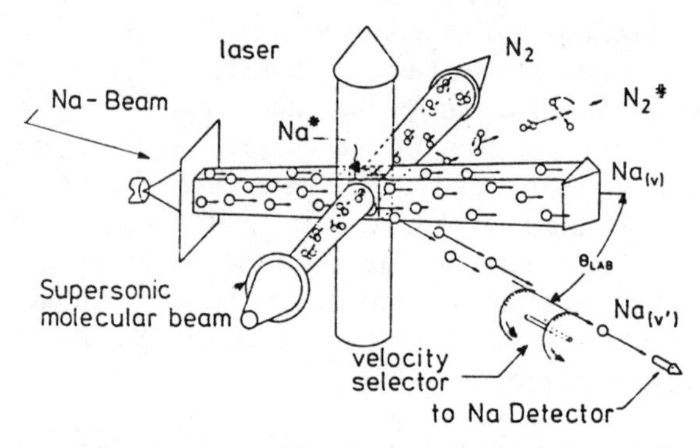

Figure 47. Schematic diagram of apparatus used to investigate electronic–vibrational energy transfer in $Na^* + N_2$ system

plane (as depicted in Fig. 47) for either circular or linear polarization or propagating in the scattering plane for linear polarization with the electric vector of the light either in plane or perpendicular to it. The experimental geometry is depicted in Fig. 48. The scattering plane is as usual defined by the relative velocities before (v_{rel}) and after (v'_{rel}) the collision. The alignment

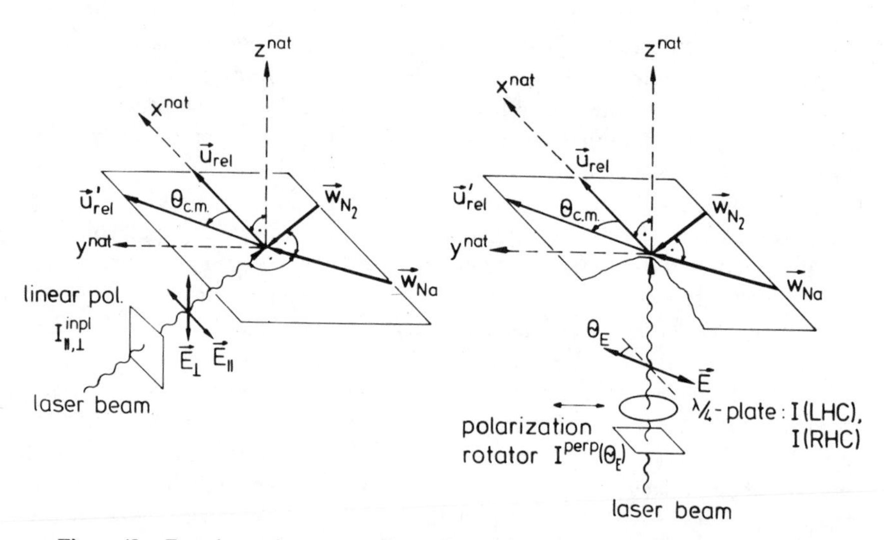

Figure 48. Experimental geometry. Rate of particles scattered under cm scattering angle θ_{cm} called $I^{perp}(\theta_E)$ if Na beam is excited by linearly polarized light propagating perpendicular to scattering plane, I(LHC) or I(RHC) if circularly polarized light used, and $I_{\parallel}^{inpl}$ or $I_{\perp}^{inpl}$ for linearly polarized light propagating in scattering plane having its electric vector in or perpendicular to scattering plane, respectively. Reproduced from ref. 14.

parameters $(\gamma, P_l^+, \rho_{00})$ are obtained from Eq. (7) by carrying out a least-squares fitting procedure on the measured intensity ratios:

$$\frac{I^{\text{perp}}(\theta_E)}{I_\perp^{\text{inpl}}} = \frac{1 - a_0/4[1 - 3\rho_{00} + 3(1-\rho_{00})P_l^+ \cos 2\gamma \cdot \cos 2(\theta_E - \gamma)]}{1 - a_0/2(1 - 3\rho_{00})} \tag{17}$$

where the observed intensity $I^{\text{perp}}(\theta_E)$ is given as

$$I^{\text{perp}}(\theta_E) = \tfrac{1}{2}(I^{\max} + I^{\min}) + \tfrac{1}{2}(I^{\max} - I^{\min}) \cos 2(\theta_E - \gamma). \tag{18}$$

A typical signal is shown in Fig. 49. This "raw" data gives the ratio $I^{\max}/I^{\min}$ and also yields a value for γ directly.

The angular momentum transfer $L_\perp^+$ is obtained by measuring the difference in scattering intensity with LHC and RHC polarized light:

$$\frac{I(\text{LHC}) - I(\text{RHC})}{I(\text{LHC}) + I(\text{RHC})} = \frac{3|O_0|L_\perp^+(1-\rho_{00})}{2 + a_0(1 - 3\rho_{00})}. \tag{19}$$

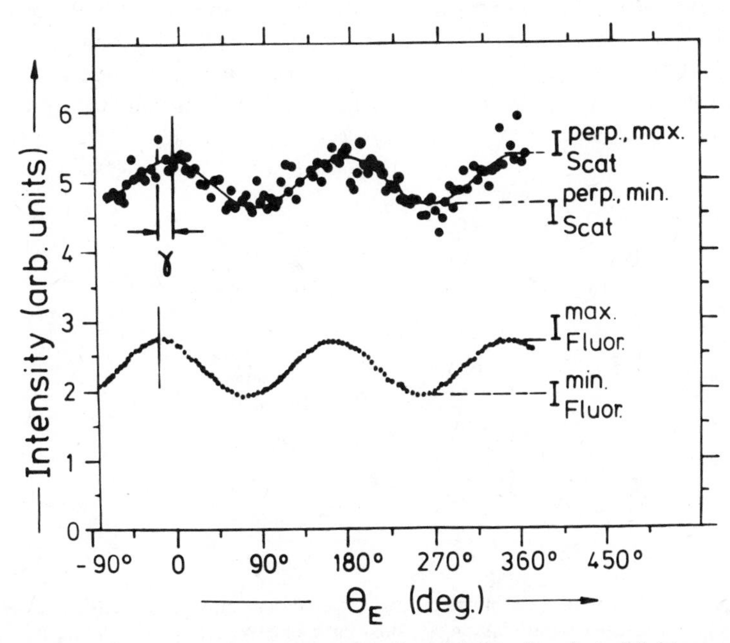

Figure 49. Upper curve: Raw data for excitation with linearly polarized light propagating perpendicular to scattering plane. Lower curve: Fluorescence intensity that monitors optical pumping process and is seen to have expected $\cos 2\theta_E$ behavior. Reproduced from ref. 14.

The present situation differs substantially from the previous systems discussed in this chapter since the molecular alignment would be expected to play a role in the outcome of the collisions. We can only control the alignment and orientation of the excited atom, the molecular alignment; that is, the relative position of the intramolecular axis with respect to the internuclear axis separating the atom and molecule can only be described by an appropriate statistical distribution. However, in spite of this inherent averaging present in the experiment, we can still observe substantial polarization effects, as Fig. 49 indicates.

The relevant potential energy surfaces for the NaH_2[112] and NaN_2[113,114] systems have been calculated and are an essential requirement for the understanding of the collision dynamics. The model based on these quantum chemical calculations will be outlined in what follows. The surfaces are, of course, dependent on three coordinates: R, the distance between the atom and the center of mass of the molecule; r, the molecular bond length; and ε, the angle between R and r. A schematic view of a cut through the potential energy surfaces of the ground ($\tilde{X}A'$) and first excited states ($\tilde{A}A'$, $\tilde{A}A''$, and $\tilde{B}A'$) is shown in Fig. 50 for a geometry close to C_{2v} ($\varepsilon \sim 90°$). The most recent calculations[114] that determined the energy surfaces by ab initio configuration interaction treatments in the framework of the multireference doubly excited configuration interaction method (MRDCI) showed that the quenching process occurs most favorably for the C_{2v} configuration and geometries close to this. A crossing can also occur for the collinear approach,

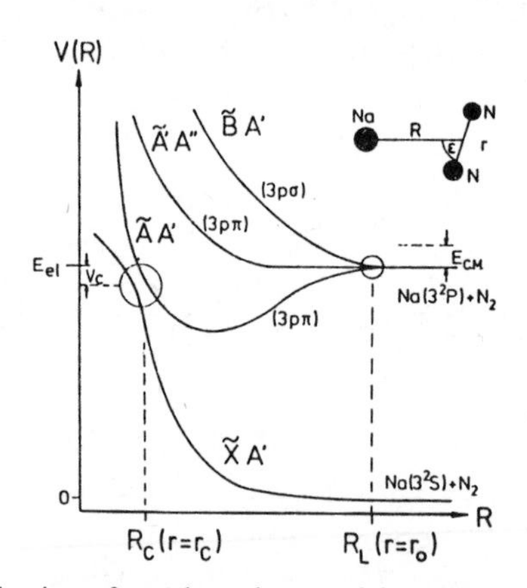

Figure 50. Schematic view of cut through potential energy surfaces of NaN_2 in ground ($\tilde{X}A'$) and first excited states ($\tilde{A}A'$, $\tilde{A}'A''$, $\tilde{B}A'$) for geometry close to C_{2v}. There is avoided crossing near R_c, r_c, where quenching process can occur. Transition region from space-fixed to body-fixed motion near locking distance R_L also indicated. Note that $r_c > r_0$, r_0 being equilibrium molecular bond distance. Reproduced from ref. 14.

but this is not so favorable energetically, and the region of geometries for which it is accessible is much narrower than for the perpendicular approach. However, at thermal initial energies, neither of these crossings can be reached when the molecular bond length r is at its equilibrium value. As the molecular bond expands, a crossing between the lowest excited state and the ground state becomes accessible. Although this bond stretching will bring the system higher up on the potential surface for large R, for small R, the net result of the *bond stretching* is an *attraction* leading to a crossing below the asymptotic $Na^*(3^2P)$ excitation energy (see Fig. 50). This lowering of the excited-state potential can be regarded as occurring via a partial transfer of the 3p valence electron to the molecule as the internuclear distance R decreases. This forms a potential well for the triatomic system by filling an antibonding orbital of the molecule. Thus, the N_2 bond distance is increased, and the avoided crossing between the excited $\tilde{A}A'$ state and the ground $\tilde{X}A'$ state can be reached. It should be stressed that this is only a *partial* charge transfer; no ionic intermediate state is formed as is the case with collisions between alkali atoms and halogen[115] or alkyl halide[116] molecules.

This crossing is responsible for the observed electronic-to-vibrational energy transfer[101] in collisions with thermal initial energies. The following picture of the collision process based on model calculations for NaH_2[117,118] and NaN_2[119] has been obtained, the first detailed theoretical studies being carried out for the NaH_2 system. Since the initial kinetic energy is low, the approaching excited three-particle system will follow the minimum energy path on the $\tilde{A}A'$ surface ($\tilde{A}\,^2B_2$ in C_{2v}). There, it reaches the crossing with the ground state along an essentially attractive path when the N–N bond is stretched. It will jump with a high probability to the ground state, $\tilde{X}\,^2A_1$, in which the molecular potential is practically unchanged. Thus, after the surface hop, the nitrogen is in a nonequilibrium position and will start to vibrate with an amplitude corresponding to the bond stretch at the crossing point. The nuclear motion is illustrated in Fig. 51*a* for the C_{2v} geometry. Here, the minimum energy path is shown on a contour map of the excited $\tilde{A}\,^2B_2$ state. The electronic transition occurs at the region marked C, where the crossing has its minimum energy. The motion of the particles on the ground-state surface, $\tilde{X}\,^2A_1$, is illustrated schematically in Fig. 51*b*. The trajectory starts at the lowest point of the crossing seam, C, where it has left the excited surface. Since r_c, the molecular bond length at the moment the crossing occurs, is substantially larger than the equilibrium bond length, the molecule finds itself in a vibrationally excited state as the atom and molecule quickly move apart on the repulsive ground-state surface.

The magnitude of the quenching cross section as well as its dependence on initial kinetic energy[80] has been explained qualitatively fairly well[112,101] by the "absorbing sphere" model.[80] This model predicts a maximum impact

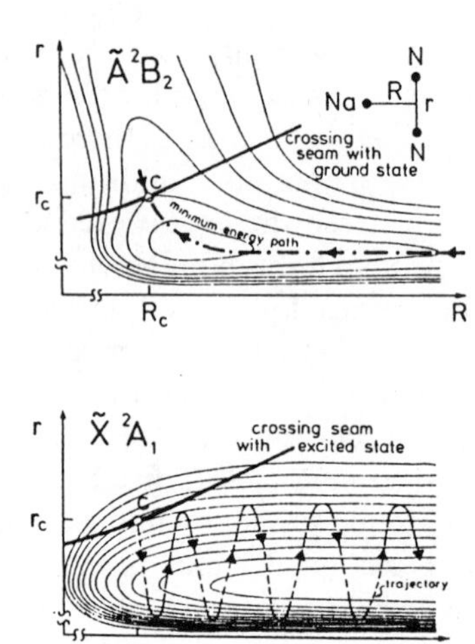

Figure 51. (*a*) Schematic of contour map for first excited NaN_2 state in C_{2v} symmetry ($\tilde{A}\,^2B_2$). Minimum energy path (–·–) and crossing seam (——) with $\tilde{X}\,^2A_1$ ground-state surface indicated. Actual particle trajectory may look as indicated by dashed line. (*b*) Schematic of contour map for ground state of NaN_2 system. Trajectories (– – –) start at crossing C with excited state.

parameter $b_{max} = 8.5$ a.u. and is very nicely confirmed in the case of Na–H_2 by surface-hopping trajectory calculations based on DIM potentials.[117] All trajectories with impact parameters up to this value may reach the crossing seam and potentially lead to quenching. In cell experiments where there is no control over the atomic alignment, it is normally assumed that only one-third of the excited atomic state distribution participates in the quenching, that is, only those atoms with a $3p\pi$ orbital correlating with the $\tilde{A}A'$ molecular state, as indicated in Fig. 50. If one calculates the surface-hopping probability from the "absorbing sphere" model, $\sigma_q \approx \pi b_{max}^2 w/3$, using the experimentally determined quenching cross section ($\sigma_q \approx 18\,\text{Å}$ for $E_{cm} = 0.16\,\text{eV}$),[80] one obtains a surface-hopping probability w of about 0.8,[14] which is larger than the maximum Landau–Zener probability (0.5) for a twofold transition through the crossing. We could thus perhaps be justified in surmising that more than one atomic alignment somehow contributes to the quenching process. However, the large probability could also be due, for example, to trajectories that pass the crossing seam many times during the collision.

The geometries we can prepare in the laboratory in the space-fixed frame are shown schematically in Fig. 52 for large internuclear separations of the atom and molecule. Here, we can see that the two molecular geometries that lead to quenching, B_2 (in C_{2v}) and Π (in $C_{\infty v}$ symmetry), can be formed

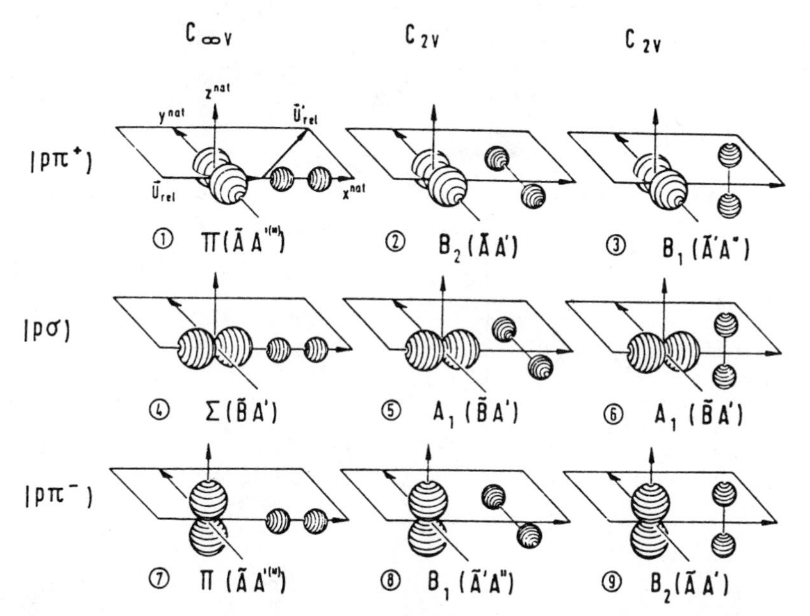

Figure 52. Schematic illustration of possible asymptotic preparations of Na 3p orbitals with respect to molecular alignment. C_{2v} and $C_{\infty v}$ configurations taken as representative of latter, although in reality any linear combination of nine geometries is possible. Experimentally, only atomic alignment can be controlled, allowing $|p\pi^+\rangle$, $|p\sigma\rangle$, and $|p\pi^-\rangle$ preparation with respect to scattering plane as defined by particles' relative velocities before and after collision. Reproduced from ref. 14.

asymptotically in two different ways: by preparation of either a $p\pi^+$ or a $p\pi^-$ atomic state.

The three alignment parameters observed experimentally $(P_l^+, \gamma, \rho_{00})$ are shown in Figs. 54 and 56 for a fixed energy transfer, $\Delta E_{\text{vibrot}} \sim 0.9$ eV, where the cross section is sharply peaked,[120] as seen in Fig. 53. This energy is approximately equivalent to the energy required for exciting the third vibrational level of the nitrogen molecule. All parameters show a significant dependence on scattering angle. The alignment angle γ (Fig. 54) shows a particularly interesting behavior. For forward scattering ($\theta_{\text{cm}} = 0°$), $\gamma = 0°$, which indicates that a $p\sigma$ preparation is most efficient for the quenching process. As the scattering angle increases, γ also increases until, for backward scattering ($\theta_{\text{cm}} = 180°$), $\gamma \sim 90°$, indicating that $p\pi^+$ preparation is most efficient. In the space-fixed frame (Fig. 52), a $p\sigma$ atomic orbital preparation corresponds to either a $^2A_1$ molecular state (C_{2v} symmetry), where the main axis of the atomic electron charge cloud is aligned perpendicular to the internuclear axis of the molecule, or a $^2\Sigma$ molecular state ($C_{\infty v}$ symmetry),

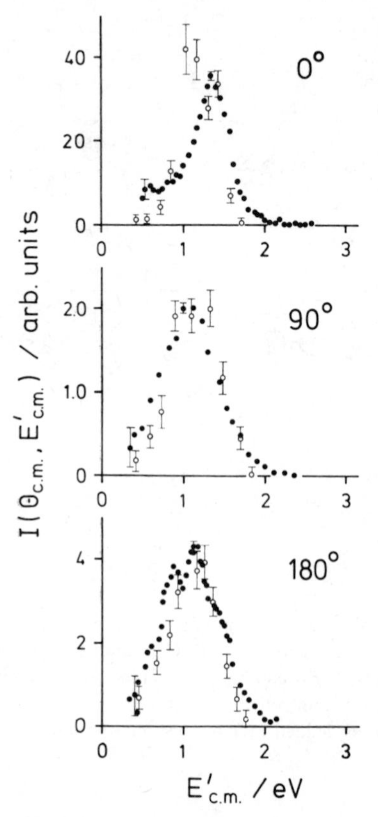

Figure 53. Double differential quenching cross sections for $Na^* + N_2$ at different cm scattering angles as function of final relative kinetic energy E'_{cm} and energy transferred to molecule $\Delta E_{vibrot} = E_0 - E'_{cm}$, with $E_0 = 2.26$ eV. Experimental points[120] given (●) together with Poppe calculations[114] (○).

which has the atomic electron charge cloud aligned along the molecular internuclear axis, neither of which can lead to quenching. Thus, it would appear that we again have a situation where, at a particular distance R_L, there is a transition from space-fixed to body-fixed motion, as indicated schematically in Fig. 50. The maximum quenching probability will be found when the excited sodium atom is aligned at an angle γ such that the space-fixed charge cloud approaches the locking radius R_L in a way that leads to formation of the B_2 molecular state (C_{2v} symmetry) with the p orbital lying in the scattering plane, parallel to the molecular internuclear axis (Fig. 52). This situation is illustrated schematically in Fig. 55 for three different trajectories. The trajectories are essentially attractive before reaching the

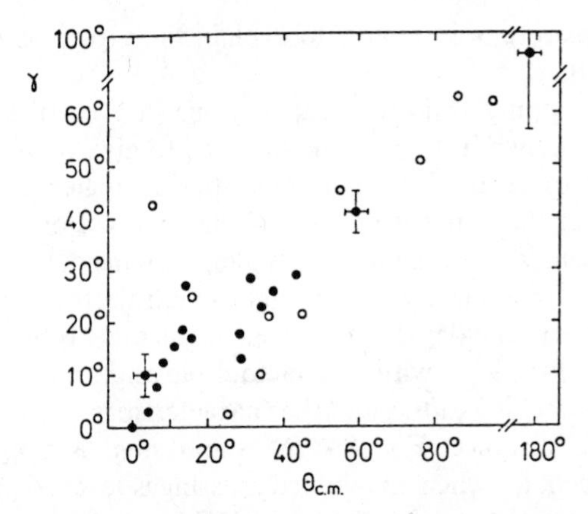

Figure 54. Alignment angle γ for quenching process $Na^* + N_2 \rightarrow Na + N_2^{\#}$ as function of cm scattering angle for final relative kinetic energy $E'_{cm} = 1.35\,eV$, which corresponds to peak in differential cross section. Experimental results (●); theoretical calculations (○).

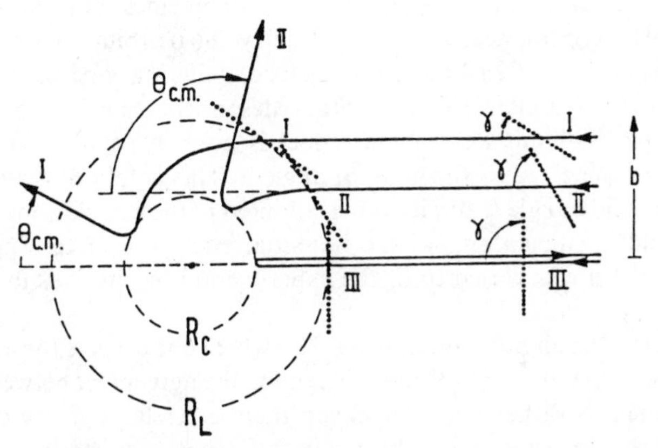

Figure 55. Schematic trajectories to explain atomic alignment angle γ for which maximum quenching cross section found. Atomic orbitals stay space fixed for $R < R_L$. Quenching occurs at R_c and leads to repulsive trajectories on way out. Reproduced from ref. 14.

crossing radius R_c and repulsive after a surface hop to the ground state at R_c. Thus, for trajectories with impact parameters greater than the locking radius, leading to small scattering angles, the system will remain space fixed, and a pσ atomic orbital preparation (i.e., $\gamma = 0$) will lead to the required molecular symmetry. As the impact parameter decreases, the scattering angle θ_{cm} increases, and the alignment angle must also increase for quenching to occur until, for backward scattering, $\theta_{cm} = 180°$, we have $\gamma = 90°$. With this very simplified picture, we are able to explain the experimental finding that $\gamma \approx \frac{1}{2}\theta_{cm}$ (Fig. 54).

The surface-hopping trajectory calculations of γ carried out by Poppe et al.[114] are also shown in Fig. 54 and are seen to give very good agreement with experiment, especially for the larger scattering angles. The model Poppe et al. use involves the concept of the locking radius. For $R > R_L$, only the geometric (Coriolis and centrifugal) coupling is considered. In this region, the sodium atom travels freely on a straight-line trajectory, and the p orbital remains space fixed. The density matrix elements are strongly dependent on these long-range forces,[121] with two factors playing an important role: the magnitude of the locking radius and the characteristics of the electronic wave functions at this distance. For $R < R_L$, the system is treated as moving adiabatically except for when an avoided crossing is reached and the p orbital is locked into the molecular frame. At the locking radius, the product states of the Na(3p) and the $N_2(^1\Sigma_g)$ are projected onto the molecular wave function responsible for the quenching. Since reflection symmetry is conserved in the collision, if the reflection symmetry of the molecule does not change, the p orbital must lie in the molecular plane, which coincides with the scattering (x^{nat}–y^{nat}) plane. In collinear and C_{2v} symmetry, the p orbital is directed along the y^{nat} axis (see Fig. 52). For the geometric configurations between those two extremes (C_s symmetry), the p orbital is described as a linear combination of pσ and pπ^+ orbitals and in the model is given by a rotation of the p orbital about the z^{nat} axis through an angle α. This rotation angle depends on the orientation angle ε, and its value, adopted in the calculations, is guided by the ab initio calculations of the potential energy surfaces. The locking radius R_L, which was varied to fit the experimental results, was found to be 9.8 a.u.

For the smallest angular range, $\theta_{cm} \lesssim 30°$, where one expects the long-range forces to dominate the σ–π coupling dynamics, the agreement between theory and experiment is rather poor. However, there are relatively few calculated points, and the statistics is unfavorable in this range of scattering angles due to the $\sin\theta_{cm}$ weighting factor.

The remaining two alignment parameters are shown in Fig. 56; again, both the experimental and theoretical results are displayed and show very good agreement.

For forward scattering ($\theta_{cm} = 0$), $P_l^+ = \frac{1}{3}$, which tells us that the pσ preparation ($\gamma = 0$) is twice as efficient as the pπ^+ preparation for quenching. This anisotropy decreases as the scattering angle increases until, for scattering at $\theta_{cm} = 90°$, P_l^+ is approximately 0.1 and the pσ and pπ^+ atomic orbitals are contributing almost equally to the quenching process. This is what we would expect from our naive locking model picture. For a scattering angle of 90° c.m. we have $\gamma \sim 45°$ (see Fig. 54), and our atomic p orbital will be a

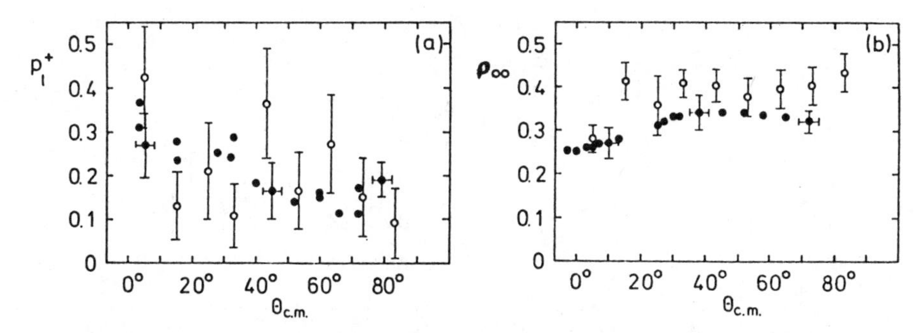

Figure 56. $P_l^+(a)$ and $\rho_{00}(b)$ for quenching process $Na^* + N_2 \rightarrow Na + N_2^\#$ as function of cm scattering angle for fixed relative kinetic energy $E'_{cm} = 1.35$ eV. Experimental results (●); theoretical calculations (○).

linear combination of approximately equally weighted σ and π^+ atomic orbitals (see Fig. 55).

The relative height of the charge cloud, ρ_{00}, is shown in Fig. 56*b*. If we neglect spin–orbit effects, the magnitude of ρ_{00} gives the relative cross section for a change in the molecular reflection symmetry. This is because in the atomic transition $|3p\pi^-\rangle \rightarrow |3s\rangle$ the reflection symmetry of the atom changes, and as the total reflection symmetry of the system has to be conserved during the collision, the reflection symmetry of the molecule must also change ($\Delta N + \Delta M_N =$ odd, where N is the angular momentum quantum number of the molecule). This parameter, ρ_{00}, has a value of ~ 0.25 for forward scattering and rises to ~ 0.33 for scattering angles $\theta_{cm} \gtrsim 30°$. Thus for scattering angles between $\sim 30°$ and 90° c.m. (where $P_l^+ \sim 0.15$), all three atomic orbital preparations, pσ, pπ^+, and pπ^-, contribute fairly equally to the quenching with the ratio pσ:pπ^+:pπ^- = 0.39:0.28:0.33. For forward scattering, the appropriate ratio is pσ:pπ^+:pπ^- = 0.5:0.25:0.25, clearly showing the predominance of the pσ preparation.

Figure 57 shows the experimental values of P_l^+ and ρ_{00} for forward scattering as a function of the final relative kinetic energy E'_{cm}. The theoretical predictions for ρ_{00} are also given. For symmetry reasons, at forward scattering, the relationship $P_l^+ = (1 - 3\rho_{00})/(1 - \rho_{00})$ must hold, and this is indeed the case here (within the bounds of the experimental error bars), which gives a nice check on the experimental results. As mentioned earlier, the differential cross-section peaks for an energy transfer corresponding to about three vibrational quanta of the N_2 molecule (i.e., for a final relative kinetic energy of 1.35 eV) (Fig. 53). At around this energy, P_l^+ has a maximum value of about 0.4, which drops for lower and higher final kinetic energies to a value of ~ 0.17. On the other hand, ρ_{00} has its minimum value of 0.25

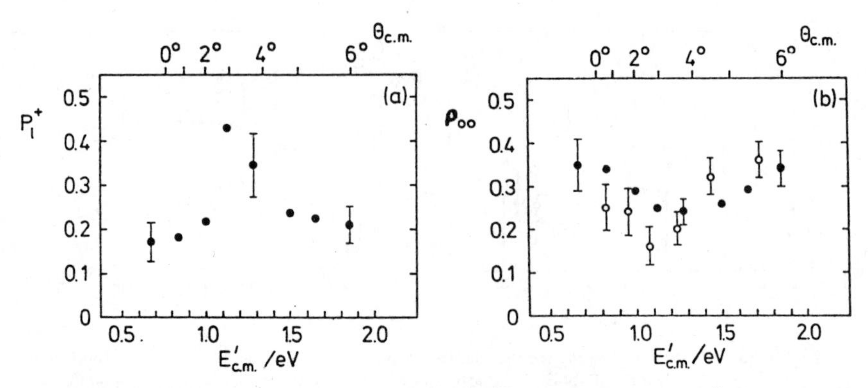

Figure 57. $P_l^+(a)$ and $\rho_{00}(b)$ for quenching process $Na^* + N_2 \rightarrow Na + N_2^\#$ as function of final relative kinetic energy $E'_{\rm cm}$ for forward scattering. Experimental results (●); theoretical calculations (○).

for this energy and increases to ~0.35 for lower and higher final kinetic energies. The differential cross section increases drastically where ρ_{00} decreases. Therefore, the cross section for reflection symmetry changing processes must increase less strongly than the others for $\theta_{\rm cm} \simeq 0°$ and $E'_{\rm cm} = 1.35$ eV. Figure 52 shows that the preparation of the different geometries at large internuclear separations is equally probable for both $p\pi^+$ and $p\pi^-$ atomic orbital preparation. The only difference is the reflection symmetry with respect to the scattering plane, and thus the fact that $\rho_{00} < \frac{1}{3}$ indicates that the coupling elements for reflection symmetry changing processes must be smaller than those for conserving both atomic and molecular reflection symmetry.

Finally, the experimental measurements of the orientation parameter $L_\perp^+$ are displayed in Fig. 58 as a function of scattering angle for $E'_{\rm cm} = 1.35$ eV. As we would expect for such a complicated triatomic system, the net transfer of angular momentum observed is small, $|L_\perp^+| \lesssim 0.1$. However, it does show some structure being positive for small scattering angles, reaching a peak at $\theta_{\rm cm} = 5°$ and crossing the axis to negative values at $\theta_{\rm cm} \simeq 15°$. As discussed in Section III, the angular momentum transfer depends on the magnitude and sign of the phase difference of the potentials involved, the symmetry of the relevant states, and whether the interaction is effectively repulsive or attractive. At small scattering angles, both attractive and repulsive trajectories contribute to the scattering in the quenching process, whereas at larger scattering angles, only repulsive trajectories will contribute. Thus, one explanation of the experimental results would be that at low scattering angles the attractive trajectories dominate, giving, in this case, a net positive angular momentum transfer. However, for the larger scattering angles with repulsive

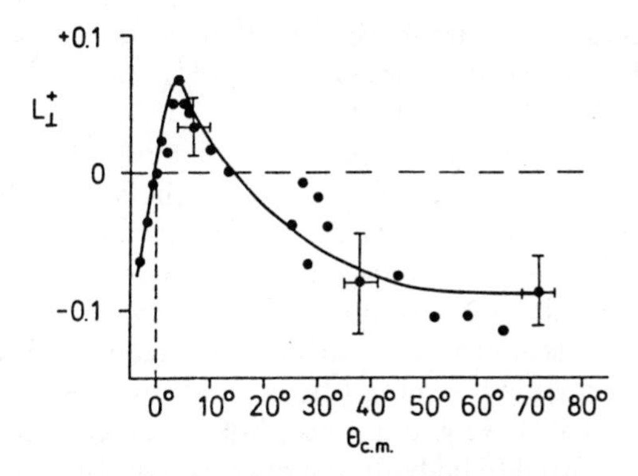

Figure 58. Angular momentum transfer observed experimentally for quenching process $Na^* + N_2 \rightarrow Na + N_2^{\#}$ as function of cm scattering angle for fixed relative kinetic energy $E'_{cm} = 1.35\,eV$.

trajectories, the angular momentum transfer will be negative. This is, of course, a very tentative picture, and in reality, the situation may be much more complicated. Unfortunately, there are no theoretical predictions of $L_\perp^+$ available at present.

In this section, we have shown that polarization effects exhibit an interesting behavior even in such complicated situations as atom–molecule collisions, where the molecular alignment can only be described statistically. The classical model developed by Poppe et al.[114] gives a nice description of the larger scattering angle part ($\theta_{cm} \gtrsim 30\,ev°$), the collisions where the long-range dynamics do not play such an important role. The concept of a locking radius is not quite so straightforward in this system as it was, for example, for the inelastic and superelastic scattering in the $Na^+ + Na^*(3p)$ system discussed in Section III since there may well be more than one locking radius involved due to the different molecular alignments in the quenching process. In spite of these complications, the classical calculations involving only two fitting parameters (R_L and α) give surprisingly good agreement with the experimental results and are a very encouraging first step along the path to a full quantum mechanical understanding of the quenching process in these collisions.

VIII. CONCLUSION

In this chapter, we have tried to illustrate the stimulating possibilities of orbital polarization studies in the investigation of elementary molecular dynamic processes. Without attempting to be comprehensive, we hope to have shown a few rather typical situations in which new insights into the underlying mechanisms could be gleaned and to have indicated where future studies along these lines may be informative. We have also indicated the limitations of these types of studies and the areas of research in which further efforts are needed.

The situation appears to be very well understood in atom–atom or atom–ion collisions at suprathermal energies in the range between a few and some hundred electronvolts. At large internuclear distances, the polarized atomic orbital stays essentially space fixed, whereas at small separations of the interacting particles, a description in terms of the quasi-molecule formed during the collision is appropriate. This may be treated rather rigorously by semiclassical calculations involving rotational coupling for the electron orbital motion. Qualitatively, the concept of a locking radius at which the change from space-fixed to body-fixed motion is thought to occur has proved to be very useful and intuitively appealing, although certainly somewhat oversimplified.

At present, the step from this model picture to something one may call orbital stereospecifity in a chemical reaction still seems somewhat uncertain. There seems hope that with some further efforts, both on the theoretical as well as on the experimental side, one may sort out what in more complicated situations may still be described in the framework of atomic orbital dynamics and what has to be ascribed to truly quantum mechanical effects. Success in understanding the polarization effects in the electronic-to-vibrational energy transfer between an excited atom and a diatomic molecule is very encouraging. For more complex systems and real chemical reactions, one may have to be content with more hand-waving models. The goal of the types of studies discussed here, focusing on small model systems, is to provide the guidelines to understanding, probing, and perhaps even manipulating individual reaction dynamics by making use of polarization methods in laser chemistry.

References

1. R. W. Wood, *Verh. d. Deutsch. Phys. Ges.* **13**, 72 (1911).
2. R. J. Donovan, *Prog. React. Kinet.* **10**, 253 (1979).
3. J. W. McGowan (ed.), *The Excited State in Chemical Physics*, Wiley, New York, 1975.
4. K. P. Lawley (ed.), *Dynamics of the Excited State*, Wiley, New York, 1982.
5. J. L. Picque, G. Spiess, and F. J. Wuilleumier (eds), *Atomic and Molecular Collisions in a Laser Field, Journal de Physique*, Colloque C1 **46** (1985).
6. Springer Proceedings in Physics Vol. 1, N. K. Rahman, C. Guidotti, and M. Allegrini (eds.), *Photons and Continuum States of Atoms and Molecules*, Springer-Verlag, Berlin, 1987.
7. C. H. Greene and R. N. Zare, *Ann. Rev. Phys. Chem.* **33**, 119 (1982).
8. C. H. Greene and R. N. Zare, *J. Chem. Phys.* **78**, 6741 (1983).
9. P. L. Houston, *J. Phys. Chem.* (to be published).
10. I. V. Hertel and W. Stoll, *J. Phys. B* **7**, 583 (1974).

11. I. V. Hertel and W. Stoll, *Adv. At. Mol. Phys.* **13**, 113 (1978).
12. R. Düren, E. Hasselbrink, and H. Tischer, *Phys. Rev. Lett.* **50**, 1983 (1983).
13. A. Bähring, I. V. Hertel, E. Meyer, and H. Schmidt, *Z. Phys. A* **312**, 293 (1983).
14. W. Reiland, G. Jamieson, H.-U. Tittes, and I. V. Hertel, *Z. Phys. A* **307**, 51 (1982).
15. C. T. Rettner and R. N. Zare, *J. Chem. Phys.* **77**, 2416 (1982).
16. P. S. Weiss, J.-M. Mestdagh, H. Schmidt, F. Vernon, M. H. Covinsky, B. A. Balko, and Y. T. Lee, Recent Advances in Molecular Reaction Dynamics, conference proceedings, R. Vetter and J. Vigue (eds.), Paris, 1986, p. 15.
17. S. R. Leone, *Ann. Rev. Phys. Chem.* **35**, 109 (1984).
18. G. Rahmat, J. Verges, R. Vetter, F. X. Gadea, M. Pelissier, and F. Spiegelmann, Recent Advances in Molecular Reaction Dynamics, conference proceedings, R. Vetter and J. Vigue (eds.), Paris, 1986, p. 225.
19. J. Lorenzen, H. Hotop, and M.-W. Ruf, *Z. Phys. D* **1**, 321 (1986).
20. W. Bussert, T. Bregel, R. J. Allan, M.-W. Ruf, and H. Hotop, *Z. Phys. A* **320**, 105 (1985).
21. W. Bussert, *Z. Phys. D* **1**, 321 (1986).
22. T. Bregel, W. Bussert, J. Ganz, H. Hotop, and M.-W. Ruf, *Electronic and Atomic Collisions*, Papers of the Fourteenth International Conference, D. C. Lorentz, W. E. Meyerhoff, and J. R. Peterson (eds.), Elsevier, New York, 1986, p. 577.
23. H. Waibel, W. Bussert, M.-W. Ruf, and H. Hotop, *Electronic and Atomic Collisions*, Abstracts of Papers from the Fourteenth International Conference, M. J. Coggiola, D. L. Huestis, and R. P. Saxon (eds.), Elsevier, New York, 1986, p. 376.
24. H. Waibel, M.-W. Ruf, H. Hotop, and W. Meyer, ICAP-X, Tokyo, 1986 (book of abstracts).
25. S. Runge, A. Pesnelle, M. Perdrix, and G. Watel, *Phys. Rev. A* **32**, 1412 (1985).
26. M. J. Verheijen and H. C. W. Beijerinck, *Chem. Phys.* **102**, 255 (1986).
27. L. Barbier, A. Pesnelle, and M. Cheret, *J. Phys. B* **20**, 1249 (1987).
28. L. Barbier and M. Cheret, *J. Phys. B* **20**, 1229 (1987).
29. J. Macek and D. H. Jaecks, *Phys. Rev. A* **4**, 2288 (1971).
30. M. Eminyan, K. B. MacAdam, J. Slevin, M. C. Standage, and H. Kleinpoppen, *J. Phys. B* **7**, 1519 (1974).
31. K. Blum and H. Kleinpoppen, *Phys. Rep.* **52**, 203 (1979); K. Blum and H. Kleinpoppen, *Phys. Rep.* **96**, 251 (1983); J. Slevin, *Rep. Prog. Phys.* **47**, 461 (1984).
32. H. W. Hermann and I. V. Hertel, *Comments At. Mol. Phys.* **12**, 61 (1982); V. Kempter, *Comments At. Mol. Phys.* **10**, 287 (1981); N. Andersen and I. V. Hertel, *Comments At. Mol. Phys.* **19**, 1 (1986).
33. N. Andersen, J. W. Gallagher, and I. V. Hertel, *Rep. Prog. Phys.*, (to be published).
34. N. Andersen, J. W. Gallagher, and I. V. Hertel, *Electronic and Atomic Collisions*, Papers of the Fourteenth International Conference, D. C. Lorentz, W. E. Meyerhoff, and J. R. Peterson, and Elsevier, New York, 1986, p. 57.
35. R. Witte, E. E. B. Campbell, C. Richter, H. Schmidt, and I. V. Hertel, *Z. Phys. D* **5**, 101 (1987).
36. I. V. Hertel, *Fundamental Processes in Energetic Atomic Collisions*, NATO Advanced Study Institute, Ser. B., H. O. Lutz, J. S. Briggs, and H. Kleinpoppen (eds.) **103**, 519 (1983).
37. R. Düren, H. O. Hoppe, and H. Pauly, *Phys. Rev. Lett.* **37**, 743 (1976).
38. L. Hüwel, J. Maier, R. K. B. Helbing, and H. Pauly, *Chem. Phys. Lett.* **74**, 459 (1980).
39. L. Hüwel, J. Maier, and H. Pauly, *J. Chem. Phys.* **76**, 4961 (1982).

40. U. Lackschewitz, J. Maier, and H. Pauly, *Chem. Phys. Lett.* **88**, 233 (1982).
41. U. Lackschewitz, J. Maier, and H. Pauly, *J. Chem. Phys.* **84**, 181 (1985).
42. R. Düren and E. Hasselbrink, *J. Chem. Phys.* **85**, 1880 (1986).
43. F. van den Berg and R. Morgenstern, *Chem. Phys.* **90**, 125 (1984); F. van den Berg, R. Morgenstern, and C. Th. J. Alkemade, *Chem. Phys.* **93**, 171 (1985).
44. W. Beyer and H. Haberland, *Phys. Rev. A* **29**, 2280 (1984); Ch. Bender, W. Beyer, H. Haberland, D. Hausamann, and H. P. Ludescher, in *Atomic and Molecular Collisions in a Laser Field*, J. L. Picque, G. Spiess, and F. J. Wuilleumier (eds.), *J. Phys., Colloque C1* **46** (1985).
45. R. D. Levine and D. R. Herschbach (eds.), *J. Phys. Chem.* (to be published).
46. I. V. Hertel, H. Schmidt, A. Bähring, and E. Meyer, *Rep. Prog. Phys.* **48**, 375 (1985).
47. N. Andersen and S. E. Nielsen, *Europhys. Lett.* **1**, 15 (1986).
48. N. Andersen and S. E. Nielsen, *Z. Phys. D* **5**, 309 (1987).
49. S. E. Nielsen and N. Andersen, *Z. Phys. D* **5**, 321 (1987).
50. G. S. Panev, N. Andersen, T. Andersen, and P. Dalby, *Z. Phys. D* **5**, 331 (1987).
51. H. Schmidt, A. Bähring, E. Meyer, and B. Miller, *Phys. Rev. Lett.* **48**, 1008 (1982).
52. A. Bähring, I. V. Hertel, E. Meyer, W. Meyer, N. Spies, and H. Schmidt, *J. Phys. B* **17**, 2859 (1984).
53. A. Bähring, E. Meyer, I. V. Hertel, and H. Schmidt, *Z. Phys. A* **320**, 141 (1985).
54. R. J. Allan and H. J. Korsch, *Z. Phys. A* **320**, 191 (1985).
55. R. Shakeshaft, *J. Phys. B* **5**, 559 (1972).
56. A. Fischer and I. V. Hertel, *Z. Phys. A* **304**, 103 (1982).
57. U. Fano and J. H. Macek, *Rev. Mod. Phys.* **45**, 553 (1973).
58. J. Grosser, *Z. Phys. B* **14**, 1449 (1981).
59. K. Burnett, *Phys. Rep.* **118**, 339 (1985).
60. H. Schmidt, A. Bähring, and R. Witte, *Z. Phys. D* **1**, 71 (1986).
61. M. O. Hale, I. V. Hertel, and S. R. Leone, *Phys. Rev. Lett.* **53**, 2296 (1984).
62. (a) D. Neuschäfer, M. O. Hale, I. V. Hertel, and S. R. Leone, *Electronic and Atomic Collisions* Papers of the Fourteenth International Conference, D. C. Lorentz, W. E. Meyerhoff, and J. R. Peterson (eds.), Elsevier, New York 1986, p. 585. (b) W. Bussert, D. Neuschäfer, and S. R. Leone, *J. Chem. Phys.*, to be published. (c) W. Bussert and S. R. Leone, *Chem. Phys. Lett.*, to be published.
63. B. Pouilly and M. Alexander, *J. Chem. Phys.*, **86**, 4790 (1987).
64. A. Z. Devdariani and A. L. Zagrebin, *Chem. Phys. Lett.* **131**, 197 (1986).
65. M. Alexander, private communication.
66. F. M. Penning, *Naturwissenschaften* **15**, 181 (1927).
67. J. G. Kircz, R. Morgenstern, and G. Nienhuis, *Phys. Rev. Lett.* **48**, 610 (1982).
68. G. Nienhuis, *Phys. Rev. A* **26**, 3137 (1982).
69. D. M. Jones and J. S. Dahler, *Phys. Rev. A* **31**, 210 (1985).
70. E. W. Rothe, R. Theyunni, G. P. Reck, and C. C. Tung, *Phys. Rev. A* **31**, 1362 (1985).
71. M.-X. Wang, M. S. de Vries, and J. Weiner, *Phys. Rev. A* **33**, 765 (1986).
72. H. A. J. Meijer, H. P. v.d. Meulen, R. Morgenstern, I. V. Hertel, E. Meyer, H. Schmidt, and R. Witte, *Phys. Rev. A* **33**, 1421 (1986).

73. E. W. Rothe, R. Theyunni, G. P. Reck, and C. C. Tung, *Phys. Rev. A* **33**, 1426 (1986).
74. M.-X. Wang, M. S. de Vries, and J. Weiner, *Phys. Rev. A* **34**, 1869 (1986).
75. M.-X. Wang, J. Keller, J. Boulmer, and J. Weiner, *Phys. Rev. A* **34**, 4497 (1986).
76. J. Keller R. Bonanno, M.-X. Wang, M. S. de Vries, and J. Weiner, *Phys. Rev. A* **33**, 1612 (1986).
77. G. Nienhuis, *Electronic and Atomic Collisions*, Papers of the Fourteenth International Conference, D. C. Lorentz, W. E. Meyerhoff, and J. R. Peterson (eds.), Elsevier, New York, 1986, p. 569.
78. D. M. Jones and J. S. Dahler, *Phys. Rev. A*, **35**, 3688 (1987).
79. H. A. J. Meijer, H. P. v.d. Meulen, and R. Morgenstern, *Z. Phys. D* **5**, 299 (1987).
80. J. R. Barker and R. E. Weston, Jr., *J. Chem. Phys.* **65**, 1427 (1976).
81. A. Henriet, F. Masnou-Seeuws, and C. Le Sech, *Chem. Phys. Lett.* **118**, 507 (1985).
82. G. F. Hanne, Cz. Szmytkowski, and M. van der Wiel, *J. Phys. B* **15**, L109 (1982).
83. I. V. Hertel, M. H. Kelley, and J. J. McClelland, *Z. Phys. D.* **6**, 163 (1987).
84. J. J. McClelland, M. H. Kelley, and R. J. Celotta, *Phys. Rev. Lett.* **55**, 688 (1985).
85. J. J. McClelland, M. H. Kelley, and R. J. Celotta, *Phys. Rev. Lett.* **56**, 1362 (1986).
86. N. D. Bhaskar, B. Jaduszliwer, and B. Bederson, *Phys. Rev. Lett.* **38**, 14 (1977).
87. B. Jaduszliwer R. Dang, P. Weiss, and B. Bederson, *Phys. Rev. A* **31**, 1157 (1985).
88. G. F. Hanne, V. Nickich, and M. Sohn, *J. Phys. B* **18**, 2037 (1985).
89. B. Jaduszliwer, G. F. Shen, J.-L. Cai, and B. Bederson, *Phys. Rev. A* **21**, 808 (1980).
90. R. Düren, *Comments At. Mol. Phys.* **14**, 127 (1984).
91. F. van den Berg, P. Bijl, and R. Morgenstern, *Z. Phys. A* **320**, 1 (1985).
92. T. Orlikowski and M. H. Alexander, *J. Phys. B* **17**, 2269 (1984).
93. P. J. Dagdigian and M. L. Campbell, *Chem. Revs.* **87**, 1 (1987).
94. See, for example, E. E. Nikitin and S. Ya. Umanskii, *Theory of Slow Atomic Collisions*, Vol. 30, Chap. 3, Springer-Verlag, Berlin, 1984.
95. R. Düren, E. Hasselbrink, and G. Moritz, *Z. Phys. A* **307**, 1 (1982).
96. R. Düren, E. Hasselbrink, and G. Hillrichs, *Chem. Phys. Lett.* **112**, 414 (1984).
97. M. Kohmohto and U. Fano, *J. Phys. B* **14** L447 (1981).
98. S. Lemont and G. W. Flynn, *Ann. Rev. Phys. Chem.* **28**, 261 (1977).
99. I. V. Hertel, *The Excited State in Chemical Physics II*, J. W. McGowan (ed.), Wiley, New York 1981, p. 341.
100. W. H. Breckenridge and H. Umemoto, *The Dynamics of the Excited State*, K. P. Lawley (ed.), Wiley, New York, 1982, p. 325.
101. I. V. Hertel, *The Dynamics of the Excited State*, K. P. Lawley (ed.), Wiley, New York, 1982, p. 475.
102. H. Horiguchi and S. Tsuchiya, *J. Chem. Phys.* **72**, 455 (1980).
103. D. S. Y. Hsu and M. C. Lin, *Chem. Phys. Lett.* **42**, 78 (1978).
104. I. V. Hertel, H. Hoffmann, and K. Rost, *Phys. Rev. Lett.* **36**, 861 (1976).
105. J. A. Silver, N. C. Blais, and G. H. Kwei, *J. Chem. Phys.* **71**, 3412 (1979).
106. W. Reiland, H.-U. Tittes, C.-P. Schulz, and I. V. Hertel, *Phys. Rev. Lett.* **48**, 1398 (1982).
107. W. Reiland, H.-U. Tittes, C.-P. Schulz, and I. V. Hertel, *Chem. Phys. Lett.* **91**, 329 (1982).

108. W. Reiland, H.-U. Tittes, I. V. Hertel, V. Bonacić-Koutecký, and M. Persico, *J. Chem. Phys.* **77**, 1908 (1982).
109. G. Jamieson, W. Reiland, C.-P. Schulz, H.-U. Tittes, and I. V. Hertel, *J. Chem. Phys.* **81**, 5805 (1984).
110. I. V. Hertel and W. Reiland, *J. Chem. Phys.* **74**, 6757 (1981).
111. C.-P. Schulz, H.-U. Tittes, and I. V. Hertel, *Z. Phys. D.* (to be published).
112. P. Botschwina, E. Meyer, W. Reiland, and I. V. Hertel, *J. Chem. Phys.* **75**, 5438 (1981).
113. P. Habitz, *Chem. Phys.* **54**, 131 (1980).
114. D. Poppe, D. Papierowska-Kaminski, and V. Bonacić-Koutecký, *J. Chem. Phys.* **86**, 822 (1987).
115. For example, J. Los and A. W. Kleyn, *Alkali Halide Vapours*, P. Davidovits and D. L. McFadden (eds.), Academic Press, New York, 1979.
116. E. E. B. Cowan, M. A. D. Fluendy, A. M. C. Moutinho, and A. J. F. Praxedes, *Mol. Phys.* **52**, 1125 (1984).
117. N. C. Blais and D. G. Truhlar, *J. Chem. Phys.* **79**, 1333 (1983).
118. N. C. Blais, D. G. Truhlar, and B. L. Garret, *J. Chem. Phys.* **78**, 2956 (1983).
119. P. Archirel and P. Habitz, *Chem. Phys.* **78**, 213 (1983).
120. W. Reiland, C.-P. Schulz, H.-U. Tittes, and I. V. Hertel, *Chem. Phys. Lett.* **91**, 329 (1982).
121. D. Poppe, *Z. Phys. D* **1**, 207 (1986).

GRAPHS IN CHEMICAL PHYSICS OF POLYMERS

SEMION I. KUCHANOV, SERGEI V. KOROLEV, AND SERGEI V. PANYUKOV

Polymer Chemistry Department, The Moscow State University, Moscow, U.S.S.R.

CONTENTS

Introduction
- I. Models and Approaches Used for Description of Branched and Network Polymers
  - A. Molecular Graphs
  - B. Ensembles of Polymeric Molecules and Random Graphs
  - C. Flory Model (Model I)
  - D. Models of Substitution Effect (Model II)
  - E. Invariance Principle
  - F. Models Allowing for Cyclization
  - G. Kirchhoff Matrix and Conformation of Molecules
  - H. Macromolecules as Graphs on Lattice and Scaling
  - I. Percolation and Other Lattice Statistical Models
- II. Subgraphs of Molecular Graphs and Microstructure of Polymers
  - A. Macromolecular Fragments and Their Subgraphs
  - B. Subgraph Stoichiometry
  - C. Probability Measure on Subgraphs
  - D. Trails and Molecular Conformations
- III. Molecules as Graphs with Coordinates and Consideration of Cyclization
  - A. Graphs Embedded in Three-Dimensional Space
  - B. Gibbs Distribution and Probability Measure on Graphs in Space
  - C. Transition to Connected Graphs
  - D. Correlation Functions and Their Generating Functionals
  - E. Equations for Generating Functionals
  - F. Case Study: Model I
  - G. One-Point Correlators and MWD
  - H. Estimation of Contributions from Cycles of Different Topologies
  - I. Perturbation Theory
  - J. Statistics of Cyclic Fragments and Possible Generalizations of Model
- IV. Diagram Technique and Field Theory
  - A. Stochastic Fields and Functional Integration
  - B. Diagram Technique

C. Spatial Physical Interactions
D. Diagram Technique in Lifshitz–Erukhimovich Model
E. Analytical Methods of Field Theory
F Self-Consistent Field Approximation
G. Multicomponent Systems and Extension of Potts Model
H. Theory of Swelling for Stochastic Polymer Networks
I. Other Field-Theoretic Approaches to Description of Branching Polymers
V. Conformations of Gaussian Molecules and Spectral Properties of Their Graphs
A. Space Metric and Graph Metric
B. Using Kirchhoff Matrix to Calculate Distributions in Radius of Gyration and Other Conformational Characteristics of Macromolecules
C. Spectrum of Kirchhoff Matrix for Regular Networks
D. Spectral Density and Dynamic Properties of Molecule
Conclusion
Appendix: Elementary Concepts of Theory of Graphs
References

INTRODUCTION

The methods of graph theory are known to have wide application in various branches of theoretical physics and chemistry. They are usually employed, for example, in considering a series of topics of stereochemistry, for the description of kinetics of complex chemical reactions, and in quantum mechanical calculations. In addition, the most familiar tool of the contemporary theory of many-particle systems, the diagram technique, is based on graph-theoretic methods as well. One of the principal advantages of the diagram technique is its visuality, which promoted wide application of the concept of graphs, to account for the interaction in many-particle problems. The most famous applications of the diagram technique are the diagram expansions of Mayer and Feynman, the first of which has long been in the statistical physics of nonideal gases, and the second finds extensive application in quantum field theory and solid-state physics.

Solution of numerous problems in the chemical physics of polymers is greatly simplified if the problem can be expressed in terms of graph theory. This approach is particularly efficient for the description of branched and net polymers, which represent the ensembles of macromolecules, differing in the number of structural units and in their conjunction. In order to account for the structural isomerism of macromolecules arising in such systems, it is convenient to put in correspondence to each of the macromolecules a molecular graph, which is similar to the structural formula in classical organic chemistry. The synthetic polymers, however, are the sets of practically infinite numbers of individual chemical compounds, and hence, the statistical ensembles of molecular graph corresponding to these polymers involve the same number of different representatives. Their distribution in the polymeric pattern is random and defined by the conditions of its synthesis. As a consequence, the theory of polymers has to deal with the ensembles of random graphs, and to find the probability measures of these graphs, it is necessary to consider the process of preparation of the polymeric pattern in the course of which the set of macromolecules corresponding to this pattern is formed. Such a combined physical and chemical consideration of the polymeric systems, as will be seen later, is one of the principal peculiarities of their theoretical description.

The results of the graph-theoretic approach to describe the branched polymers were initially arranged and summed up in the review by Gordon and Temple,[1] the first, to make a significant contribution in the field. However, this review, as well as the subsequently published monograph,[2] discussed merely treelike molecular graphs regardless of the arrangement of their elements in space. The probabilities of such different graphs in the statistical ensemble, defining the fractions of the corresponding structural isomers in the polymeric pattern, can be established analytically within the framework of a simple quasi-chemical model of its formation with prohibited intramolecular reactions. Section I of this Chapter, which is of introductory character, formulates this and some other models used for the description of branched and network polymers as well as gives a brief account of papers available on the application of graph-theoretic methods in this field. The definitions of the main concepts of the physicochemistry of polymers are outlined in the first part of Section I so that no preliminary preparation in the field is necessary for its comprehension. The other sections contain mainly original results obtained by the present authors in which new graph-theoretic approaches were employed as compared to the traditional ones.[1,2] This enabled the solution of a series of important problems in the chemical physics of branched polymers.

The traditional approach to describe branched polymers implies that the probabilities of all macromolecules of the polymeric pattern are given. Section II gives an account of a new method to describe the molecular structure of branched polymers by setting relative fractions of different fragments of molecules. To such fragments, there correspond certain subgraphs of the molecular graphs. In order to particularize the description of the macromolecular structure, one should consider the subgraphs containing ever-increasing numbers of vertices representing the monomeric units and the edges connecting them, which correspond to chemical linkages. By analogy with hydrocarbons, many so-called structure-additive properties of polymers can be calculated on the basis of average numbers of different fragments of small size in macromolecules. Using graph theory, it is possible to find certain relationships of topological stoichiometry connecting the numbers of various subgraphs. Since the fractions of every possible fragment of branched molecules can now be measured by molecular spectroscopy techniques with high accuracy, such relationships prove to be essential in interpreting the spectroscopic data.

Many macroscopic features of branched polymers are determined not only by their primary structure but also by the spatial arrangement of their constituent units. Thus, for example, in calculating the average sizes of macromolecules, their hydrodynamic radius, or the intensity of light scattering, it is necessary to perform averaging over a probability measure that accounts not only for the modes of chemical bonding of fragments but also for their mutual arrangement in space. Such a measure is also required in the theory of the formation of polymeric networks allowing for the intramolecular reactions of cycle formation. Section III describes the Gibbs distribution method of constructing the probability measure on the set of labeled graphs, each fragment of which has a definite value of its spatial coordinates. In calculating the macroscopic features of the polymeric pattern, it is occasionally necessary to perform averaging over the probability distribution in the ensembles of graphs embedded in space. In this context, the introduction of correlation functions and their generating functional allows us, as shown in Section III, to obtain, numerous general results in a relatively simple and elegant way. With such a description of the spatial probability measure, it is possible to account, in a natural way, for the formation of cyclic fragments in the molecules of finite dimensions, that is, to go beyond the framework of average field approximation. The latter approximation is known to describe adequately the observed data in the system where the branched polymers were formed in melts or concentrated solutions. However, as the solution becomes more dilute, there are observed ever-increasing deviations from the results of the mean field theory since the intramolecular reactions begin to play the essential role in the formation of the ensemble of macromolecules. This effect can be accounted for by the presentation in Section III of perturbation theory in small parameter ε, the value of which is the reciprocal concentration of units in solution. The zero-order approximation in this parameter, when only tree graphs are involved, gives the results of the mean field approximation, while each subsequent order of perturbation theory accounts for the cycles of ever more complex topology.

A field-theoretic consideration of the ensembles of branched macromolecules[3] is an alternative approach to that described in Section III based on the theory of branching random processes. Application of the field-theoretic methods is related to the generating functional of the Gibbs distribution since the probabilities of state of such statistical ensembles can be represented as a continual integral over a random field, which is proportional to the fluctuating density of units or chemically reacting functional groups.

The calculation of this integral by the method of stationary phase for $\varepsilon \to 0$ results in the thermodynamic potentials of the mean field theory, and the calculation of high-order corrections accounts for the field fluctuation using specific methods of perturbation theory adapted for functional integrals. For this purpose, the diagram technique is developed in Section IV, which was also employed for the calculation of pair correlation functions. This method proved to be most efficient in the statistical theory of branched polymers taking into account the physical (spatial) interactions of molecules besides chemical ones. This version of the theory accounts for the thermodynamic affinity of the polymer toward the solvent and thus describes phase transitions in the process of the formation of polymeric networks.

Section V is devoted to another aspect of the application of graph-theoretic methods to the chemical physics of polymers. The probability distribution for the distance between the endpoints of a sufficiently long linear polymeric molecule is known to be described (neglecting the spatial interactions) by the Gauss function. The Gaussian branched and network polymers, the molecules of which consist of linear Gaussian chains connected at the branching points, are treated in a similar way. Averaging over the probabilities of their different arrangement in space, arising in the calculation of the macroscopic features of molecules, is evidently reduced to the calculation of multidimensional Gaussian integrals. In this way, the analytical formulas are obtained, which involved eigenvalues of the Kirchhoff matrix of the molecular graph, characterizing its topological structure. Thus, it is possible to express many features of the molecule, which can be determined by experiment, via the spectrum of the Kirchhoff matrix of the molecular graph. Evaluation of the eigenvalues of this matrix is still more important since the relaxation times of the molecule are connected in a simple way with these eigenvalues. This allows us to express the elasticity parameters of the polymeric network via the spectral density of the Kirchhoff matrix of its molecular graph. Since the spectrum of a graph of a Gaussian molecule determines its statistical sum, and, consequently, its probability in the equilibrium statistical ensemble, the solution of the problem of evaluating the eigenvalues allows us to obtain automatically the topological distribution of the products of equilibrium intramolecular cyclization. An example of the calculation of such a system is given in Section V.

To conclude this brief introduction, we emphasize that no specific knowledge, neither in polymer science nor in graph theory, is required of the reader. All the necessary preliminary information is presented in the first part of Section I and the Appendix.

I. MODELS AND APPROACHES USED FOR DESCRIPTION OF BRANCHED AND NETWORK POLYMERS

A. Molecular Graphs

The molecules of polymers consist of different numbers of structural units—the *monomer units*. If all structural units are identical, the resulting polymer is called a *homopolymer*, as opposed to the *copolymers*, consisting of different units, in which the number of types is usually not large. The mode of conjunction of units defines the topological structure of the molecule. In the simplest case the molecule can represent a linear chain (Fig. 1.1*a*). Besides these linear molecules, there are a great number of branched and network polymers characterized by different topologies of molecules. The latter except for the monomer units also involve the *functional groups*, which enter the chemical reactions to form the chemical bond that joins two molecules. The topology of the resulting compound depends on the position of the reacting functionalities in the initial molecules. Hence, the functional groups, together with monomer units, also define the structural formula of the polymeric molecule, which can be visualized most clearly as a *molecular graph*.

The molecular graph can be represented with different levels of particularization (Fig. 1.1). In the most simple version the monomer units can be represented by the vertices of a graph, where the edges of the graph correspond to the chemical linkages between the units (Figs. 1.1*b*, *d*). The functional groups are depicted on the graph by pendant vertices (Fig. 1.1*e*), denoted by open (white) circles, as opposed to solid (black) circles denoting the monomer units. If the polymer consists of units or functionalities of different types, there must be enough colors to distinguish the vertices of the molecular graph.

It should be noted that the molecular graph does not necessarily define the individual chemical compound in an unambiguous way. Besides the stereoisomery exemplified in Fig. 1.2, there exist some other types of spatial *isomery*.[4] In what follows, however, we will not distinguish between these isomers, exhibited by the same molecular graph, suggesting them all to have identical graphical configuration by definition. The separation of the ensemble of polymeric molecules into the sets of such isomers (Fig. 1.3) is determined by the choice of the law of correspondence between the graph vertices and molecular fragments. Thus, if individual atoms are taken as molecular fragments, the graphical configuration corresponds unambiguously to a definite structural isomer.[4] When monomer units are chosen as molecular fragments, different structural isomers may have identical graphical configuration. The peculiarities of the theoretical description of branched polymers arising in this case are discussed in Section I.E.

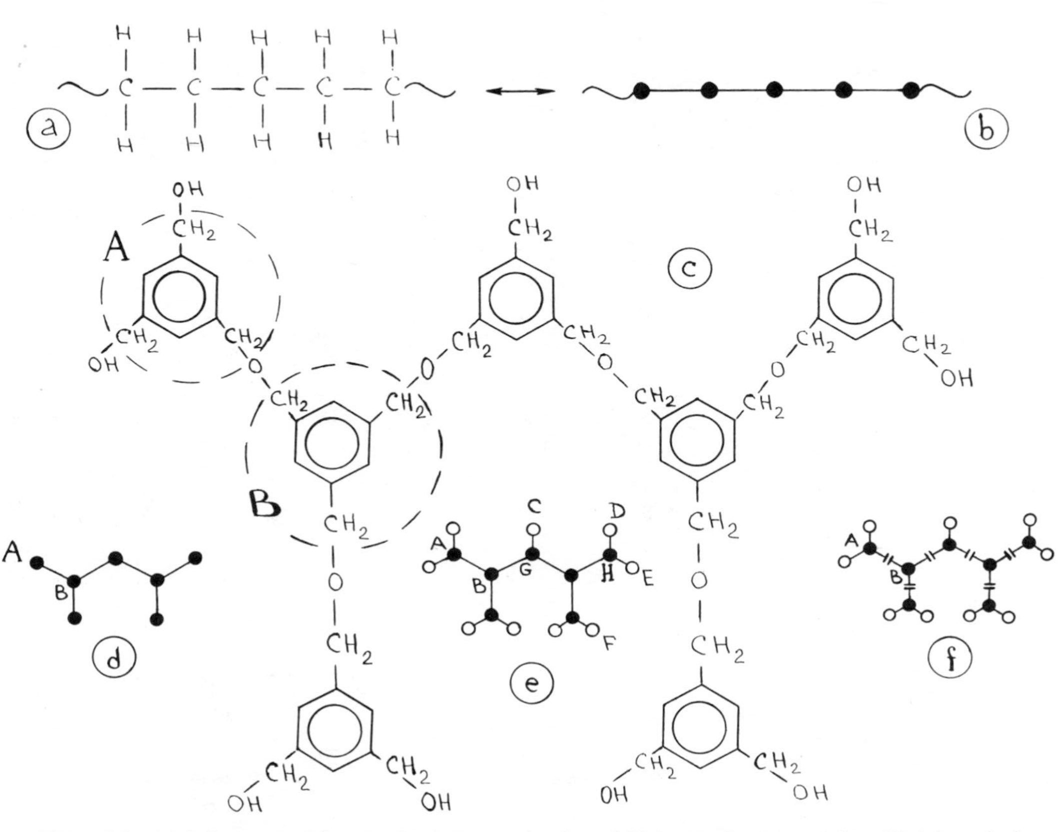

Figure 1.1. (*a*) A fragment of linear polyethylene molecule and (*b*) its graph representation. (*c*) A branched molecule of seven monomer units of 1, 3, 5-trimethylolbenzene and its pictures (*d*)–(*f*), three molecular graphs representing it in various details. Dashed lines encircle fragments of adjacent monomer units **A** and **B**.

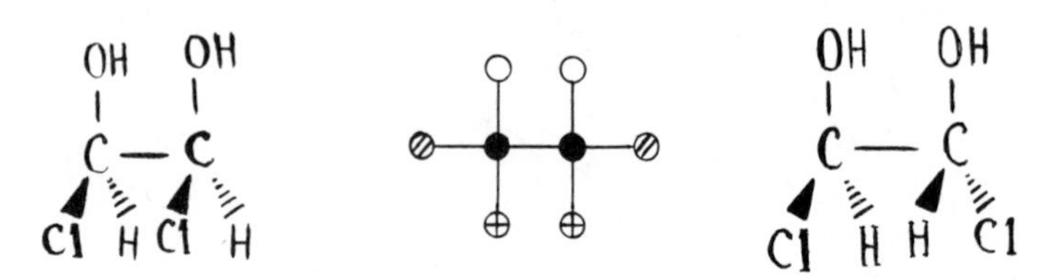

Figure 1.2. Two stereoisomers represented by the same molecular graph.

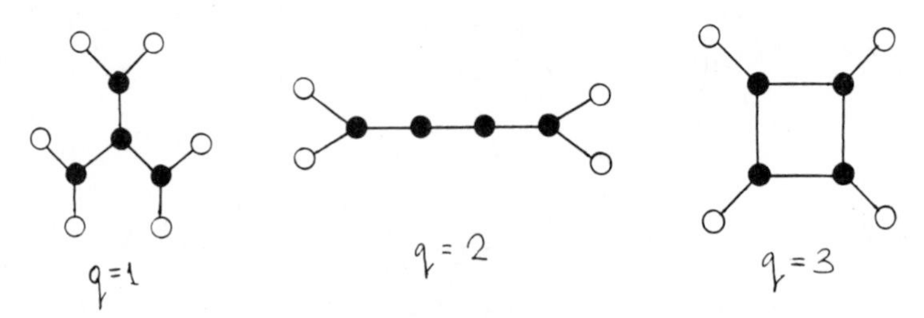

Figure 1.3. Different isomers of polymerization degree $l = 4$ (tetramers).

In some processes of polymer formation the reaction of bond formation can be reversible. For example, the polymer in Fig. 1.1*c* is formed as a result of the condensation reaction

$$—\mathrm{OH} + \mathrm{HO}— \rightarrow —\mathrm{O}— + \mathrm{H_2O}, \tag{1.1}$$

where the ether linkage —O— is capable of breaking under the action of the molecule H_2O to give two hydroxyl functional groups —OH. Hence, such a linkage can be considered as a reactive internal functional group, which is convenient to denote as a pair of cuts on the edge of the graph, connecting the monomer units (Fig. 1.1*f*). In systems with several types of functionalities the cuts bear the information on the colors of vertices corresponding to the reacted groups. Then each pair on the edge characterizes its color, so that edges in this case can be regarded as colored. When some linkage is broken, each of the cuts is substituted by the corresponding pendant vertex. The convenience of such notation is revealed in considering the fragments of molecular graphs (Section II) as well as when the elements of the graph are marked with labels, the coordinates of the corresponding fragments of molecules (Section III).

The mode of representation of the molecular graph is defined by the nature of the problem to be solved. Graphs depicting units and groups, but not linkages, are used most frequently (Fig. 1.1*e*). Each of the three methods allows (taking into account the color of vertices and edges) to recover unambiguously the other two, since the number and types of functionalities of the initial monomers are known from their structure.

One of the principal features of the polymeric molecule is its molecular mass. When all units are identical, the molar mass can be easily calculated provided the degree of polymerization, that is, the number of units l contained in the molecule, is known. Molecules of identical degree of polymerization can differ by their topology. Isomers of the same degree of polymerization l will be distinguished in the following by numbering them in an arbitrary way with some index q and will be called (l, q)-mers. The properties of different isomers can differ markedly depending on the topological structure of their molecular graphs.

A complete description of these graphs by depicting them in a figure, or with the help of topological matrices, is hardly effective in the case of molecules of high degree of polymerization ($l \gg 1$). Hence, we have to confine ourselves to less detailed characteristics, which reveal to some extent the structure of the graph. For example, the local structure can be defined by specifying the number of vertices of different color in the graph, different pairs of adjacent vertices, triples, and so on. (Fig. 1.4). As the cyclic rank of a graph and the spectra of topological matrices are considered its global characteristics. Such characteristics can be connected with some measurable parameters of the isomer depicted by this graph.

For example, the so-called structure-additive properties[5,6] depend on the number of subgraphs of a certain kind with a small number of vertices contained in the molecular graph. In the simplest case a property such as, say, the enthalpy of formation of the molecule is specified only by the number

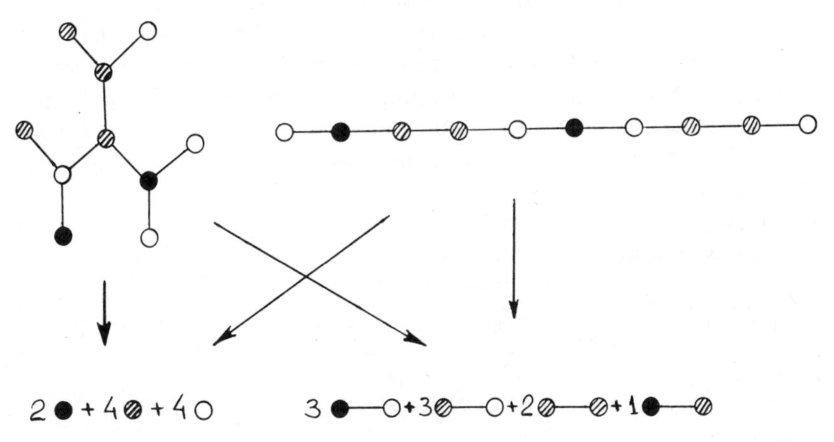

Figure 1.4. Two graphs having identical local topologies given by numbers of different small subgraphs.

of chemical bonds of different types in the molecule.[5,6] Hence, to calculate the enthalpy, it is sufficient to determine in the colored graph the number of subgraphs consisting of pairs of vertices of all possible combinations of colors (Fig. 1.4).

Among the molecular properties, which are not structure additive, there are those defined by the global structure of the graph. Generally, such properties depend on the set of conformations of the molecule. Each of the *conformations* of some isomer is characterized by the mutual arrangement of its fragments in space. In the theoretical calculation of such properties for the particular configuration of the molecule, it is necessary within the framework of some model of the polymeric chain[7,8] to carry out averaging over the whole set of possible conformations taking into account their probabilities. It is evident that the knowledge of only the local structure of the molecular graph is insufficient for this purpose, since, in principle, any molecular fragments, even those separated in the graph by a path of any length, may be arranged in space in close proximity. The probabilities of different mutual arrangements of molecular fragments in space can be accounted for by using the correlation functions, similar to those considered in Section III.

B. Ensembles of Polymeric Molecules and Random Graphs

Any real macroscopic polymeric system can, in principle, be represented as a graph. However, the number of its connected components, each corresponding to one molecule, is comparable to the Avogadro number; that is, it is practically infinite. In addition, the molecules change their configuration continuously as a result of chemical reactions. Hence, the only efficient method to describe such a system is the statistical method. In this case, on the set of all possible molecular graphs a probability measure is defined in such a way that the probability of some molecular graph is proportional to the concentration of molecules denoted by this graph in the polymeric system.

Such a measure on the nonrooted graphs is convenient for the description of the ensemble of molecules comparable to each other in sizes. However, the process of formation of branched polymers is characterized by the specific effect, referred to as gelation, when a huge macromolecule (*gel*) is formed that occupies the whole reaction volume. The number of units forming this molecule is comparable in magnitude to the total number of monomer units in the system, although the probability of such a molecule, specified in the preceding, is practically negligible since it happens to be a single molecule of a vast number of finite ones comprising the *sol-fraction*. The probability of an arbitrary vertex in the set of random connected graphs to belong to its infinite element (gel) is equal to the fraction of units entering the gel. In this case the probability measure is constructed on the set of rooted graphs. This measure is related in a straightforward manner to the distribution of nonrooted graphs but appears to be more favorable for the description of the polymeric systems. When one of the vertices of a nonrooted graph is incidentally chosen as a root, the graph becomes a rooted one. Sorting out all the vertices of an l-mer in this way, one obtains l-rooted graphs (Fig. 1.5), so that the probability of the graph representing a definite molecule varies in proportion to the number of its units, that is, to its molar mass. When a pendant vertex is chosen as a root, the resulting graph is called a planted graph. The transition from the nonrooted graphs to those planted is also reasonable and defines a measure proportional to the fraction of functional groups of a given isomer among all unreacted groups of the system.

In experimental studies of the individual chemical compounds consisting of identical molecules, any measured property can be regarded as relating to any particular molecule. In contrast to this, all various (l, q)-isomers composing the polymeric system contribute to its physicochemical properties; thus, it necessitates averaging over the probability measure on the set of molecular graphs. If the averaged characteristics depend on the spatial arrangement of the molecular fragments, then the first step, as in the case of the ensemble of identical molecules (Section I.A), must include averaging over all possible conformations of each molecule. Next, a subsystem can be conditionally separated, containing the molecules of equal degree of polymerization l, and this subsystem is averaged over different configurations of isomers. As a rule, the value of any property for such a set of molecules is given as the arithmetic mean of contributions of isomers of this set. The final averaging over l's can be performed with different weights depending on the method of experimental evaluation of a calculated property. For example, for

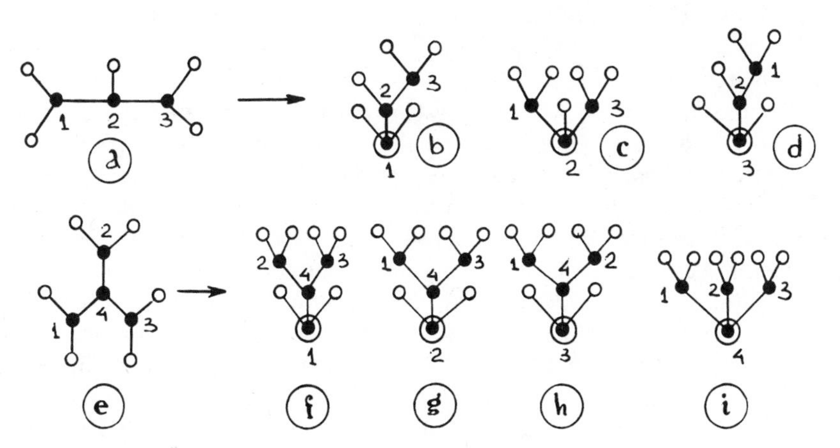

Figure 1.5. Transition to rooted graphs associated with change in graph probability in proportion to number of nodes. Graph roots are circled.

osmometric measurements the contribution of the l-mers is proportional to their concentration $c(l)$, while in light-scattering measurements it is proportional to $l^2c(l)$. Therefore, the final averaging can be performed both over the number *molecular weight distribution* (*MWD*) $f_N(l)$ and over weight MWD $f_W(l)$ as well as over MWD $f_Z(l)$:

$$f_N(l) \equiv \frac{c(l)}{\sum c(l)}, \qquad f_W(l) \equiv \frac{lc(l)}{\sum lc(l)}, \qquad f_Z(l) \equiv \frac{l^2c(l)}{\sum l^2c(l)}. \tag{1.2}$$

The measure on the set of nonrooted graphs, constructed in the preceding, corresponds evidently to the number MWD, while that on the set of rooted graphs is associated with the weight MWD. In a similar way, the distribution $f_Z(l)$ required for the calculation of the Z-average characteristics can be associated with random two-rooted graphs.

The probability measure on graphs not only contains information on the distribution of molecules by the number of units in them but also allows us to find the fractions of various isomers. By analogy with (1.2), there can be defined different *molecular structure distributions* (*MSDs*) $f_N(l, q)$, $f_W(l, q)$, and $f_Z(l, q)$ that distinguish the molecules in accordance with their molecular graphs.

These distributions depend naturally on the chemical peculiarities of monomers and on polymer formation conditions. In polymer chemistry there are various methods for the synthesis of branched and networks polymers that lead to different ensembles of polymeric molecules. One of the most convenient methods is the reaction of polycondensation,[9,10] and we will exemplify the applicability of graph theory for the description of polymers mainly with this method. In the framework of the clearly defined model of polymer formation, MSD can be calculated on the basis of main physical and chemical principles.

For the equilibrium polymeric systems the probability measure is defined by the Gibbs distribution, whereas for the nonequilibrium systems it is calculated on the basis of kinetic equations of the reactions between the macromolecules. These stages, which are required to derive the statistical characteristics of polymers, are occasionally omitted, being substituted by frequently elegant but speculative considerations appealing to common sense.

It should be remembered that the conditions of polymer processing essentially differ, as a rule, from those of synthesis. The MSD formed in the course of polymer formation is usually retained in the product, defining its properties in the subsequent operation. Properties, defined by feasible conformations of molecules, will naturally differ with varying external conditions, for example, temperature and solvent. The limits of these variations are evidently defined by the nature of MSD, so that the problem of calculating operation parameters for the polymer cannot be solved while disregarding the process of its formation. In the theoretical calculation of MSDs the conformational distribution of molecules should also be taken into account in some cases, since it determines the feasible reaction channels of their functional groups, for example, the competition between the intra- and intermolecular reactions in different isomers. In these cases the conformational statistics of molecules under the conditions of their synthesis has to be considered, although with a known MSD of products this information is quite useless.

Some properties of network polymers (e.g., their elasticity) are determined, besides the configurational structure of the network, also by its topological limitations, connected with the mutual impermeability of the polymeric chains. These limitations can affect essentially the conformational set of network polymers. Thus, occasionally, it is necessary to distinguish the topological isomers. A simple example is shown in Fig. 1.6. The compounds formed by topological linkages are called "cathenans" and are quite familiar in organic chemistry.[11,12] Such topological linkages can arise only for the molecular graphs embedded in three-dimensional space. This spatial topology should be distinguished from that of the graph defined by its homeomorphism.[13] In what follows we will consider the "topology" only in its graph-theoretic sense, since the concept of spatial topological isomerism is beyond the scope of this chapter. This is explained by the fact that we consider primarily the equilibrium processes of the formation of branched and network polymers, and in this case no topological limitations arise, since in virtue of the reversibility of the reaction of chemical bond formation the transitions between different topological isomers become possible.

If the polymer is prepared under nonequilibrium conditions, then for the calculation of the statistical characteristics of its topological structure, it is necessary to solve the corresponding kinetic equations describing the evolution of the molecular ensemble in the course of the chemical reactions between the molecules. The methods for working out and solving such equations are

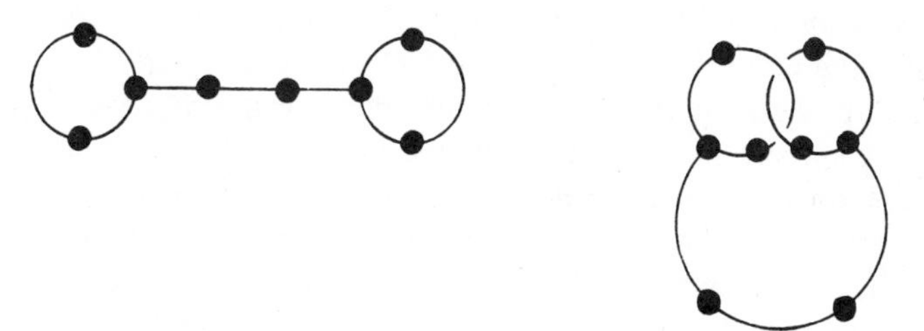

Figure 1.6. Pair of topological isomers having identical configurations and different conformational sets.

currently well developed even for the branched polymers.[2] However, since these methods do not employ graph-theoretic concepts, we will not discuss them here.

Equilibrium processes are widely used in the polycondensation method of the synthesis of polymers, and hence, this method is appropriate to demonstrate general approaches. When the low-molecular-mass product [e.g., water, liberating in the etherification reaction (1.1)] is removed from the reaction zone sufficiently slowly, the reaction system at each instant of time will be in a state that is sufficiently close to equilibrium. In this case the MSD of the polymer formed is defined by the Gibbs distribution, which changes over time as the by-product is removed.

The characteristic peculiarities in the formation of branched polymers are manifested in a simple and sufficiently universal system, the molecules of which are built of units of a single type. The initial monomer bears f identical functionalities, so that the degree of any node of the molecular graph equals to f if the graph involves the pendant vertices corresponding to the functionalities. An example of the molecule formed from a three-functional monomer is shown in Fig. 1.1*c*. Different suggestions on the nature of the interaction between the functionalities and molecules lead to the physico-chemical models of the f-functional polycondensation of different levels of complexity. The simplest model was proposed in the pioneer work of Flory.[14]

Model I is based on two main postulates:

1. The reactivity of all functionalities is invariable in the course of polycondensation.
2. There are no intramolecular reactions of cyclization; hence, all the molecular graphs are trees.

In real systems each of the postulates of this model of *ideal* polycondensation can be violated to some extent. Thus, the activity of some functionality can vary due to inductive, steric, or any other effects as a result of bond formation between the units. For example, the unit G in Fig. 1.1*e* is linked with two neighboring units, whereas the unit H is linked with only with one. Thus, the activity of groups C and D may differ, whereas the groups D, E, and F possess equal reactivity. Such kinetic or thermodynamic *substitution effects*, caused by the change in the local topology of the molecular graphs, are considered in *model II*, proposed for the first time in ref. 15.

The second postulate by Flory holds true rather well for the polymers with rigid chains in concentrated solutions or melts. However, in dilute solutions the frequency of molecular collisions decreases, whereas the probability of interaction between the functionalities belonging to a single molecule is usually weakly dependent on dilution. As a result, the fraction of intramolecular reactions is so increased that it is necessary to account for them in the construction of the theory, thus resulting in *model III*.

In some cases the process of formation of branched and network polymers is essentially affected by the difference between the interaction of molecular fragments with each other and with a solvent. Depending on the polymer–solvent thermodynamic affinity, a phase separation may occur in the course of polycondensation. To calculate such systems, it is necessary to take into account, besides the chemical reaction between the functionalities, the physical interactions between the monomer units, thus arriving at *model IV* (Section IV).

Model I differs qualitatively from the others in that the MSD of products of nonequilibrium polycondensation obtained in this model proves to be the same as that of a polymer obtained under equilibrium conditions.[2] It is evident that the concentrations and conversions of functional groups, which are the parameters of this distribution, will be determined in the first case by kinetic factors and in the second by thermodynamic ones. However, the analytical dependencies of MWD on these parameters in both cases are quite the same for any number of types of monomers and their functionalities.

C. Flory Model (Model I)

As the most simple case, it is reasonable to consider within the framework of this model the equilibrium conditions of polycondensation, the product MWD of which was calculated for the first time by Stockmayer.[16] Instead of using the combinatorial calculations of this author, the generalization of which on the complex multicomponent systems is quite troublesome, it is more favorable to make use of graph-theoretic methods. To this effect, in the following,[1,2] we consider the microcanonical ensemble composed of N monomer units combined in a fixed number of molecules Π occupying the volume V. The state of such a system can be represented by a disconnected graph G, that is, a forest consisting of trees, each describing a single molecule. The energy of all states is identical, since in model I it depends only on the number of linkages $N-\Pi$. Hence, the probability of any state is defined merely by the multiplicity of its degeneration, that is, by the number of different ways of conjunction of N monomers into the given configuration G. As was shown by Gordon and Temple,[1,17] the change in the combinatorial (topological) entropy in turning from sate G_1 to state G_2 is related to the orders of the automorphism groups $\mathscr{S}(G_1)$ and $\mathscr{S}(G_2)$ of the graphs:

$$\Delta S = S_2 - S_1 = R \ln [\mathscr{S}(G_1)/\mathscr{S}(G_2)], \tag{1.3}$$

where R is the universal gas constant. If $\mathscr{S}(l,q)$ is the order of the automorphism group of the (l,q)-isomer and $m_{l,q}$ is the number of such isomers in the configuration G_i, then $\mathscr{S}(G_i)$ is the product of $m_{l,q}$ factors $\mathscr{S}(l, q)$ (independent transpositions of vertices of each graph) and the factorial $m_{l,q}!$ (transpositions of identical connected components of the graph G_i).

Let us take N separate monomers as an initial state and a final state G characterized by the numbers $m_{l,q}$. Then according to the foregoing considerations, the multiplicity of degeneration of the state G is equal to

$$W(G) = \exp\frac{\Delta S}{R} = \frac{N![\mathscr{S}(1)]^N}{\prod_{l,q}\{m_{l,q}![\mathscr{S}(l,q)]^{m_{l,q}}\}} \tag{1.4}$$

where $\mathscr{S}(1) = f!$ is the number of ways to transform the molecular graph of the monomer into itself. In particular, for the ensemble consisting of units combined in a single treelike molecule involving the $(l-1)$th bond, the formula (1.4) gives the number of different ways to obtain the (l, q)-isomer from the monomers:

$$W(l, q) = l!(f!)^l/\mathscr{S}(l, q). \tag{1.5}$$

The most probable distribution[16] of molecules maximizes expression (1.4). In this case the concentration $c(l, q)$ of the (l, q)-isomer turns out to be proportional to the multiplicity of its degeneration (1.5). The proportionality factor can be found from the normalization conditions,[16] which require the total numbers of units and molecules to be N and Π, respectively. This factor can be expressed via the *conversion*

$$p = 2(N - \Pi)/fN, \tag{1.6}$$

that is, the fraction of the reacted functional groups, and thus the concentration of the (l, q)-mers assumes the form

$$c(l, q) = \frac{N}{V}\frac{(f!)^l}{\mathscr{S}(l, q)}\frac{f(1-p)^2}{p}\left[p\frac{(1-p)^{f-2}}{f}\right]^l. \tag{1.7}$$

The reciprocal dependence (1.5) between the number $W(l, q)$ of different ways of "assemblage" of the isomer and the order of its automorphism group can be established also by another graph-theoretic approach based on the labeling of graphs. For this purpose, let us enumerate l monomer units with numbers from 1 to l and number the groups from 1 to f, separately for each monomer (Fig. 1.7). To each method of formation of the certain (l, q)-isomers from these monomers there corresponds some numeration of vertices and cuts of the molecular graph. The total number of ways are identical, leading to the

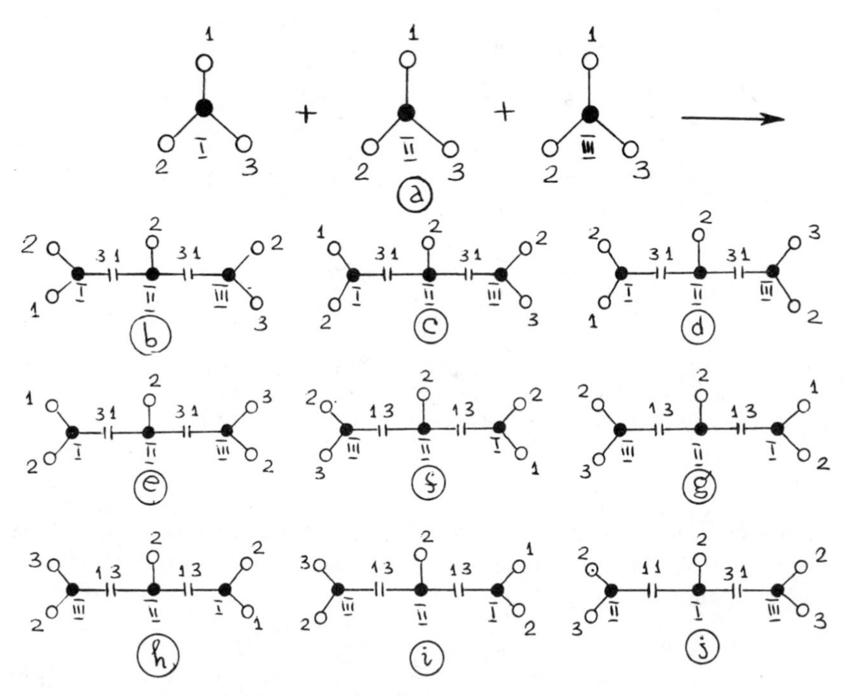

Figure 1.7. (*a*) Labeled isomers and their nine combinations [(*b*)–(*j*)] in trimer; eight of them [(*b*)–(*i*)] yield isomorphic graphs ($\mathscr{S}(3) = 2 \times 2 \times 2$).

isomorphic numbered graphs (Fig. 1.7). The number of copies of any numbered graph coincides with the number of ways it can be transformed into itself with the preserved adjacency of vertices; that is, it coincides with the order of the automorphism group $\mathscr{S}(l, q)$ of this graph. In this way, $W(l, q)$ again is expressed as (1.5).

To obtain the concentration $c(l)$ of all molecules of a given degree of polymerization l (i.e., the MWD), it is necessary to sum items (1.7) with all possible q's. Summing the reciprocal orders of automorphic groups arising in this case can be reduced to the known problem of the enumeration of ordered rooted trees.

Let us embed an ordered rooted tree into the half-plane (Fig. 1.8), and assign the order (e.g., from left to right) of edges incident to each vertex. The f branches, outgoing from a root, can be transposed in $f!$ ways. For each of the remaining vertices the number of transpositions of branches is $(f-1)!$. Since any of l vertices of the tree can be selected as its root, the total number of ordered trees associated with the labeled molecular graph is equal to $lf[(f-1)!]^l$. After elimination of labels, some trees become identical

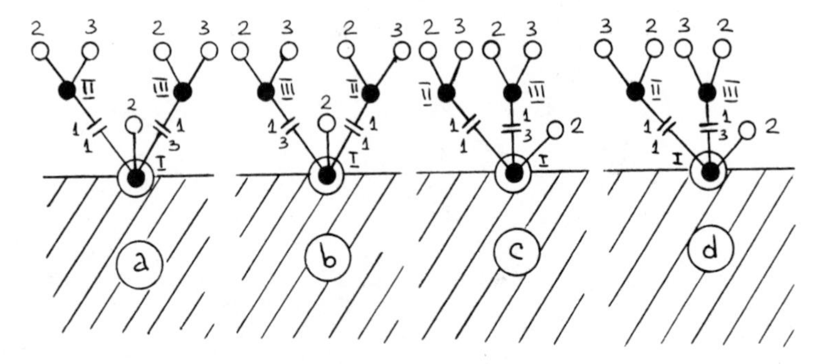

Figure 1.8. Four ordered rooted trees obtained from labeled graph of Fig. 1.7*j*, where the middle unit is chosen as the root. After marks are erased, two trees [(*a*) and (*b*)] become identical (isomorphic), but they are still different from other trees [(*c*) and (*d*)].

(isomorphic), with the number of copies of each tree being again $\mathscr{S}(l, q)$. As a result, we obtain the formula for the number of different rooted ordered trees representing a given (l, q)-isomer[1,17]:

$$D(l, q) = l\frac{f[(f-1)!]^l}{\mathscr{S}(l, q)} = \frac{W(l, q)}{(l-1)!\, f^{l-1}}. \tag{1.8}$$

The calculation of the concentration $c(l)$ is now reduced to the enumeration of all rooted ordered trees with l f-functional nodes. An algorithm for the solution of this problem was given by Good.[18] It is based on the graphical representation of the Newtonian binomial as the simplest rooted labeled trees (Fig. 1.9). In a similar way, to each term in the expansion of the enumerating *generating function* (*g.f.*) $g^{\mathrm{e}}(y)$, there corresponds a single ordered tree (Fig. 1.10)

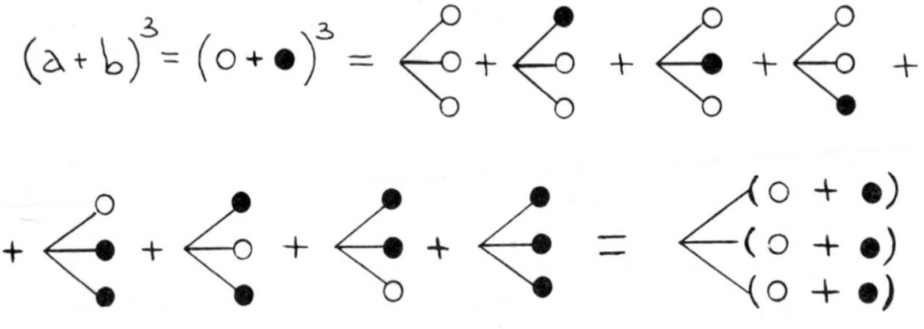

Figure 1.9. Graphical representation for Newton binomial formula.

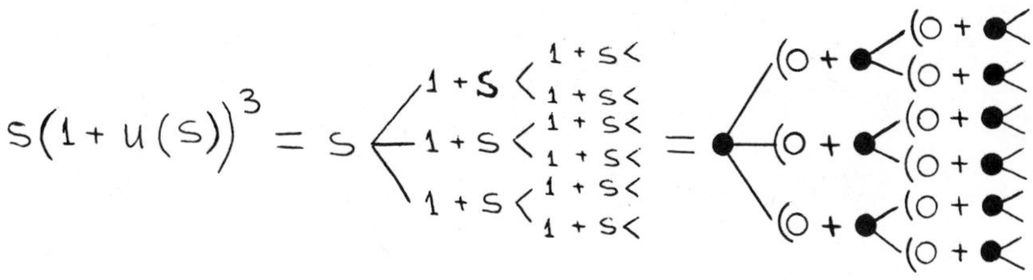

Figure 1.10. Graphical representation for Eq. (1.9).

$$g^{e}(s) = \sum_{l} D(l)s^{l} = s(1 + s(1 + s(1 + s(\cdots)^{f-1})^{f-1})^{f-1})^{f} = s(1 + u^{e}(s))^{f}, \quad (1.9)$$

$$u^{e}(s) = s(1 + u^{e}(s))^{f-1}, \qquad D(l) = \frac{f[(f-1)l]!}{(l-1)![(f-2)l+2]!}. \quad (1.10)$$

The coefficient at the lth power of the "counter" s (which is a dummy variable) turns out to be exactly the number $D(l)$ of all different ordered trees with l vertices. These numbers can be calculated in the explicit form by using the Lagrange expansion or its generalization,[19] and then the concentration of l-mers and MWDs can be expressed via (1.7) and (1.8):

$$\frac{lc(l)}{N} = f_{W}(l) = D(l)\{p^{l-1}(1-p)^{(f-2)l+2}\}. \quad (1.11)$$

The interpretation of the equation is quite clear: The probability of finding an arbitrarily chosen unit of the system as belonging to an l-mer coincides with the total probability of all rooted ordered trees with l vertices, the probability of any such tree being $p^{l-1}(1-p)^{(f-2)l+2}$. It should be noted that the number $D(l)$ of different ordered trees appearing in Eq. (1.11) is not related directly to the number of all possible isomers of the type q of a given degree of polymerization, the evaluation of which has resulted in a number of classical graph-theoretic papers (cf. ref. 20).

The description of the MWD is greatly simplified by using the probability g.f., which was obtained[2] by substituting $sp(1-p)^{f-2}$ for s in the enumerating g.f. $g^{e}(s)$. Subsequent substitution of the function $u(s)$ by $(1-p)u^{e}(s)/p$ reduces the g.f. to the standard form:

$$G_{W}(s) \equiv \sum_{l} f_{W}(l)s^{l} = s(1-p+pu(s))^{f}, \qquad u(s) = s(1-p+pu(s))^{f-1}. \quad (1.12)$$

The use of the g.f. allows us to perform MWD averaging in a simple way, without explicit knowledge of this distribution. For example, the calculation of the weight-average degree of polymerization is reduced to differentiating the g.f.:

$$P_W \equiv \sum_l l f_W(l) = \left.\frac{dG_W(s)}{ds}\right|_{s=1} = \frac{1+p}{1-(f-1)p}. \tag{1.13}$$

The expression for the g.f. of the number MWD can be obtained from (1.2):

$$\begin{aligned} G_N(s) &\equiv \sum_l f_N(l)s^l = \left[\sum_l l f_N(l)\right]\sum_l \frac{f_W(l)}{l} s^l = P_N \int_0^s \frac{G_W(y)}{y}\,dy \\ &= P_N\left[G_W(s) - \frac{fpu^2(s)}{2}\right]. \end{aligned} \tag{1.14}$$

The latter expression was obtained by integrating by parts, and the number-average degree of polymerization P_N, which is equal to the mean number of units in the molecule, is, according to (1.6),

$$P_N \equiv \sum_l l f_N(l) = \frac{N}{\Pi} = \frac{2}{2-fp}. \tag{1.15}$$

Comparing the enumerating g.f. (1.9) and (1.10) and the probability g.f. (1.12) shows that the latter can also be represented by the right part of Fig. 1.10; it is only necessary to assign to each white pendant vertex of the tree a weight $1-p$ and to the black nonrooted vertex a weight p. This corresponds to the construction of the probability measure on the set of rooted ordered trees. Each depicts a definite realization of a random branching process[21] evolving in the following way.

Every realization of the process starts with a single black particle, which generates f descendants (Fig. 1.11). Each descendant turns out to be either white with a probability of $1-p$ or black with a probability of p, which coincides with the fraction of the reacted functionalities. The white individuals are incapable of subsequent reproduction, whereas those that are black nonrooted always generate $f-1$ descendants. Then history is reiterated to degeneration, when in some generation there happen to occur only white individuals incapable of further reproduction (Fig. 1.11). Since any tree with l vertices involves exactly $l-1$ edges between them and $(f-2)l+2$ terminal vertices, the probability of its appearance among the realizations of the branching process is equal to the term in braces on the right-hand side (r.h.s.) of (1.11); that is, it coincides with the measure constructed in the preceding section.

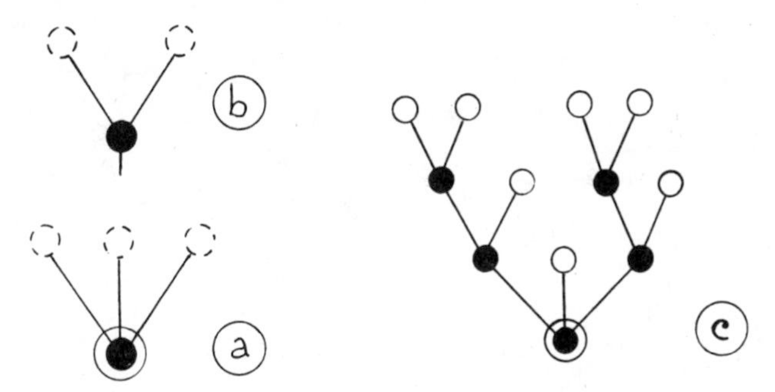

Figure 1.11. A particle, which is at the root of the branching process, has always f descendants (dashed lines): each of them may be (*a*) black or white. In turn, every black descendant yields $(f-1)$ new ones [(*b*)]. The process terminates when in a generation all the particles are white, which do not yield offspring [(*c*)].

The use of the well-developed tools of the theory of branching processes in polymer chemistry was initiated in the pioneer work of Gordon[22] and currently is generally accepted.[2,23] The phenomenon of gelation in the branched polymeric system means, in terms of branching processes theory, that the probability of degeneration is not equal to unity; that is, among the realizations there appears one with infinite lifetime. This can occur if the mean number of black descendants of an arbitrarily reproducing particle is not less than unity, that is, $p(f-1) \geqslant 1$. The minimal value of the conversion $p^* = (f-1)^{-1}$ satisfying this inequality corresponds to the gel point when among the molecular graphs there appears, with nonzero probability, an infinite one. The weight-average degree of polymerization (1.13) in the vicinity of the gel point grows to infinity. When the gel has been formed, the branching process describes the history of finite realizations, the formulas being the same (1.12) but the conversion p being replaced by some other parameter. As a result, the probability measure on the set of molecular graphs is renormalized and thereby describes a conditional probability to find an arbitrarily chosen vertex as belonging to the molecular graph of the l-mer with the condition that this vertex does not belong to the infinite graph representing the gel.

All these considerations, which are relevant to graphs containing white vertices, can be applied to the graphs with only black vertices, which represent the units, such as those shown in Fig. 1.1*d*. In this case the realizations of the branching process involve only black individuals (Fig. 1.12*a*), and the number of descendants of each individual is now random. The probability that k of f groups of a unit have been reacted independently is $C_f^k p^k (1-p)^{f-k} = pfC_{f-1}^{k-1} p^{k-1} (1-p)^{f-k}/k$. This probabiity describes the appearance of k descendants for an individual of zero generation. In all other

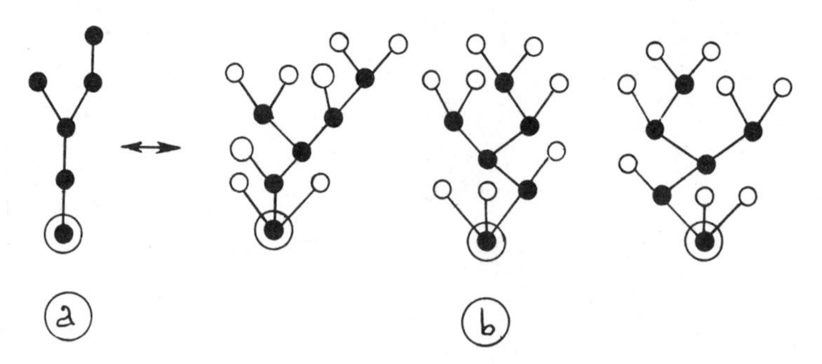

Figure 1.12. (*a*) Realization of branching process with no white particles (*b*) corresponding to several rooted trees with the functional groups indicated.

generations an individual is capable of generating no more than $f-1$ descendants, and the vertex corresponding to this individual is of order k if the number of its descendants is $k-1$, the probability of which is $C_{f-1}^{k-1}p^{k-1}(1-p)^{f-k}$. As a consequence, the probability of an ordered rooted tree with l_k vertices of order k $(k=1,\ldots,,f)$, including the root of order i, is equal to

$$\mathscr{P}_i(l_1,\ldots,l_f)=\frac{f}{i}\prod_{k=0}^{f}(C_{f-1}^{k-1})^{l_k}p^{l-1}(1-p)^{(f-2)l+2}. \tag{1.16}$$

In this case the probability $\mathscr{P}_i$ (3, 2, 1) of the tree shown in Fig. 1.12*a* naturally differs from the probabilities of trees in Fig. 1.12*b*, the latter expressed by the term in braces in Eq. (1.11). The weight fraction of a definite isomer, as in the foregoing, coincides with the total probability of all rooted ordered trees representing it but now with no white vertices. Although the probability of trees has now changed, at the same time their number decreases (Fig. 1.12). To establish the multiplicity of such degeneration, we observe that to a nonrooted vertex there can be "attached" k white vertices in C_{f-1}^{k-1} different ways, whereas to a rooted one the same can be done in $C_f^i=fC_{f-1}^{i-1}/i$ ways. Thus, the change in the number of trees exactly makes up for the combinatorial factor in Eq. (1.16).

To the molecular graph in Fig. 1.1*e* there corresponds its own branching process, the individuals of which are the same as in the first process considered in the preceding. In contrast to this latter process, the reproducing individuals generate white descendants directly, whereas the black individuals pass a stage of 'germ," depicted by cuts, with each germ transforming unambiguously into a black individual capable of further reproduction.

D. Models of Substitution Effect (Model II)

From the viewpoint of graph theory, the account of the substitution effect of functional groups does not lead to principal distinctions from the Flory model, since the combinatorial entropy of formation of the system is determined, as before, by the number of ways the system can be built up of monomers. However, the energy of a state is no longer determined merely by the number of linkages, as was the case in model I. As a consequence, we should make use of the canonical ensemble instead of the microcanonical one.

To write the energy factor for the Gibbs distribution, it is convenient to distinguish the units by the number i of reacted groups.[24–26] The unit of the ith kind is depicted on the molecular graph as a vertex directly connected with i black and $f-i$ white vertices (Fig. 1.13). The number of ordered trees representing some isomer is determined, as before, by Eq. (1.8), but now the summing of the reciprocal orders of automorphism groups is performed only over the isomers of identical energy, (i.e., with identical distribution of vertices by their kind). Now the trees are enumerated in such a way that the counter s_i corresponds to the vertex of the ith kind. The enumerating g.f. with such an arrangement of counters takes the form[24]

$$g^{e}(\mathbf{s}) \equiv \sum_{l} D(\mathbf{l}) \prod_{i} (s_i)^{l_i} = \sum_{i=0}^{f} s_i C_f^i (u^{e}(\mathbf{s}))^i, \tag{1.17}$$

$$u^{e}(\mathbf{s}) = \sum_{i=1}^{f} s_i C_{f-1}^{i-1} (u^{e}(\mathbf{s}))^{i-1}, \tag{1.18}$$

where the component l_i of a vector $\mathbf{l}$ denotes the number of units of the ith kind in the molecule.

The enumerating g.f. can be transformed into the probability g.f. by simply renormalizing the argument $\mathbf{s}$, just as in model I. In this case each component s_i acquires a factor connected with the formation energy of the unit of the ith kind. Instead of the conversion p in the renormalization procedure, use is made of the fractions λ_i of units of the ith kind to all units of the system. With these

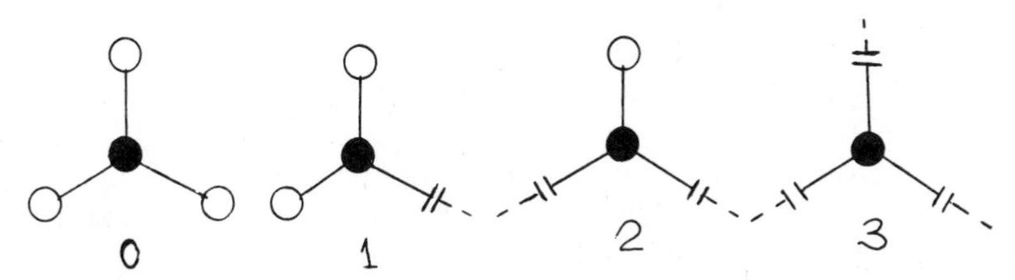

Figure 1.13. Units of various kinds that differ in number of adjacent black vertices.

specifications the expression for the g.f. of the MWD takes the form

$$G_W(\mathbf{s}) \equiv \sum_{\mathbf{l}} f_W(\mathbf{l}) \prod_i (s_i)^{l_i} = \sum_{i=0}^{f} s_i \lambda_i (u(\mathbf{s}))^i, \tag{1.19}$$

$$u(\mathbf{s}) = \sum_{i=1}^{f} s_i d_i [u(\mathbf{s})]^{i-1}, \qquad d_i = \frac{i\lambda_i}{pf}, \qquad pf = \sum_i i\lambda_i, \tag{1.20}$$

which can be interpreted in terms of the theory of branching processes[2,15] and defines the probability measure on the set of trees.

The relation between the probability parameters of the branching process and the equilibrium constants of the elementary reactions for the models of substitution effects was established for the first time elsewhere[27] on the basis of main properties of equilibrium reactions.[28]

In model I each f group reacts in an independent manner, so that the probability λ_i of finding, among the f groups of a single unit, exactly i that have been reacted is $\lambda_i = C^i_f p^i (1-p)^{f-i}$. With these values for the fractions Eqs. (1.19) and (1.20) transform into (1.12) provided all counters s_i are set equal, $s_i = s$.

In the preceding model for the first-shell substitution effect, the distribution parameters on the set of trees are the fractions λ_i of different subgraphs consisting of a single unit. More complex versions of model II can be formulated where the activity of the group is determined not only by the kind of the adjacent unit but also by its further topological environment. The probabilities of different molecular graphs in this case will be expressed via the fraction of subgraphs, where the maximal diameter depends on the assumptions adopted in the model.

The theory of such random graphs is discussed in the papers by Matula and co-workers.[29,30] In these papers a probability measure is constructed on the set of all rooted disordered subgraphs composed in a random way from some base set of subgraphs of small sizes. With this random construction each of several possible "extensions" of a subgraph is chosen with the probability proportional to the fraction of new base subgraphs arising in this way. For example, when the rooted subgraphs of Figs. 1.14*a*, *b*, corresponding to the vertices of different kind, are taken as the base subgraphs, to the broken linkage in Fig. 1.14*a* there can be attached one of the rooted subgraphs of Fig. 1.45*b*. According to the algorithm,[29,30] the probabilities of the resulting random subgraphs (Fig. 1.14*c*) should be proportional to the relative fractions of the fragments added. The probability of a subgraph of any size can be obtained by reiterating this procedure. However, in this case, all possible extensions have to be sorted out at each step, so that application of this algorithm for sufficiently large subgraphs is rather troublesome. This problem

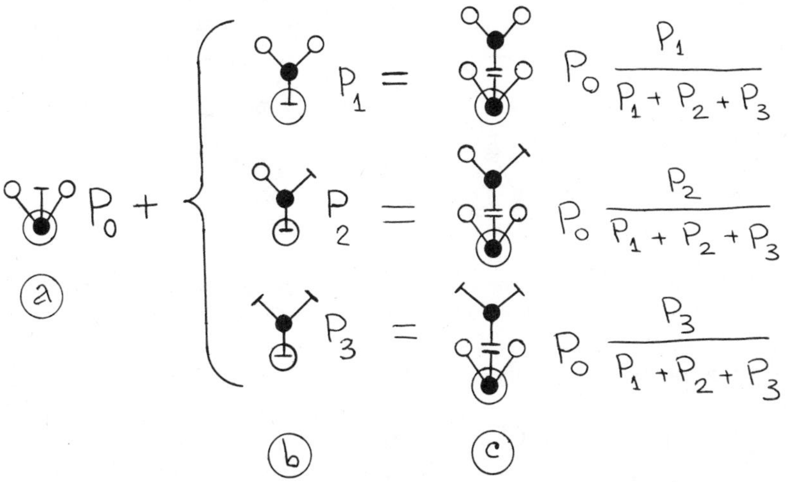

Figure 1.14. Example of simplest "continuations" of subgraph, according to algorithm of refs. 29 and 30. Subgraph probability indicated on right.

can be solved entirely only for complete molecular graphs (such as the upper graph in Fig. 1.14*c*). The expression for the concentrations of different *l*-mers obtained in this way[29] can be reduced to the form obtained later[31] by the method of enumeration of rooted trees with a given distribution of the kind of vertex. An equivalent result can be obtained by expanding the g.f. (1.19) of the branching process in powers of counters. It is quite natural since a random extension of a graph (Fig. 1.14) can be regarded as an elementary act of the reproduction of individuals of the branching process. The theory of such processes allows us to compute the probabilities of different subgraphs in a more efficient manner (see Section II).

Application of the method of random graphs[29,30] is most reasonable for the calculation of probabilities of subgraphs of small sizes, that is, sufficient, for example, for the calculation of the conditions of gelation. In this case more complex models of model II type can be considered with the base subgraphs containing some black vertices.[29,30] Random extension of rooted subgraphs again can be interpreted as reproductions of individuals, which, however, become overlapping (Fig. 1.15), that is, inconvenient in the theory of random branching processes. The latter theory deals, as a rule, with the individuals reproducing in an independent manner. The only correlation in the creation act that is taken into account is related to the prehistory of the individual, that is, to some information about its "father" and "grandfather" (Fig. 1.15). However, from the viewpoint of chemistry, the kind of a "brother" D affects the reactivity of groups of the particle C in the same way, as does

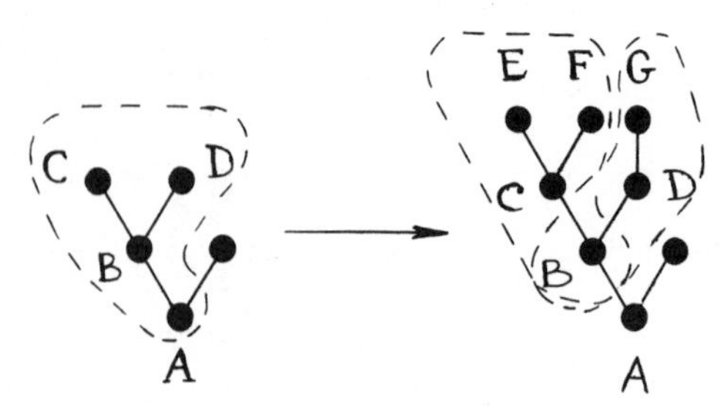

Figure 1.15. Choice of proliferating superparticles (circled with dashed lines) of branching process for higher order substitution effect model.

the genus of its "grandfather" A. Hence, it is reasonable to combine the "brothers" C and D together with their ancestors A and B into one larger "superindividual." When C and D generate the "children" E, F, and G, B becomes a "grandfather," giving rise to new superindividuals, one of which involves the units B, C, E, and F and another that involves the units B, D, and G (Fig. 1.15).

The transition from the combinatorial entropy to the number of ordered trees described can be easily generalized for the systems with several types of monomers. To this effect, it is sufficient to color the graph vertices with a greater number of colors and to take into account the coloring under automorphism transformations of this graph. If the degree of a vertex of some νth color is f_ν, the branches outgoing from this vertex can be transposed in $(f_\nu - 1)!$ (for the root in $f_\nu!$) ways. Let us specify a molecule by a composition vector l with components l_ν ($\nu = 1, 2, \ldots$) the numbers of graph vertices of different colors. In this case, for the qth isomer of such an l-mer, the number of ordered rooted trees can be obtained by transposing all vertices of its molecular graph in the form[1,17]

$$D(\mathbf{l}, q) = \sum_{\nu} l_\nu f_\nu \prod_{\mu} [(f_\mu - 1)!]^{l_\mu} / \mathscr{S}(\mathbf{l}, q), \tag{1.21}$$

which is a generalization of Eq. (1.8).

Some authors[31-34] calculated the total number of rooted trees with a given composition vector $\mathbf{l}$ and found the function of the weight size composition distribution (SCD) via formulas similar to (1.11). From the viewpoint of practical calculations, such an approach is hardly efficient, since in calculating the physicochemical properties of polymers, it becomes necessary to average over this distribution. It is much more efficient for this purpose to employ the generating functions from the very beginning, whereas the values of $D(\mathbf{l})$ and the respective probabilities $f_W(\mathbf{l})$ can be recovered, if needed, from the g.f. using the generalization of the Lagrange expansion.[18,19]

The types of trees to be enumerated with the help of the g.f. are determined by the model adopted, namely, by the dependence of the energy of formation of the molecule on its configuration. Thus, in model I in the most simple case the energies of all bonds are assumed identical, and consequently, the total concentration $c(\mathbf{l})$ of molecules of a given composition, which is proportional to the sum of reciprocal orders of the automorphic groups $(\mathcal{S}(\mathbf{l}, q))^{-1}$, is reduced, according to (1.21), to enumerating the ordered trees with a preassigned distribution of the vertex types. The enumerating g.f. for the trees with l_ν vertices of type ν is expressed as (Fig. 1.16)

$$g^{e}(\mathbf{s}) \equiv \sum_{\mathbf{l}} D(\mathbf{l}) \prod_{\nu} (s_\nu)^{l_\nu} = \sum_{\nu}' s_\nu \left(\sum_{\mu} u_\mu^{e}(\mathbf{s}) \right)^{f_\nu},$$
$$u_\nu(\mathbf{s}) = s_\nu \left(\sum_{\mu} u_\mu^{e}(\mathbf{s}) \right)^{f_\nu - 1}. \tag{1.22}$$

When the pendant vertices can be discarded, the respective counters should be set at $s_\nu = 1$; the functions $u_\nu^{e}(\mathbf{s})$ corresponding to these vertices also become equal to unity [cf. (1.9)]. The primed summation symbol in (1.22) denotes the summation exclusively over those colors ν in which the root can be colored. For example, when the pendant vertices in Fig. 1.16 are assumed to be nonrooted, the last item in this figure is omitted. Renormalizing the arguments in the enumerating g.f. gives the probability g.f. for the weight MWD, obtained formerly in terms of the theory of branching processes.[35]

The substitution effect for the functionalities complicates the enumeration problem quite insignificantly. Thus, in this case, it is merely sufficient to distinguish, besides the types of units, (as in the preceding section), their kinds on which the energy of molecules in model II depends. More essential alterations are required for the systems, in which each of the monomers carries an arbitrary number of various functionalities. In this case the energy of a bond, depicted by some edge of the graph, is no longer defined by the colors of the adjacent vertices. These complex multicomponent systems were considered for the first time by Whittle.[36–41]

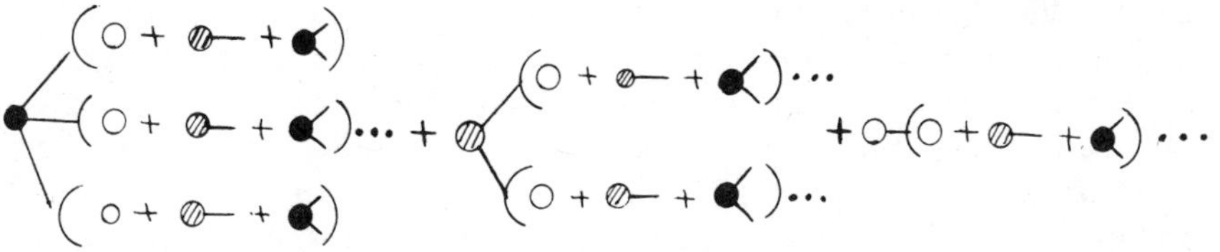

Figure 1.16. Graphical representation for enumerating generating function of colored ordered trees; generalization of Fig. 1.10.

The most general case of multicomponent systems is discussed in refs. 24 and 42, where the g.f. of the MWD is expressed in the form traditional in the theory of branching processes, whereas the probability parameters, coinciding with fractions of units of different kinds, are expressed via the equilibrium constants of elementary reactions. The probability measure on the set of molecular graphs can be constructed only on the basis of branching processes, since the g.f. of the SCD merely defines the total concentrations of isomers with either identical composition or identical distribution of units by their kinds.

The energy of the molecule with bonds of different types depends not only on the composition $\mathbf{l}$ but also on the numbers b_{ij} of bonds formed in the reaction between the functionalities of the ith and jth types; thus, the calculation of the MWD and SCD is reduced to enumerating the ordered trees with given distribution $\mathbf{l}$ of colored vertices and the chemical bond matrix $\mathbf{B} = \{b_{ij}\}$. In this case, besides the units' counters s_ν, it is necessary to introduce the counters x_{ij} for bonds of different types. The enumerating g.f. for such trees is of the form[42]

$$g^{\mathrm{e}}(\mathbf{s}, \mathbf{X}) \equiv \sum_{\mathbf{l},\mathbf{B}} D(\mathbf{l}, \mathbf{B}) \prod_{\nu} (s_\nu)^{l_\nu} \prod_{i \leqslant j} (x_{ij})^{b_{ij}}$$
$$= \sum_{\nu} s_\nu f_\nu! \prod_i \{(\xi_i^{\mathrm{e}}(\mathbf{s}, \mathbf{X}))^{f_{i\nu}} / f_{i\nu}!\},$$

$$\xi_i^{\mathrm{e}}(\mathbf{s}, \mathbf{X}) = 1 + \sum_{j\mu} x_{ij} s_\mu (f_\mu - 1)! \frac{f_{j\mu}}{\xi_i^{\mathrm{e}}} \prod_{\kappa} \{(\xi_k^{\mathrm{e}}(\mathbf{s}, \mathbf{X}))^{f_{k\mu}} / f_{k\mu}!\}. \tag{1.23}$$

Here $f_{i\nu}$ denotes the number of groups of the ith type in the monomer of the νth type. Likewise, by introducing some additional counters, there can be enumerated the trees, differing not only in the types but also in the kinds of units,[24] that are required for the complex systems in model II. Renormalizing the arguments in these enumerating g.f.'s gives, in the usual way, the probability g.f. of the weight SCD.

E. Invariance Principle

Up to this point it was implied that the molecular configuration is completely described by its molecular graph. This holds true for the molecules consisting of those monomers with functionalities that can be rearranged freely in space, for example, when they are connected with the monomer unit by flexible polymeric chains (Fig. 1.17*a*). In such starlike molecules, with sizes that are much greater than those of their central fragments, for any set of coordinate values of this fragment and functionalities, any pair of those functionalities can be interchanged without affecting the others. During such interchanging, rigid monomers (Fig.1.17*b*) are rearranged in space exclusively as a whole. The

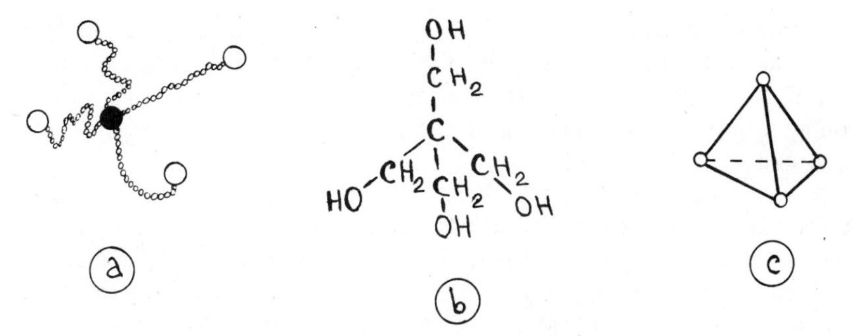

Figure 1.17. (*a*) Model of monomer with flexible chains between unit and the groups. (*b*) A rigid monomer and (*c*) its representation as tetrahedron.

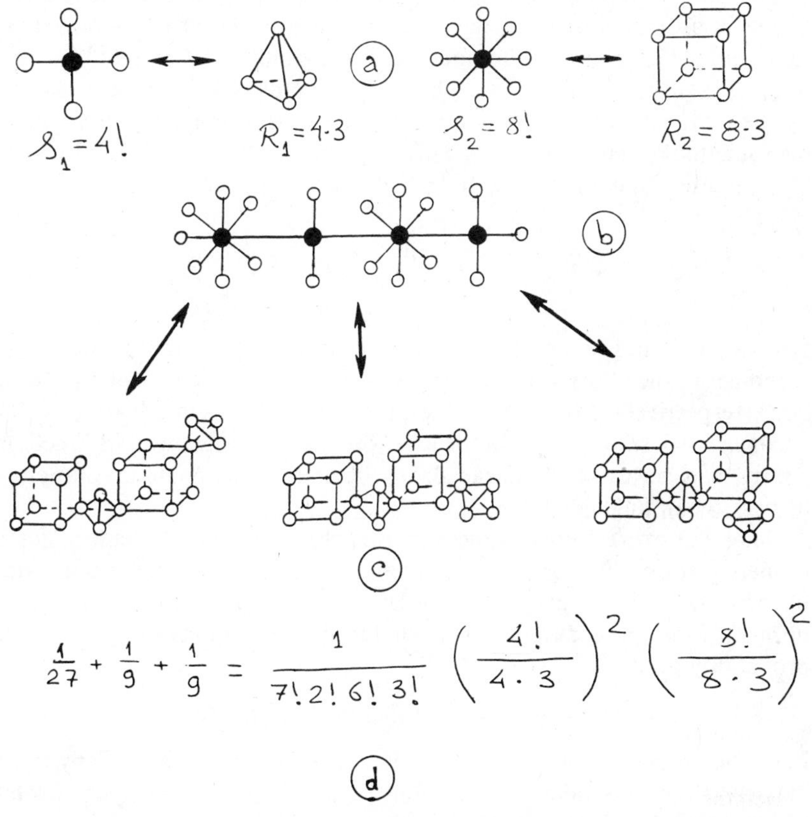

Figure 1.18. (*a*) Models of rigid monomer given by tetrahedron and cube. Molecular graph [(*b*)] depicts three isomers [(*c*)]. (*d*) Relation between orders of isomorphism groups for monomers and molecules given by invariance principle, Eq. (1.24).

macromolecule in this case can be represented as a set of rigid fragments connected with the joints, the chemical bonds. Here, to a single molecular graph there generally corresponds, several stereoisomers with different configurations that cannot be transformed into each other in space[1] (Fig. 1.18).

At this level of description there arises the problem of calculating the molecular configurational distribution, which defines the fractions of different stereoisomers in the polymeric pattern. However, when all stereoisomers with identical graphical configuration $(\mathbf{l}, q)$ have identical energy, the theoretical results based on the concept of molecular graphs remain valid provided the concentration $c(\mathbf{l}, q)$ denotes the total content of all such stereoisomers. Validity of this suggestion follows from the invariance principle established by Gordon and Temple.[1,43]

To elucidate this point, we observe that the group of all possible rotations of a stereoisomer r in space, the order of which is denoted as $\mathscr{R}(\mathbf{l}, q, r)$, is a subgroup of the group of automorphisms of the molecular graph.[44] Inserting $\mathscr{R}(\mathbf{l}, q, r)$ in place of $\mathscr{S}(\mathbf{l}, q)$ in (1.5) leads, in a standard way, to the expression for the sum of concentrations $c(\mathbf{l}, q, r)$ over all stereoisomers r, represented by the same molecular graph $(\mathbf{l}, q)$. This sum can be reduced to (1.7) by virtue of the mathematical formulation of the invariance principle[1,43] (Fig. 1.18):

$$\sum_r \frac{1}{\mathscr{R}(\mathbf{l}, q, r)} = \frac{1}{\mathscr{S}(\mathbf{l}, q)} \left\{ \prod_\nu \left(\frac{\mathscr{S}_\nu}{\mathscr{R}_\nu} \right)^{l_\nu} \right\}. \tag{1.24}$$

Here $\mathscr{S}_\nu = f_\nu!$ and $\mathscr{R}_\nu$ are the orders of the groups of automorphisms for the monomer of the νth type for its graphical and three-dimensional representations, respectively. For example, for pentaerythritol (Figs. 1.17*b*, *c*) $f_{PE} = 4$ and $\mathscr{R}_{PE} = 4 \cdot 3$, since for any of the four ways of choosing one of its functional groups as a vertex of the tetrahedron, there are three possible rotations of its base by an angle of 120°.

Since the expression in braces in the r.h.s. of (1.24) is identical for all isomers q, it does not prevent the summation over q's of the concentrations of molecules with a given composition vector $\mathbf{l}$ and vanishes in the renormalization procedure when going from the enumerating g.f. to the probability g.f.

F. Models Allowing for Cyclization

Even the groups, separated on the molecular graph by a path of any length, may enter the intramolecular reactions. It is essentially troublesome to take into account such long-range correlations, compared to those of short range considered in model II. A direct application of the Stockmayer's method described in Section I.C to the graphs involving cycles leads to the physically senseless result that gelation can occur for any low conversion,[45] which is related to the dramatically rapid growth of the number of ways in which l monomers can be assembled into the molecule of an l-mer. The paradox is explained by the fact that not all of these ways are equally probable since the number of possible conformations (methods of arrangement in space) of the region of molecules that enters the essentially decreases on cycle formation. To date, no solution is available for such configurational–conformational problem in its general formulation. A detailed analysis of this problem is presented in Section III. Here we confine ourselves to some approximate methods for its solution.

Analysis of cyclization in linear molecules shows[46] that the formation of a single cycle leads to a decrease in conformational entropy. Since for nonoverlapping cycles the conformational entropy is an additive characteristic, the entropy for the isomers with any unit belonging to one cycle only can be regarded as known. In this case the molecular graphs are cacti[19,47] (Fig. 1.19) and are characterized by a vector of cycles $\mathbf{m} = \{m_0, m_1, \ldots, m_n, \ldots\}$, where the nth component m_n is the number of cycles of size n, and m_0 is the number of nodes that enter no cycle. The isomers characterized by the

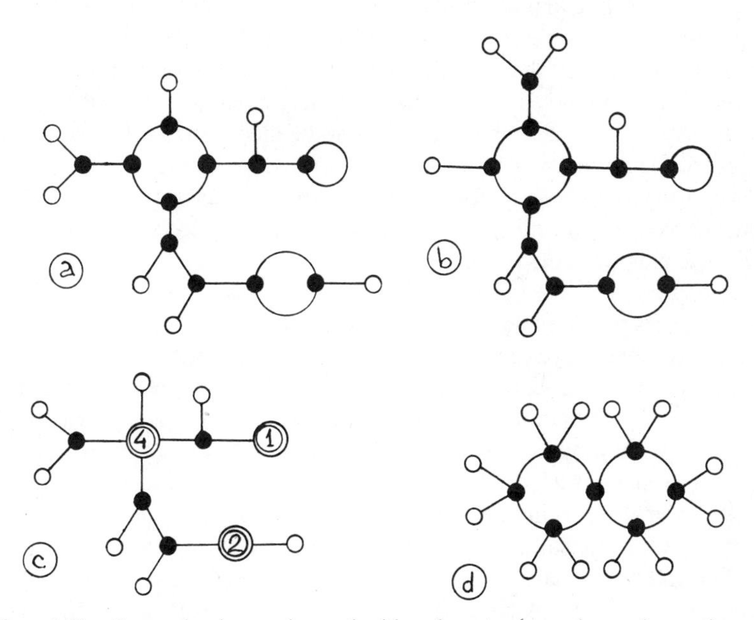

Figure 1.19. Two molecular graphs, cacti, with cycle vector $\{m_0 = 4, m_1 = 1, m_2 = 1, m_4 = 1\}$ [(a) and (b)] and their common leaf composition, contracted tree [(c)]; numbers at the nodes indicate the colours. (d) Cactus is discarded since it contains a node, which belongs to two cycles simultaneously.

same vector of cycles have identical formation energy and conformational entropy, and by using the invariance principle (Section I.E), calculation of the number of ways in which such graphs can be assembled is reduced to the problem of enumerating the ordered trees, which has already been solved. For this purpose let us depict a cactus as a tree by contracting each of cycles of size $n \geqslant 1$ into one vertex and, to distinguish it from others, painting it in the nth color (Fig. 1.19). The functionality of such a quasi-monomer is $f_n = (f-2)n$, so that the total concentration of all isomers represented by a certain contracted tree $(\mathbf{m}, q)$ with m_n vertices of the nth color is proportional, according to (1.24), to

$$\sum_i \frac{1}{\mathscr{S}_i} = \frac{1}{\mathscr{S}(m, q)} \prod_{n \geqslant 1} \left\{ \frac{[(f-2)n]!}{2n[(f-2)!]^n} \right\}^{m_n}. \tag{1.25}$$

Now $\mathscr{S}_i$ denotes the order of the automorphism group of the noncontracted molecular graph of the ith isomer, and $2n[(f-2)!]^n$ is the order of the automorphism group of the separate cycle. The factors 2, n, and $[(f-2)!]^n$ correspond, respectively, to the mirror reflection of the cycle, its rotations, and transpositions of unreacted functionalities of each of n units of the cycle. Next, Eq. (1.21) leads to enumerating the ordered trees of a definite composition. Renormalizing the enumerating g.f. (1.22) for these trees affords the probability g.f. for the distribution of molecules by the number of cycles of various sizes involved, which can be interpreted in terms of branching processes.[48,49] This distribution was established initially[50] via a direct combinatorial calculation of the number of ways the molecules with a given vector of cycles can be assembled. This method, in contrast to that described in the preceding, is more cumbersome, and its generalization to multicomponent systems presents a problem.

The degree of correctness of the preceding assumption on the lack of complex cycles in molecules can be estimated after solution of a more general problem with these cycles being accounted for (cf. Section III). Further, the applicability of the model to some real polymeric system can be established. To date, no a priori methods exist to specify the class of systems described by other approximate methods to consider cycle formation. For example, the random graph theory[51] makes extensive use of the unrestrictedness of the degree of vertex, precluding the use of this theory for the description of polymers. Whittle assumes,[36–41] as does Stepanov,[51] that the intramolecular bonds between any pair of functionalities in the molecule are formed with identical probability regardless of its configuration and size. As exemplified in ref. 45, this assumption may drastically distort the behavior of the system. Bruneau,[52–54] neglecting the conformational peculiarities of cyclization, is also subject to similar criticism. Nevertheless, the approach of critics,[55–57] which is known as the approximation of spanning trees, suffers from similar drawbacks, as does its subsequent extension.[58,59]

To obtain the spanning tree of the graph, it suffices to break some linkages and to label the functionalities thus derived as specific (i.e., cycleforming). Next, selecting one of the units as a root, these trees are regarded as the realization of the branching process generating occasionally white cycle-forming individuals (labeled in Fig. 1.20 with a cross). However, the selection of the spanning tree still remains ambiguous. At the same time, to any realization of the branching process there may correspond different isomers,

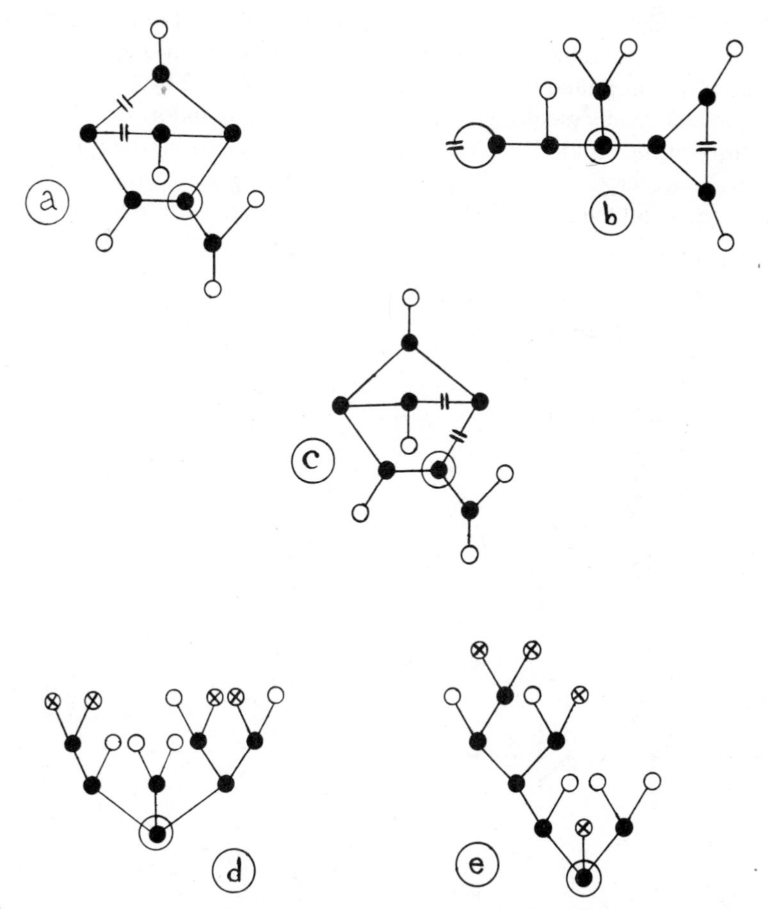

Figure 1.20. Two different molecular graphs [(*a*) and (*b*)] may have same spanning tree [(*d*)], while breaking different bonds in two identical graphs [(*a*) and (*c*)], (indicated by cuts) one may get different spanning trees [(*d*) and (*e*)].

where the conformational entropy and, hence, the formation probability may be essentially different (Fig. 1.20). Besides these principal disadvantages, which have to do with the lack of one-to-one correspondence between molecular graphs and rooted ordered trees, there arise some complications in constructing the branching process that describes such frame trees. A deficiency of this method was recognized by the authors,[55] who, in fact, did not claim to calculate the MWD by this method but rather confined themselves to such characteristics of the local structure of graphs as the distribution of units by kinds.

Among the papers taking into account the conformational statistics of molecules, an original approach should be pointed out,[60] in which the authors elucidate the observed inhomogeneity of the network polymers. It was shown that for a fixed number of nodes and bonds of the molecular graph the isomers represented by a tree with attached groups of cycles of small size are characterized by greater probability, in contrast to uniformly cross-linked structures of the regular lattice type. The proof is based on the operation of sticking together the two adjacent edges belonging to a single cycle, with simultaneous attachment of another (noncyclic) edge with a pendant vertex

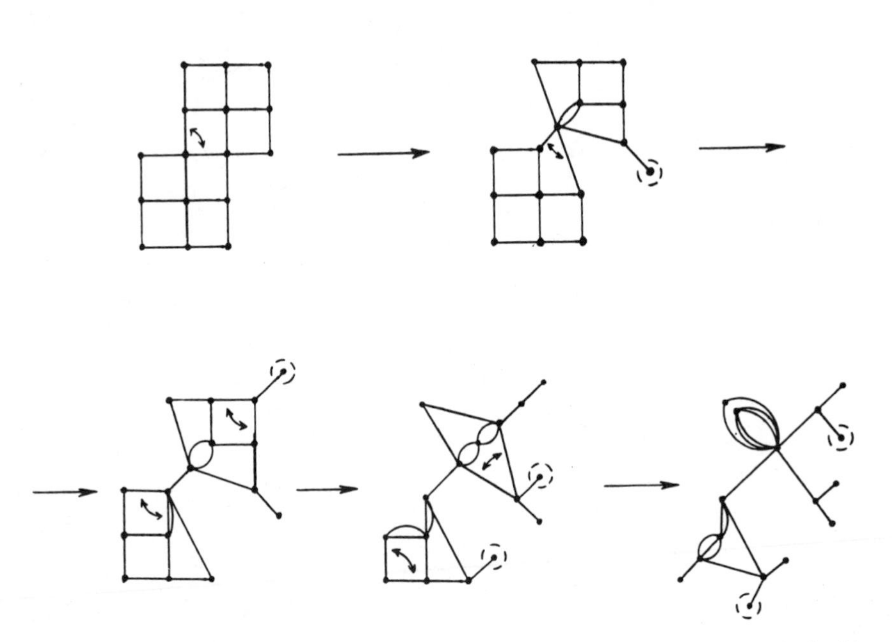

Figure 1.21. Appearance of inhomogeneities in network. Attaching edges marked with arrows and adding pendant vertex (circled with dashed line), one gets graph corresponding to molecule having larger conformational entropy than original one.

(Fig. 1.21). This operation does not alter the number of vertices and edges of the graph, that is, the energy of the molecule, whereas the number of its possible conformations in space was shown[60] to have increased. Sequential reiteration of this operation, accompanied by the increase in entropy with unaffected enthalpy, leads to the inhomogeneity of the network mentioned previously, since the free energy for such configurations is less than that for the homogeneous state. However, these considerations neglect the combinatorial component of the entropy related to the number of ways in which the molecule can be assembled. With due regard for this factor, the validity of the results of ref. 60 is open to discussion. The probabilities of cycles of different topology and size can be estimated in a quantitative way by using the field-theoretic methods,[61,62] although many of the results established by these methods can also be obtained by generalizing the method of contraction of a cactus into a tree, described in the foregoing, to the cycles of complex topology, followed by the calculation of the conformational entropy losses on the formation of the complex cycle.[63] A detailed account of these methods will be given in Section III.

The description of a huge macroscopic molecule, the gelfraction of the polymer, also necessitates consideration of the cycles. This follow from the fact that it is impossible to embed an infinite tree with a uniform density of nodes in space, whereas in the presence of cycles, such a "Malthusian paradox" does not occur.[23] That is why even the Flory model allows intramolecular reactions in gel. In this model a sol fraction consisting of molecules of finite size is represented by a forest, which, as formerly, can be described by the methods considered in the preceding. In particular, the theory of branching processes allows one to establish the fraction ω_s of units entering the sol and to determine the conversion $\hat{p}$ of functionalities of finite trees.[64,65] These can be expressed via the probability of degeneration $u^* = u(1)$, which is the maximal positive root of the second equation of Eq. (1.12) for $s = 1$:

$$u^* = (1 - p + pu^*)^{f-1}. \tag{1.26}$$

The latter probability is the fraction of finite branches of the tree representing the branching process. For the gelation point $[p < p^* = 1/(f-1)]$ all realizations are of finite character and $u^* = 1$. After the gel has been formed $(p > p^*)$, we have $u^* < 1$, and

$$\omega_s = (1 - p + pu^*)^f = u^*(1 - p + pu^*), \qquad \hat{p} = pu^*/(1 - p + pu^*). \tag{1.27}$$

When $\omega_g N = (1 - \omega_s)N$ units of gel are combined into a single molecule, there appears $(1 - \omega_s)N - 1 \simeq (1 - \omega_s)N$ bonds. Hence, the conversion of functionalities of the gel p_g and the number of bonds formed as a result of intramolecular reactions, that is, the cyclic rank r_g of the molecular graph of the gel, can easily be calculated

$$p_g = 1 - \frac{fN(1-p) - f\omega_s N(1-\hat{p})}{fN(1-\omega_s)} = p\frac{1+(u^*)}{1+pu^*}, \tag{1.28}$$

$$r_g = \tfrac{1}{2}(fN)p_g(1-\omega_s) - (1-\omega_s)N = (N(1-u^*)\left[\frac{fp}{2}(1+u^*) - (1-pu^*)\right] \tag{1.29}$$

The parameter r_g plays an important role in the theory of high elasticity since in one of its contemporary versions[66,67] the elastic free energy and, consequently, the force required for the deformation of the network is proportional to its cyclic rank.

In other theories the front factor incorporates, instead of the cyclic rank, the number of elastically active chains or nodes.[69,70] The difference between these two values, which are calculated by the theory of branching random processes,[71] is exactly the cyclic rank of the network.[67] Numerous attempts have been made[72] to elucidate the influence of different network defects, such as inactive and small-size rings and pendant vertices, on the elastic energy. Graph theory may prove very helpful in solving these questions. However, even for the perfect nets, no generally accepted model of high elasticity is currently available, which would relate unambiguously the strain and deformation to the topological structure of the network.[68,72–74] Thus, the problem of the correct description of the network polymers is at present far from being settled.

G. Kirchhoff Matrix and Conformation of Molecules

The topological matrices of graphs contain complete information on the global and local structure of the graph, as does its graphical representation. Graph theory makes extensive use of the adjacency **A** and incidence **B** matrices and occasionally of the Kirchhoff matrix **K**, which is obtained from $-\mathbf{A}$ when the ith element of its principal diagonal is substituted by the degree of the ith vertex. The Kirchhoff matrix can also be obtained by assigning arbitrary orientation to the edges of the graph and then multiplying the incidence matrix of the resulting orgraph **B** by its transpose $\mathbf{B}^\mathrm{T}$:

$$\mathbf{K} = \mathbf{B}\mathbf{B}^\mathrm{T}. \tag{1.30}$$

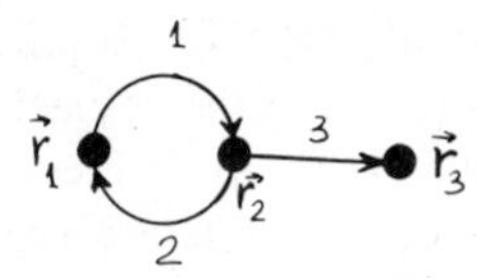

Figure 1.22. Labeled orgraph.

For example, for the orgraph shown in Fig. 1.22,

$$\mathbf{A} = \begin{pmatrix} 0 & 2 & 0 \\ 2 & 0 & 1 \\ 0 & 1 & 0 \end{pmatrix}, \qquad \mathbf{B} = \begin{pmatrix} 1 & -1 & 0 \\ -1 & 1 & 0 \\ 0 & 0 & -1 \end{pmatrix},$$

$$\mathbf{K} = \begin{pmatrix} 2 & -2 & 0 \\ -2 & 3 & -1 \\ 0 & -1 & 1 \end{pmatrix} = \begin{pmatrix} 1 & -1 & 0 \\ -1 & 1 & 1 \\ 0 & 0 & -1 \end{pmatrix} \begin{pmatrix} 1 & -1 & 0 \\ -1 & 1 & 0 \\ 0 & 1 & -1 \end{pmatrix} = \mathbf{BB}^{\mathrm{T}}. \tag{1.31}$$

In fact, the scalar square of the ith row of the matrix $\mathbf{B}$ is the degree of the ith vertex, and the scalar product of the ith row and the jth row is the number of edges connecting the vertices i and j taken with opposite sign. In polymer science the matrix $\mathbf{K}$ plays the significant role, since the conformational statistics of the molecule is directly related to the characteristics of this matrix.[75]

When monomeric units are connected by sufficiently long polymeric chain, as for the starlike monomers of the type shown in Fig. 1.17*a*, the distance between these units in the absence of physical interactions is distributed according to Gauss's law.[7,8] Hence, such molecules can be represented as bead units connected by springs with the force constant $k_B T\gamma$, where k_B is the Boltzmann constant, T is the absolute temperature, and $2/(3\gamma)$ is the mean-square extension of the spring. The macromolecular conformation in this model is characterized by the coordinates $\mathbf{r}_i$ of units, and the energy $U_{\mathrm{el}}\{\mathbf{r}_i\}$ of some conformation is the sum of potential energies of all springs:

$$U_{\mathrm{el}}\{\mathbf{r}_i\} = k_B T\gamma \sum_{i-j} |\mathbf{r}_i - \mathbf{r}_j|^2. \tag{1.32}$$

This sum over all pairs of adjacent units i–j can be expressed[75] in matrix form by using the Kirchhoff matrix. For this purpose let us introduce a 3×1 matrix **R**, where the elements $r_{\alpha i}$ are the projections of the radius vectors of molecular units onto Cartesian coordinate axes ($\alpha = 1, 2, 3$). The vectors $\mathbf{r}_i - \mathbf{r}_j$ of distances between the adjacent units are represented by the columns of the matrix **RB**, for example, for the molecules in Fig. 1.22,

$$\begin{pmatrix} r_{11}-r_{12} & r_{12}-r_{11} & r_{12}-r_{13} \\ r_{21}-r_{22} & r_{22}-r_{21} & r_{22}-r_{23} \\ r_{31}-r_{32} & r_{32}-r_{31} & r_{32}-r_{33} \end{pmatrix} = \begin{pmatrix} r_{11} & r_{12} & r_{13} \\ r_{21} & r_{22} & r_{23} \\ r_{31} & r_{32} & r_{33} \end{pmatrix} \begin{pmatrix} 1 & -1 & 0 \\ -1 & 1 & 1 \\ 0 & 0 & -1 \end{pmatrix} \tag{1.33}$$

In this case, the molecular energy (1.32) is expressed in a simple way via the Kirchhoff matrix of its graph:

$$\sum_{i-j} |\mathbf{r}_i - \mathbf{r}_j|^2 = \mathrm{Tr}(\mathbf{RB}(\mathbf{RB})^{\mathrm{T}}) = \mathrm{Tr}(\mathbf{RBB}^{\mathrm{T}}\mathbf{R}^{\mathrm{T}}) = \mathrm{Tr}(\mathbf{RKR}^{\mathrm{T}}), \tag{1.34}$$

where Tr is the trace of the matrix. This equation allows one to rewrite the configurational integral for the molecular states in the form

$$\int \exp\{-U_{\mathrm{el}}\{\mathbf{r}_i\}/k_{\mathrm{B}}T\} \prod_{i \neq p} \mathbf{dr}_i = \int \exp\{-\gamma \mathrm{Tr}(\mathbf{RKR}^{\mathrm{T}})\} \prod_{i \neq p} \mathbf{dr}_i \tag{1.35}$$

where integration is performed over all units apart from one singled out (e.g., the pth). The multidimensional Gaussian integral (1.35) is calculated by standard methods. The latter integral is expressed via the determinant of the pth principal minor of the matrix **K**, which, according to the matrix theorem about trees,[47] is equal to the number of spanning trees of the graph.

Thus, even such simple information on the Kirchhoff matrix as the value of any of its principal minors allows us to find the free energy of the polymeric molecule. The degree of particularization in the description of the conformational statistics increases with the information content on the matrix **K**. Thus, the knowledge of its spectrum allows us to establish the mean dimensions of the molecule and the distribution for its radius of gyration.[75] The generalization of the theories by Rouse[76] and Zimm[77] allows us, on the basis of the same information, to calculate the dynamic properties of the Gaussian molecule, that is, the spectrum of its relaxation times.[75,78] To this effect, Forsman[78,79] makes use of the analogue $\mathbf{B}^{\mathrm{T}}\mathbf{B}$ for the Rouse matrix[76] instead of the matrix $\mathbf{K} = \mathbf{BB}^{\mathrm{T}}$, which is the generalization of the Zimm matrix[77] for branched molecules. Since the nonzero eigenvalues of the Kirchhoff matrix coincide with those of the Rouse matrix, these two different approaches lead to identical results.

In some problems of the conformational statistics of branched molecules, use is made of the inverse (in a generalized sense[80]) matrix to **K**. The inverse matrix is employed for the calculation of the mean value for the hydrodynamical radius of the molecule as well as the joint distribution of different linear combinations of radius vectors of monomeric units.[75] A detailed account of these and related topics[81–83] is presented in Section V.

H. Macromolecules as Graphs on Lattice and Scalling

The theoretical description of the condensation phenomenon is often based on the well-known model of a *lattice gas*, according to which the particles can occupy only the sites of some regular lattice.[84] To take into account qualitatively the interaction between particles (strong repulsion, i.e., the "effect of excluded volume" at small distances r, attraction at intermediate $r \sim a$, and absence of interaction at large distances $r \gg a$), the following must be considered: First, the lattice gas model prohibits two particles from occupying the same site; second, to each pair of neighboring particles on the lattice with characteristic dimension a, there is assigned identical negative value of interaction energy (Fig. 1.23).

In a similar way the statistical physics of polymers consider their lattice models, in which the molecular graphs are embedded in a regular spatial lattice (cf. Fig. 1.23). In this case the vertices and edges of these graphs can be arranged only in the sites and bonds of the lattice, respectively, the coordination number f of which coincides with the functionality of the monomer. The obvious advantage of these lattice polymeric models is their

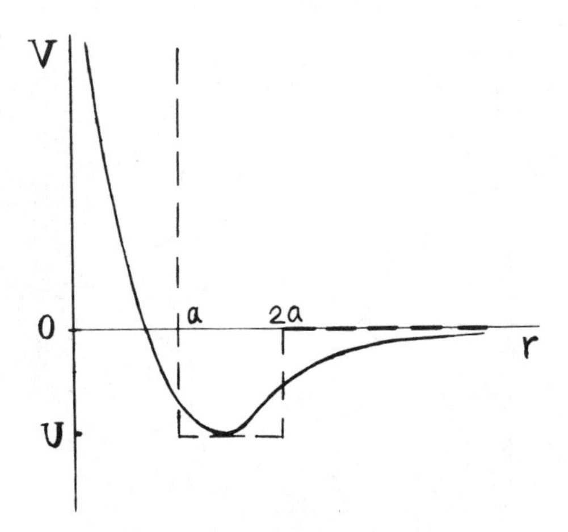

Figure 1.23. Actual form of typical pair interaction potential (solid line) and its lattice model approximation (dashed line).

ability to account for cycle formation and, in addition, to incorporate in a natural way (as in a lattice gas) the physical interactions between monomeric units. These models, however, cannot be employed for the calculation of some significant characteristics of the polymeric system, such as, for example, the conversion at the gelation point p^*, which depends on, besides f, the lattice geometry. It should be remembered that the lattice models claim to describe merely asymptotic values of their statistical parameters, which are determined by the behavior of real polymeric systems on a scale much larger than the characteristic dimension of the lattice.

A striking feature of such behavior is its universal character. In the region of universality near the threshold of gelation, mean molecular weights and sizes of macromolecules, weight of gel fraction, and some other quantities are characterized by the power dependence on the proximity of the conversion p to its critical value p^*. The power exponents for these relationships are referred to as the *critical exponents.* According to the contemporary theory of phase transitions,[85,86] these are determined merely by dimensionality of space d and symmetry of the model, which specify its class of universality.

A particular class of models constitute those on the *Bethe lattice*, where all subgraphs are trees (cf. Fig. 1.24). These models are especially interesting since they admit exact solution of the problems considered. The critical exponents obtained in this way coincide with those found within the framework of the well-known *mean (self-consistent) field* approximation, when polymeric systems are regarded as continual. The Bethe lattice is peculiar in that it cannot be embedded into a space of finite dimensionality d, and thus, it is considered as corresponding to some infinite-dimensional space. In fact,

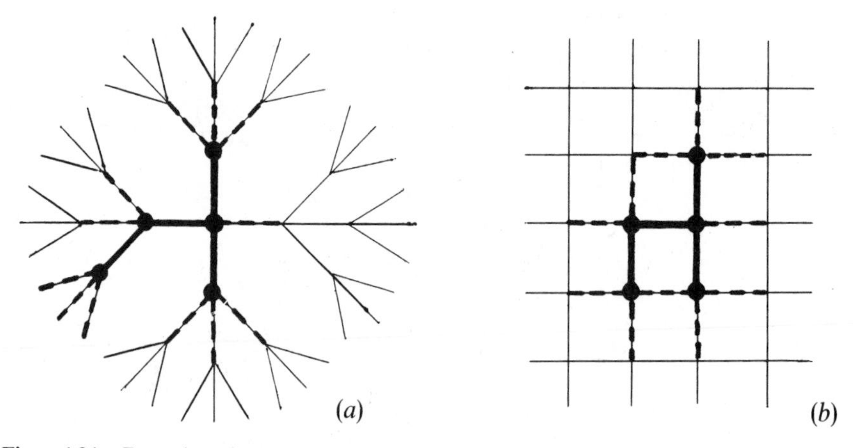

Figure 1.24. Examples of graph arrangements is (*a*) Bethe lattice and (*b*) square lattice. Perimetric bonds indicated----

the threshold of gelation on hyperlattices (which are arranged as the square, $d=2$, and cubic, $d=3$, lattices but are embedded in a space of greater dimensionality) increases monotonically with d, and in the limit $d \to \infty$ it gives the asymptotic value corresponding to the Bethe lattice. The contemporary theory of phase transitions not only deals with integer dimensionality $d>3$ but also makes extensive use of the concept of continuous dimensionality. The critical indices calculated for the Bethe lattice were found to be exact in the space of dimensionality $d>d_c$.

The upper critical dimensionality d_c proves to be an important parameter characterizing the class of universality of the model at hand. The concept of continuous dimensionality enables one to employ the well-developed tools of the "renormgroup" to calculate the critical indices as expansions in the parameter $\varepsilon = d_c - d$ (ref. 86). These methods are based on the "scaling" ideas,[85-87] which employ the similarity in the behavior of features of the system in the region of universality on changing the spatial scale. The scaling consideration of branched macromolecules and gelation, enabling the calculation of critical indices for some lattice models of polymers, have found wide application in numerous theoretical works.[87-89]

The aim of the scaling approach in the description of such polymeric systems is to establish the asymptotic law for their characteristics in the proximity of the gelation point $|p-p^*| \to 0$ and in the limit $l \to \infty$ for the degree of polymerization of macromolecules. The first dependence determines completely the mean-weight degree of polymerization for finite molecules up to (P_W) and after ($\hat{P}_W$) the gel point, the fraction of units ω_g entering the gel, and some other MWD-averaged characteristics of the system. Within the region of universality (i.e., in the proximity of the gel point), the following asymptotic formulas are valid:

$$P_W = c_+(p^*-p)^{-\gamma}, \qquad \hat{P}_W = c_-(p-p^*)^{-\gamma}, \qquad \omega_g = c'(p-p^*)^{\beta}. \quad (1.36)$$

The values of c_+, c_-, and c' as well as the critical conversion in the gel point p^* depend on lattice type and are thus not universal. Scaling theories, which aim to establish the critical indices, fail to determine these parameters. In addition to β and γ, there is a whole set of such indices, the most significant of which, ν, is determined by the $f_Z(l)$-averaged [Eq. (1.2)] radius of gyration squared for the l-mer R_l^2:

$$\xi^2 \equiv \langle R_l^2 \rangle = c'' a^2 (p^*-p)^{-\nu}. \quad (1.37)$$

From the scaling theory it follows[85,86] that for any set of critical indices only two are independent. All other indices are simple functions of the form

$$\nu d = 2\beta + \gamma. \quad (1.38)$$

Other exponents (e.g., τ and ρ) characterize the MWD of large enough clusters and their gel point sizes:

$$f_N(l) = cl^{-\tau}, \qquad R_l^2 = c'''a^2 l^{2\rho} \qquad (p = p^*), \tag{1.39}$$

the values of which are functions of β and γ:

$$\tau = \frac{3\beta + 2\gamma}{\beta + \gamma}, \qquad \rho = \frac{\tau - 1}{d} = \frac{2\beta + \gamma}{(\beta + \gamma)d}. \tag{1.40}$$

A detailed list of critical exponents and their scaling relations can be found elsewhere.[88]

To summarize this section we consider the molecular graphs arranged on a lattice; these graphs are called *lattice animals.*[89,90] The number of lattice animals is equal to the number of such arrangements referred to a single node of lattice. The simplest characteristics of the lattice animal (l.a.) as a graph are the numbers l and b of its vertices and edges, respectively, which determine its cyclic rank $r = h - l + 1$. Together with l and b, which bear information on the configuration (topology) of the molecular graph, the arrangement of its vertices on the lattice is occasionally characterized by the number t of perimetric linkages (cf. Fig. 1.24). Each of the perimetric linkages is, by definition, adjacent at least to one vertex of the l.a., coinciding, at the same time, with none of its edges. The triple (l, b, t) is sufficient to identify the l.a. in all main lattice models for branched polymers known to date. The stoichiometric relations

$$n_1 = fl - b - t, \qquad n_2 = n_1 - b, \tag{1.41}$$

provide the number n_1 of adjacent pairs of vertices of the l.a. and the number n_2 of those disconnected. For example, for l.a. (5, 4, 11) formulas (1.41) yield $n_1 = 5$ and $n_2 = 1$. Among t perimetric linkages of the l.a. there are n_2 internal and $t - n_2$ external ones, the former being adjacent to two of its vertices, whereas the latter is adjacent to only one vertex. The total number of unreacted functionalities of the polymeric molecule, represented by the l.a. (l, b, t), is obviously $a = 2n_2 + 1(t - n_2) = fl - 2b$.

The lattice animals, being the subgraphs of the lattice graph, are distinguished by their "spatial types."[90] Figure 1.25 shows that among the six types, corresponding to the molecule of a tetramer, the first four types characterize different conformations of its linear configuration. The statistical weights of these conformations, listed in brackets in Fig. 1.25, allow us to infer their relative probabilities. Types 5 and 6 in Fig. 1.25 correspond to the branched and cyclic configurations of the tetramer, each having a unique conformation.

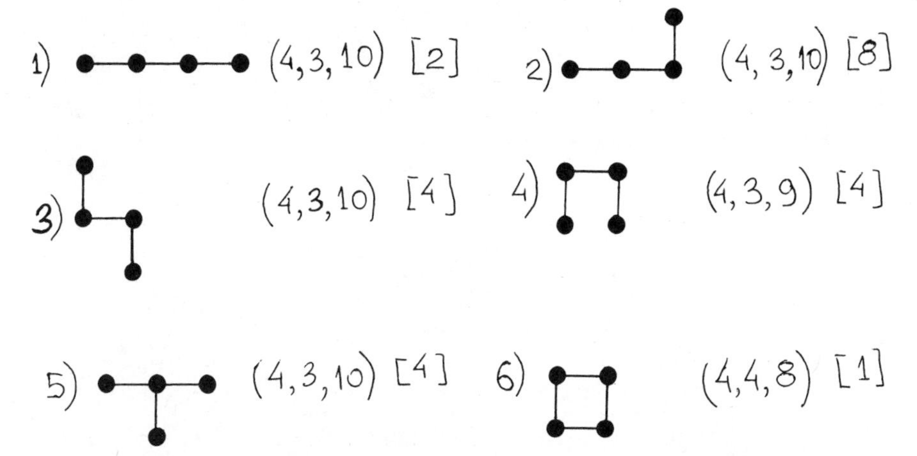

Figure 1.25. Total set of "spatial types" of lattice animals with four vertices in square lattice. Characteristics of each lattice animal given in parentheses; number of corresponding arrangements, which are obtained by rotations and reflections, given in brackets.[90]

The calculation of number of l.a. $A(l, b, t)$ is a purely combinatorial problem, its analytical solution currently unknown. For l.a.'s with small numbers of vertices computer-aided methods for the calculation of A have been developed.[89] For analytical considerations, it is convenient to make use of the enumerating generating functions

$$G^e(s, y, v) \equiv \sum_{l,b,t} A(l, b, t) s^l y^b v^t, \qquad G^e_W = s\frac{\partial G^e}{\partial s}, \tag{1.42}$$

the second of which is the g.f. for the number $lA(l, b, t)$ of different l.a. (l, b, t) containing a fixed node of the lattice (e.g., the origin). For the Bethe lattice, because of the stoichiometric relations for trees,

$$b = l - 1, \qquad t = (f - 2)l + 2, \tag{1.43}$$

this g.f. is reduced to the generating function (g)

$$G^e_W(s, y, v) = (v^2/y) g^e(syv^{f-2}), \qquad G^e_W(s, 1, 1) = g^e(s). \tag{1.44}$$

The total number of tree like l.a.'s (equal to 22) in Fig 1.25 (types 1–5) is naturally identical to that obtained by Eq. (1.9) for $D(l)/l$ for $l = f = 4$. This result is valid for any lattice but only for $l < l_m$. For certain value of l_m, which depends on the lattice type and dimensionality of space, there arise the

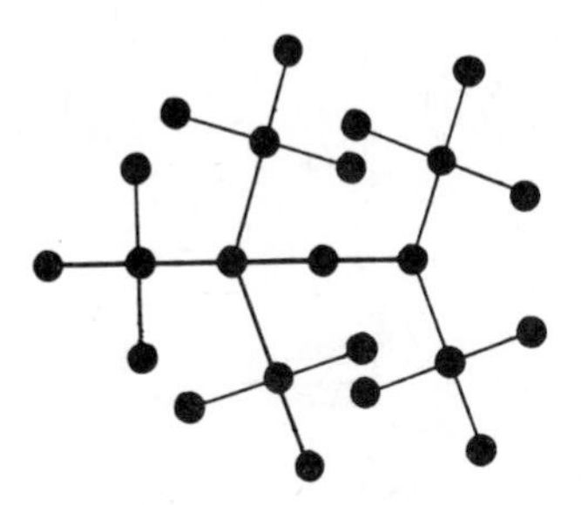

Figure 1.26. A tree ($f = 4$) with minimum number of nodes, $l = l_m = 23$, which cannot be embedded in tetrahedron lattice.[91]

topological structures of graphs, which cannot be embedded in the given lattice[91] (cf. Fig. 1.26). Thus, the g.f. (1.9) enumerates superfluous nonphysical configurations sterically prohibited due to the excluded volume effect and at the same time fails to account for the cyclic structures of the type 6 l.a. in Fig. 1.25.

I. Percolation and Other Lattice Statistical Models

The most simple of lattice models for branched polymers, at the same time allowing for the cycle formation and excluded volume effect, is that of *random percolation through bonds.* This model is intended for the description of statistics of the ensemble of macromolecules formed on equilibrium homopolycondensation of f-functional monomers. This polymeric system is represented by the lattice with coordination number f, where monomeric units occupy each N sites. Hence, all sites of the lattice turn out to be occupied by vertices of molecular graphs, whereas the bonds on lattices can be either occupied or not by the edges of these graphs, depending on the chemical reaction occurring between functionalities of adjacent monomeric units. This model assumes the probability of formation of the chemical bond p between these functionalities to be independent of their environment; that is, there are no substitution effects. Therefore, each bond on the lattice, regardless of the others present, can either be occupied by the edge of a molecular graph with a probability p or be free with a probability $1 - p$. As a consequence, in the percolation model considered, which is known as *bond problem,*[92,93] the probability $\mathscr{P}\{G_N\}$ of any realization $\{G_N\}$ of the system is determined only by the total number of edges of all its molecular graphs:

$$\mathscr{P}\{G_N\} = p^{N_b}(1-p)^{N_b^0 - N_b} = (1-p)^{N_b^0}\omega^{N_b}, \qquad \omega = p/(1-p), \tag{1.45}$$

where the total number of all lattice bonds N_b^0, neglecting the surface effects,

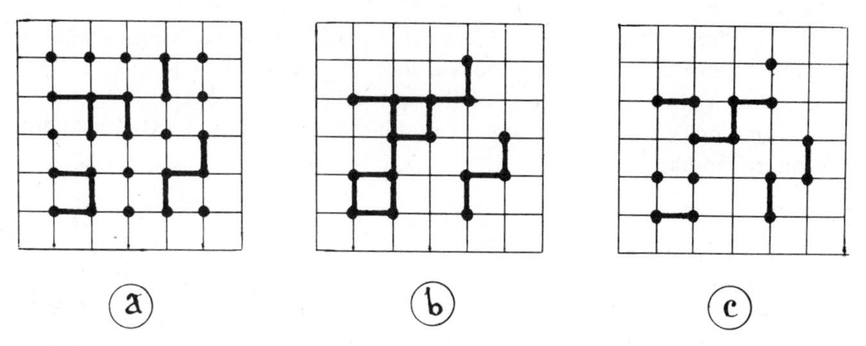

Figure 1.27. Various percolation models: (*a*) by bonds; (*b*) by sites; (*c*) by sites and bonds.

is assumed to be $\frac{1}{2}fN$. Equation (1.45) defines the probability measure on the set $\{G_N\}$ of disconnected subgraphs of the lattice graph, each corresponding to certain realizations of the percolation statistical ensemble (cf. Fig. 1.27*a*). Summing up $\mathscr{P}\{G_N\}$ [Eq. (1.45)] over all 2^N possible realizations $\{G_N\}$ naturally yields unity, since the number of equiprobable realizations of all realizations is the number of combinations of N_b^0 in N_b's.

The mean value $\langle W \rangle$ of any random quantity $W\{G_N\}$, which characterizes the realization $\{G_N\}$, can be calculated by averaging over the probability measure (1.45):

$$\langle W \rangle = \sum_{G_N} \mathscr{P}\{G_N\} W\{G_N\}, \tag{1.46}$$

where the summation is performed over all lattice subgraphs. Usually, the values w of intensive (i.e., referred to a single lattice node) characteristics of the percolation ensemble are of essential interest. These are obtained by taking the "thermodynamic" limit

$$w = \lim_{N \to \infty} (\langle W \rangle / N). \tag{1.47}$$

In particular, the number of connected components of the graph G_N, which are called *clusters*, can be taken as $W\{G_N\}$. In this case, the value w, calculated by (1.46) and (1.47), is the mean number Π of polymeric molecules in the system related to a single monomeric unit. To obtain the generating function for the MWD of these molecules, one should assume

$$W\{G_N\} = \sum s^l y^b v^t, \tag{1.48}$$

where the summation is performed over all connected components of the graph G_N. The statistical weight of each cluster is defined, according to (1.48), merely by the numbers of its nodes l, links b, and perimetric links t. The generating function for the distribution $C(l, b, t)$ of clusters by their characteristics (l, b, t) can be easily obtained by inserting (1.48) into (1.46) and interchanging the summations:

$$w(s, y, v \mid p) = \sum_{l,b,t} A(l, b, t) p^b (1-p)^t s^l y^b v^t, \tag{1.49}$$

where $A(l, b, t)$ is the number of l.a.'s (l, b, t). The g.f. $G_W(s)$ for the MWD is a partial derivative of w with respect to ln s at $y = v = 1$, that is,

$$f_W(l) = \sum_{b,t} l A(l, b, t) p^b (1-p)^t, \tag{1.50}$$

which is a generalization of Eq. (1.11) (valid for percolation on the Bethe lattice only) for arbitrary lattice. In a similar way the algorithm for the probability g.f. for cluster distribution (1.49) can be generalized by renormalizing the arguments of the enumerating g.f. for l.a.'s (1.42):

$$w(s, y, v \mid p) = G^e(s, py, (1-p)v). \tag{1.51}$$

Also, the formulas

$$\omega_s(p) = \left. \frac{\partial w}{\partial (\ln s)} \right|_{s=y=v=1}, \qquad P_W(p) = \left. \frac{\partial^2 w}{\partial (\ln s)^2} \right|_{s=y=\delta=1} \tag{1.52}$$

allow us to calculate, according to (1.46), the critical indices β and γ by taking the limit $p \to p^*$.

A particular feature that distinguishes the Bethe lattice from the others is that no formulas of the type (1.9) and (1.10), which would allow exact solution of the enumeration problem, can be derived for them. Nevertheless, some effective methods for the evolution of asymptotic dependencies have been developed in percolation theory, a rapidly developing branch of the physics of disordered systems. Thus, reduction of the polymer problem to any solved in percolation theory enables one automatically to take advantage of the results obtained in this field.

As early as 1961 Fisher and Essam[94] studied the analogy between the gelation in polymeric systems and percolation. In particular, they derived expression (1.11) by considering percolation on Bethe lattices and pointed out the relationship between this result and the theory of branching processes. These authors also compared percolation transition, when in the ensemble there appears for the first time an infinite cluster, with gelation point. However, it was not until the work of Stauffer[95] that the characteristics and notions of the ensemble of branched polymers, formed by polycondensation, were formulated in detail in terms of percolation theory. In the same paper the distinctions between critical exponents in percolation and classical theories of gelation were underlined. Almost simultaneously, De Gennes[96] proposed the process of cross-linking of linear macromolecules as a specific percolation problem. Starting from these papers,[95,96] the scaling approach to gelation as well as to melts and solutions of branched polymers has been essentially developed in a series of works.[87,88,90,97–101] In particular, more sophisticated percolation models were analyzed that took into account factors neglected in the simplest version of percolation.

One of these factors, dilution, arises when going from melts to solutions of polymers. In describing solutions the lattice framework, some fraction φ of all lattice nodes turn out to be occupied by monomeric units, whereas the remaining fraction, $1 - \varphi$, corresponds to the solvent molecules. The ideal systems, that is, those with pairwise identical energy of physical interaction between the solvent molecules and monomeric units, can be calculated[88] using the model of *random percolation through sites and bonds*.[102–105] According to this model, any lattice site can be either occupied by a vertex of the molecular graph with a probability of φ or free with a probability of $1 - \varphi$. In addition, each bond on the lattice connecting a pair of adjacent occupied sites also can be either occupied by an edge of the molecular graph or free. The probabilities of these two random events are p and $1 - p$, respectively. Hence, the clusters in this model consist of randomly distributed monomeric units connected accidentally by chemical bonds (cf. Fig. 1.27*c*). The limiting case $p = 1$, with any two occupied adjacent sites being necessarily connected by an occupied bond, corresponds to a well-known site problem[93] in percolation theory (Fig. 1.27*b*). Although in this one-parameter model the parameter value in the transition point φ^* differs from that obtained in the bond problem, p^*, still the critical exponents in both models happen to be identical.[92] A more general two-parameter model of random percolation through sites and bonds[103–105] that possesses a line of gelation rather than a point (cf. Fig. 1.28) seems to belong to the same class of equivalence. The critical dimensionality of space, d_c, for all these models is 6, and hence, it is not unexpected that in the real three-dimensional space the values of critical exponents for percolation and classical theories are markedly different (Table I). Therefore, once the asymptotic dependence of the characteristics of real polymeric systems in the region of universality has been determined experimentally, the applicability of either of these theories can be inferred.[88,97]

An essential drawback of the percolation model considered in the preceding consists in the neglect of correlations in the position of monomers due to their interaction with each other and with solvent molecules. To account for

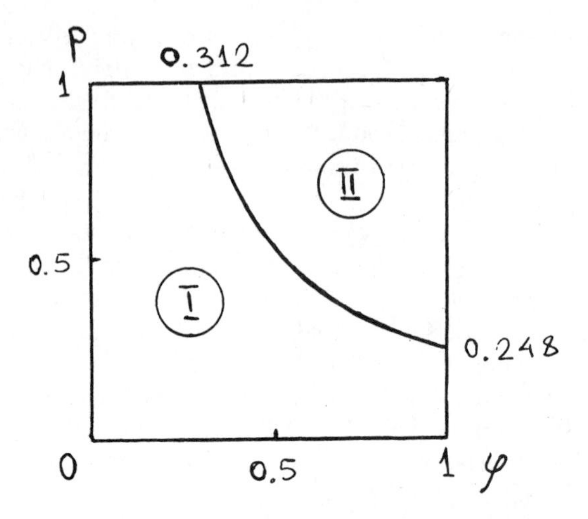

Figure 1.28. Phase diagram for random percolation by sites and bonds in simple cubic lattice as obtained by Monte Carlo simulation.[88] Domains I and II correspond to absence and presence of gel, respectively.

this, there was proposed a model of *correlated percolation through sites and bonds*.[106,107] This lattice model for the solution of branched polymers, called henceforth the general model, admits interaction between solvent molecules and monomeric units occupying adjacent sites only. This model[106] involves, together with three parameters characterizing the energy of physical interaction (unit–unit, U_{uu}; solvent–solvent, U_{ss}; and unit–solvent, U_{us}), the formation energy of a chemical bond, U_b. When the latter energy or the effective energy of physical interactions, $U = U_{ss} + U_{uu} - 2U_{us}$, is large enough as compared with the thermal energy $\beta^{-1} = k_B T$ (k_B is the Boltzmann constant, and T is absolute temperature), the position of units on the lattice can be essentially correlated. However, the chemical bonds between the units in this model are formed quite casually with a probability p, which depends in a fixed manner on the temperature and other parameters of the system. The

TABLE I Some Critical Exponents for Random Percolation on Three-Dimensional (I) and Bethe (II) Lattices

Lattice	β	γ	ν	τ	ρ
I	0.45	1.74	0.88	2.20	0.40
II	1	1	1/2	5/2	1/4

probability measure on the set of different equilibrium configurations of arrangement of monomeric units and solvent molecules on the lattice was shown[108] to be described by the Ising model. This means that the probabilities of such configurations are determined by the Gibbs distribution for the model of lattice gas, the parameters of which are expressed in an established way via those of the general polymeric model. Hence, the consideration within this model of chemical bonding of monomeric units, which is equivalent to some additional attraction between the units,[87] results merely in renormalization of energies of physical interaction in lattice gas. Thus, the gelpoint in the general polymer model[106,107] corresponds to a threshold in the model of site bond percolation with sites being correlated as in a lattice gas as opposed to,[103–105] where no correlation has been considered. The only percolation problem for lattice gas considered to date[106] was only the site problem, when bonds between adjacent sites are occupied with the probability $p = 1$.

For the calculation of molecular weight characteristics, the authors[108] have used the method essentially equivalent to that of the theory of branching

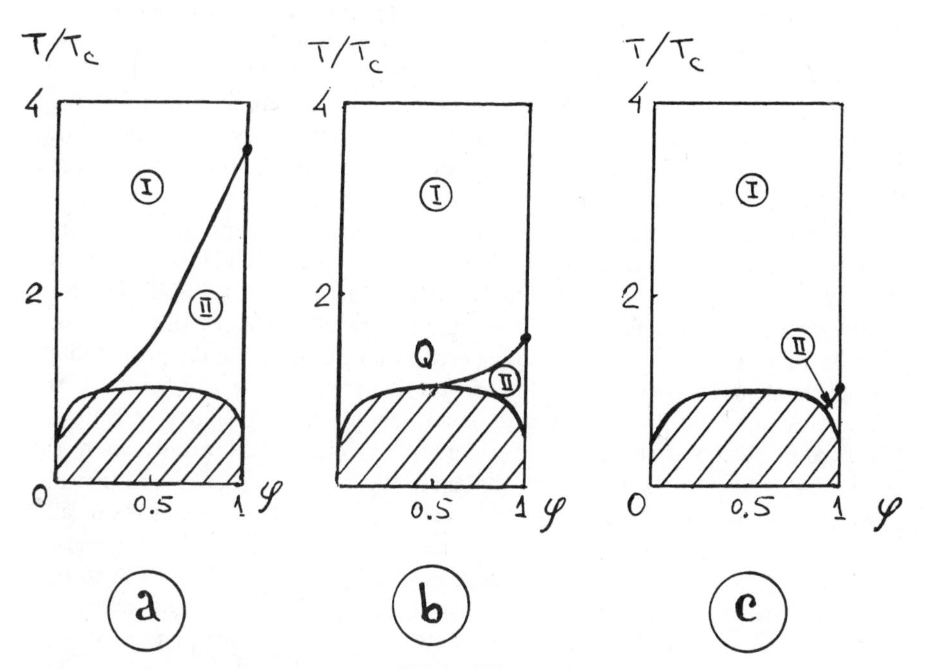

Figure 1.29. Qualitatively different types of phase diagrams for equilibrium polycondensation system in three solvents, for which critical temperature of mixing with monomer, T_c, is in (*a*) gel, at (*b*) gel formation line, and (*c*) in sol.

processes. Its applicability, however, is restricted just to Bethe lattices, for which were calculated (a) the exact value of the statistical sum and phase coexistence curve and (b) the weight-average degree of polymerization and a boundary of gelation region. A characteristic feature of the latter region, as can be seen in Fig. 1.29, is the occurrence of a maximal temperature T_{max} above which gelation does not take place, even for $\varphi = 1$, because there are few chemical bonds. For all types of solvent (i.e., for all values of energy U) there exists a temperature T_s (below the critical temperature of miscibility T_m) for which the curves of phase coexistence and gelation intersect. If in the temperature region $T_m < T < T_{max}$ the system represents a single phase (although for large enough φ an infinite network of gel can be formed), then at $T_s < T < T_m$, according to Fig. 1.29, separation in two phases takes place. Depending on the type of phase diagram, either both phases (Fig. 1.29*c*) or none of the phases (Fig. 1.29*a*) may contain a gel fraction. At $T < T_s$ only one of two phases, namely that impoverished with solvent, contains a polymeric network of gel. Phase diagrams qualitatively similar to those shown in Fig. 1.29 were obtained by Monte Carlo calculations on the three-dimensional cubic lattice[109] for the polymeric system in the same model.

By varying solvent parameters, an interesting case, $T_s = T_m$, can be realized, for which the gelation curve passes through the critical point on the phase coexistence curve (point Q in Fig. 1.29*b*). The critical behavior of the system along the gelation curve in this model of correlated percolation is just the same as in the case of random percolation. The only exception is the point Q, for which the critical indices correspond to the critical point of the lattice gas.[107,110]

In percolation models considered, intended for the description of gelation, each individual cluster is surrounded by other clusters, which is accounted for by the absence of chemical bonds along its perimeter (cf. Fig. 1.24). This intermolecular effect of excluded volume contributes a factor of $(1-p)^t$ to the statistical weight $p^b(1-p)^t$ of the percolation cluster, since the probability of each of its independent perimetric bonds is $1-p$. For the statistics of branched macromolecules in dilute solutions, when molecular interactions can be neglected, it suffices to consider solitary clusters. Naturally, there is no need to account for their environment in this case, and thus, the statistical weight of the cluster involves no factor $(1-p)^t$. This means that all conformations of any configuration arranged on a lattice of the polymeric molecule are assumed in the model of *random lattice animals* (r.l.a.'s) to be equally probable. This model, accounting for the intramolecular effect of excluded volume, is usually employed for the statistical description of the ensemble of sufficiently dilute solutions of branched polymers in a good solvent.[98,99,111–115]

For the calculation of thermodynamic features in this model it would have been sufficient to have the solution of a purely enumeration problem on the number $A(l, b)$ of ways to accommodate on the lattice the graphs with a given number of vertices l and edges b. Exact solution of this, or an even more simple problem about the number $A(l)$ or $A(b)$ of accommodations on real lattices of graphs characterized by only l or b, still was not obtained. However, the asymptotic behavior of these functions (e.g., $A(l)$ for $l \to \infty$) is available, as is the geometric dimensions R_l of l-mer molecules[88,90]:

$$A(l) \sim f_N(l) \sim l^{-\theta} \exp(-cl^{\zeta}), \qquad R_l \sim al^{\rho}. \tag{1.53}$$

The values of the universal exponents in these formulas have been found exactly[116]:

$$\zeta = 1, \qquad \theta = \tfrac{3}{2}, \qquad \rho = \tfrac{1}{2}, \tag{1.54}$$

with each index being identical for all three-dimensional lattices except the nonuniversal constant c. The last two exponents are markedly different from those obtained in the mean-field theory:

$$\zeta = 1, \qquad \theta = \tfrac{5}{2}, \qquad \rho = \tfrac{1}{4}, \tag{1.55}$$

which can be applied here in the space of dimensionality $d > d_c = 8$.

Expressions (1.53) also describe the statistics of clusters in the problem of random percolation for small values of p, since for $p \to 0$ the factor $(1-p)^t$ in the statistical weight of clusters tends to 1. Parameter c in Eq. (1.53) depends on p, and in the proximity of the transition point p^*, this dependence becomes universal:

$$c \sim (\bar{p}^* - p)^{1/\sigma}, \qquad 1/\sigma = \beta + \gamma. \tag{1.56}$$

The quantity $c^{-1} \sim l_T$ determines the number of monomeric units l_T in a typical percolation cluster with characteristic spatial size ξ (1.37). For p close to the critical value p^* two different asymptotic relations determining the distribution of large clusters exist. The clusters containing $l \gg l_T$ units are distributed according to formula (1.53) with exponents (1.54). On the other hand, clusters smaller than typical, $l_T \gg l \gg 1$, are described by expressions (1.39), which result from the general formula (1.53) for the following values of exponents:

$$\zeta = 0, \qquad \theta = 2.20, \qquad \rho = 0.40. \tag{1.57}$$

These two types of asymptotic behaviors are associated with two different fixed points, $p = 0$ and $p = p^*$, for the renorm group transformation in the random percolation model. In contrast to the first point, the second one is an unstable point of this transformation, and hence, in turning to large enough spatial scales (as compared with ξ), the system will represent an ensemble of solitary clusters. Besides these fixed points, the model considered involves another point, $p = 1$, with the following exact values of exponents

$$\zeta = \tfrac{2}{3}, \qquad \theta = -\tfrac{1}{9}, \qquad \rho = \tfrac{1}{3}. \tag{1.58}$$

This fixed point proves to be stable, and thus, enlargement of spatial scale by the renorm group transformation leads to universal behavior of the system corresponding to this point. According to this picture, finite clusters with the number of units $l \gg l_{\mathrm{T}}$ are found to be droplike, compact formations whose density is maximal and independent of their size. These droplike formations are sufficiently separated in space from each other and located within the gel network. The interaction between these formations and the network is the reason the clusters are compact. In fact, their sizes R_l are essentially greater than the characteristic dimension ξ of the network cells, and thus, the clusters cause distortion of its homogeneous structures (on a scale much larger than ξ!), accompanied by breaking a considerable number of chemical bonds. This process results in an increase of the free energy of the network, which naturally tends to recover its homogeneity. In response of the network to alien inclusion, the cluster is compressed uniformly, thus becoming compact. For p just above p^* together with drops ($l \gg l_{\mathrm{T}}$) there will occur a fraction of less compact percolation clusters ($l_{\mathrm{T}} \gg l \gg 1$) with indices (1.57). The statistics of these clusters is the same as that prior to the transition point, since the indices are identical at both sides, $p < p^*$ and $p > p^*$, of the transition point.

As opposed to the preceding one-parameter model of random percolation, a model of interacting l.a.'s was proposed[117] with two independent parameters p and q, where each cluster was assigned a statistical weight of $p^b q^t$. When parameter q tends to 1, 0, or $1 - p$, we arrive at the results relating to the statistics of random l.a.'s, drops, or percolation clusters, respectively. Accordingly, in the first case, the surface tension tends to zero; in the second it grows to infinity; and in the third it is finite.

A most detailed account of statistics for l.a.'s was carried out by Coniglio within the general lattice model of polymers,[115] who, besides the excluded volume effect, considers the difference U in the energies of interaction between monomeric units and solvent molecules. Lattice animals in this model of "correlated l.a.'s" are distinguished, in addition to by l and b, by the number n_1 of adjacent sites, pairs (i.e., adjacent monomeric units with the interaction energy U). The generating function G^A for the numbers $A(l, b, n_1)$ of these l.a.'s can be obtained,[115] with due regard for (1.41), from the g.f. w (1.49) for the model of random percolation or from the enumerating g.f. (1.42):

$$G^A(M, L, \Lambda) \equiv \sum_{l,b,n_1} A(l, b, n_1) M^l L^b \Lambda^{n_1} = w(M\Lambda^f, \quad L(\Lambda - 1)^{-1}, \quad 1|\Lambda^{-1} - 1)$$
$$= G^e(M\Lambda^f, L\Lambda^{-1}, \Lambda^{-1}), \qquad (1.59)$$

where $\Lambda = \exp(-U/k_B T)$. Thus, the description of a solitary macromolecule in dilute solution is reduced to the problem of a separate molecule in the presence of other polymeric molecules interacting with the former merely due to the factor of excluded volume.

The work of Fortuin and Kasteleyn[118,119] are conceptually important for percolation theory because it established for the first time the connection between the *geometric* and familiar *thermal* critical phenomena. The latter are properly described by the well-developed scaling theory,[85,86] and the results obtained in this theory can be directly used for the evaluation of asymptotic laws in percolation problems, thus avoiding combinatorial considerations.

The thermodynamic behavior of any equilibrium system in statistical physics is determined completely by its Hamiltonian $\mathscr{H}$, which for the three-parameter Potts model with $q = 1 + n$ states is of the form

$$-\beta\mathscr{H} = \sum_{(ij)} [J\delta(\sigma_i\sigma_j) + \mathscr{L}\delta(\sigma_i 0)\delta(\sigma_j 0)] + H\sum_i \delta(\sigma_i 0), \qquad (1.60)$$

where δ is the Kronecker delta, and the first summation is performed over all pairs (ij) of adjacent lattice sites. Each site i in this model is assumed to be occupied by an individual of $q = 1 + n$ types, numbered by the index $\sigma_i = 0, 1, \ldots, n$.

The interaction energy for two lattice adjacent individuals is nonzero only for identical individuals. In this case, the energy for the zero-type individuals, $-(J + \mathscr{L})k_B T$, differs from that for individuals of other types, $-Jk_B T$. The well-known Ising model describing the lattice gas is a particular case of the Potts model for $q = 2$. By analogy with the concept of continuous dimensionality of space, it is convenient to regard the quantity $q = 1 + n$ as a nonnegative continuous parameter. As a consequence, the partition function Z_N of the canonical ensemble on the lattice with N sites can be considered as a continuous function of q or n. The g.f. (1.49) for the distribution of clusters in the model of random percolation was shown[120] to be expressed via Z_N as

$$w(s, y, 1|p) = \lim_{q\to 1} \lim_{N\to\infty} \left(\frac{1}{N}\frac{d\ln Z_N}{dq}\right), \qquad Z_N \equiv \sum_{\{\sigma_i\}} \exp(-\beta\mathscr{H}\{\sigma_i\}), \qquad (1.61)$$

where the summation is performed over all possible realizations of the ensemble. The parameters $J, \mathscr{L}, H$ in the Potts model are related to the arguments of the g.f. (1.49) by

$$s = e^{-H}, \qquad y = (e^{J} - 1)/(e^{J+\mathscr{L}} - 1), \qquad p = 1 - e^{-(J+\mathscr{L})}. \tag{1.62}$$

Similar relations for the arguments of g.f.'s for the distribution of correlated l.a.'s,

$$M = e^{-H - f(J+\mathscr{L})}, \qquad L = e^{J} - 1, \qquad \Lambda = e^{J+\mathscr{L}}, \tag{1.63}$$

and enumerating g.f.'s,

$$s = e^{-H}, \qquad y = e^{-\mathscr{L}}(1 - e^{-J}), \qquad V = e^{-(J+\mathscr{L})}, \tag{1.64}$$

can be deduced starting from (1.59).

Thus, the statistics of clusters in the model of random percolation and correlated l.a.'s is described by the Potts model (1.60) with a single state. This allows us to take advantage of the methods of the renorm group for the description of asymptotic behavior of their statistical characteristics.

A renorm group transformation induces extension of spatial scale. The Hamiltonian (1.60) is invariant under this transformation, although its parameters undergo renormalization. Sequential extension of scale induces the sequences of the parameter values converging for systems occurring in the transition point to the limiting values $(J^*, \mathscr{L}^*, H^*)$. Depending on the initial values of the Hamiltonian parameters (1.60), the sequences will converge to one of four sets of parameters.[115] Each of the sets is associated with its own fixed point in the space of parameters $(J, \mathscr{L}, H)$, where the character of convergence determines the values of critical exponents in this point[115]:

$$\begin{array}{lll}
1. & J^* + \mathscr{L}^* = 0, \quad H^* = 0 & \text{random l.a. } (d_c = 8), \\
2. & J^*, \mathscr{L}^*, H^* < \infty & \theta\text{-point } (d_c = 6), \\
3. & \mathscr{L}^* = H^* = 0, \quad J^* < \infty & \text{percolation clusters } (d_c = 6), \\
4. & \mathscr{L}^* = H^* = 0, \quad J^* = \infty & \text{drops } (d_c = 1).
\end{array} \tag{1.65}$$

The nature of the universal behavior of the system in the neighborhood of fixed points 1, 3, and 4 has been discussed in the preceding. Point 2, where the contribution of pair interactions between monomeric units vanishes, is characterized by the critical exponent value $\rho = \frac{7}{16}$.

For the majority of fixed points in all the models considered, the critical dimensionality $d_c > 3$. Hence, it could be suggested that in the real three-dimensional space no region of applicability of the mean-field theory does exist, and measurements of critical exponents would always yield their scaling values. This, however, is not the case. Two different suits are possible, depending on the parameter Gi (the Ginzburg number).[85,86] For Gi $\ll 1$ the formulas of the mean-field theory are valid everywhere except for a small neighborhood of the fixed point. However, fluctuations of physical quantities in this neighborhood become so essential that they can be elucidated only within the framework of scaling theory. In essence, the Gi may be so negligible (as, e.g., in the theory of superconductivity[85]) that measurements in the fluctuation region are, out of reach of present-day experiment, which thus yields the exponents of mean-field theory. In another case, for Gi $\gtrsim 1$, the mean-field theory fails to be used altogether. Thus, in order to judge the region of applicability of the mean-field theory, it is necessary to express the Ginzburg number for the model through its parameters.

For cross-linking in melts and concentrated solutions of identical linear macromolecules with the number of units l, De Gennes has established theoretically[121] that the classical Flory theory is a reasonable approximation for the description of such a vulcanization, since it is characterized by Gi $\sim l^{-1/3} \ll 1$. However, others[122] question this conclusion, claiming that the mean-field theory cannot describe gelation adequately in any system. On the basis of scaling considerations of chain cross-linking in both concentrated and semidilute solutions, Daud[123] arrived at a different conclusion. According to Coniglio and Daud,[124] statistical description of the ensemble of cross-linked linear macromolecules can be performed, as for polycondensation products, by thermodynamic consideration of some lattice model. However, in contrast to polycondensation, the Hamiltonian for this model is the combination of Hamiltonians for the Potts ($q \to 1$) and n-vectorial spin ($n \to 0$) models. The latter describes the statistics of linear macromolecules with due regard for the excluded volume effect.

For the simulation of cross-linking, a procedure of lattice decoration was proposed[125,126] that consists in the substitution of each bond by the lattice two-rooted graph chosen at random from some set. This procedure leads to an increase in the critical value p^* and to diminution of the fluctuation region in the neighborhood of the transition point without affecting the exponents values. Beyond this region but still in the domain of universal behavior $(p^* - p)/p^* \ll 1$, the system is described within the mean-field theory.

Various lattice models for the branched polymers considered turned to be highly efficient for their statistical description since these models allow us to take into account in a simple way the excluded volume effect and occurrence of cyclic fragments in macromolecules. Nevertheless, applicability of results obtained within these models to real polymeric systems still remains obscure. Thus, in describing gelation, it is usually *postulated* on the basis of speculative analogies that the configurational–conformational distribution of macromolecules in the reaction system is specified by the probability measure on the set of percolation clusters. This measure may be different for various *mathematical* versions of the percolation model, and each should correspond to a quite concrete *physicochemical model* of gelation[127] (cf., e.g., Section I.B). In principle, this correspondence can be established in quite rigorous terms since in any accurately formulated physicochemical model the configurational–conformational probability measure is unambiguously defined either by the Gibbs distribution (in case of equilibrium systems) or by solutions of appropriate kinetic equations (for nonequilibrium systems). To date, no rigorous studies have been carried out to decide which of the physicochemical models of gelation are adequately described by the percolation approach.

This circumstance seems to be the main reason for the conceptual discussion that took place at the conference on polymeric networks in 1980.[128,129] Stauffer has neglected the applicability of the classical Flory theory in the proximity of the gelpoint, since, in his opinion, it yields incorrect values for critical indices.[128] Resting on the similarity between gelation and phase transition, ref. 128 suggested the use of percolation theory for the description of gelation. On the other hand, supporters of the classical theory[129] claimed it to be of universal character, suggesting that appropriate modification of this mean-field theory would allow us to describe gelation in any homogeneous system. These two contrasting viewpoints on gelation would hardly be conceivable if a correct microscopic theory of this phenomenon was available. Such a theory is expected to rest on accurate formulation of a physicochemical model of gelation to enable calculations starting from ab initio principles of statistical mechanics or kinetic theory for equilibrium and non equilibrium systems, respectively.

For systems in thermodynamic equilibrium the field-theoretic method provide such an opportunity. Once it has been possible to formulate the problem at hand in field-theoretic terms, one can employ well-developed tools to obtain in a standard way the mean-field approximations and calculate critical indices in the fluctuation region via the ε-expansion. For the lattice models described by the Potts Hamiltonian (1.60), this schedule has been successively carried.[130,131] However, applicability of field-theoretic methods in calculating the statistics of branched polymers is not necessarily restricted to lattice models; continual description of these polymeric systems can be carried out by these methods as well.[132–135] The results obtained by this approach and opportunities for its application are presented in Section IV.

II. SUBGRAPHS OF MOLECULAR GRAPHS AND MICROSTRUCTURE OF POLYMERS

A. Macromolecular Fragments and Their Subgraphs

A generally accepted polymer chemistry description of polymers by MWD bears no information on the relationship between different topological isomers, that is, on molecular configurations. Even though in some models it has been possible to construct a probability measure on the set of molecular graphs, specification of the fractions for each isomer is senseless due to an enormous number of graphs.

A similar problem arises in describing linear copolymers consisting of units of several types. In this case the configurational description of the copolymer is obtained by specifying the probabilities $P(U_k)$ of various sample sequences (U_k) of k units.[2,136–138] With increasing k, the information on the configurational structure of copolymers becomes increasingly detailed, and an exhaustive description of such a structure implies designing algorithms for calculating the probabilities of any sequences.

Extending this approach to branched polymers, we shall associate a sample sequence (U_k), referred to as a k-ad, with a connected subgraph that consists of k nodes with all outgoing edges. Some of these, the internal edges, connect the nodes of the sequence (U_k) between themselves and the pendants, whereas the remaining (external) connecting edges[13] connect these nodes with others not included in (U_k). Figure 2.1 shows that a monad $[(U_1)]$ is equivalent to a unit of a certain kind (Fig. 1.12). The most simple way to describe the topological structure of a polymer is to specify their fraction $\lambda_i = P(U_{1,i})$. Even those contain some information on the degree of branching of the polymer. Further levels of detailing the description are obtained by specifying the fractions $P(U_{2,q})$, $P(U_{3,q}), \ldots$ of various dyads $[(U_{2,q})]$, triads $[(U_{3,q})]$, and so on, which are enumerated in an arbitrary way by the subscript q separately for each set of sequences $(U_{k,q})$ with a given k. The idea that such sequences can be used to describe the topology of branched polymers has been suggested by Ziabicki and Walasek.[139] However, no new useful results in calculating the configurational statistics of macromolecules had been obtained using this idea before the works of the present authors appeared.[25,26,140]

The relation between the local topology of a molecule and its physicochemical features was discussed in Section I.A. In the simplest case a structure-additive property S of chemical individual compounds (i.e., consisting of identical molecules) is represented as

$$S = \sum_q n_q S_q, \tag{2.1}$$

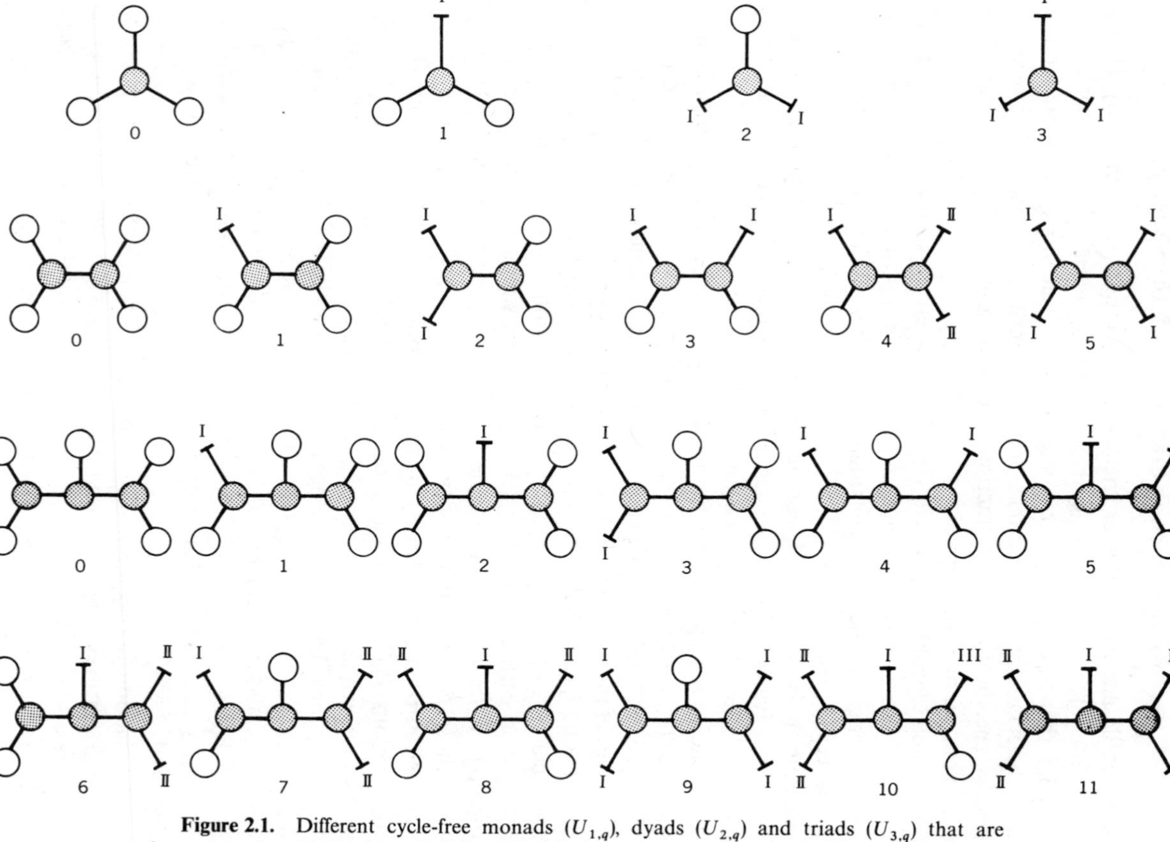

Figure 2.1. Different cycle-free monads ($U_{1,q}$), dyads ($U_{2,q}$) and triads ($U_{3,q}$) that are fragments of molecules formed of three-functional momomers. Their numbers given in Arabic numerals; equivalence classes of external edges given in Roman numerals.

where n_q is the number of different fragments of the qth kind, and S_q is the specific contribution of the qth fragment to the property S.[4,5] For a fixed number of molecular units the quantity n_q is obviously determined only by the topology of the molecular graph. Gordon and co-workers[141,142] generalized a series of such theoretical methods for the calculation of experimentally measurable properties and called them schemes of "linear combination of graph-theoretical invariants." Investigation of these schemes have induced a series of works[141–144] devoted to enumerating and coding the subgraphs of a given diameter, which, by definition, is the length of a maximal simple trail in the subgraph.[13,47] Also, the relationship between the number of various subgraphs of a given graph have been established.[140,141] These quantities are required for determination of the number of independent parameters S_q and their measurements, which can be carried out for individual low-molecular-weight compounds. However, as observed in Section I.C, the specificity of the polymeric system consists in averaging over a random set of macromolecules that is completely neglected in the enumerating approach[141–144] adapted for individual compounds.

For polymers formula (2.1) remains valid but the n_q's should be replaced by the numbers $\bar{n}_q$ of such fragments averaged over the whole set of isomers. A typical application is exemplified by the calculation of the glass transition temperature T_g for the polymer. The expression suggested by Becker[145,146] involves, in place of n_q, the number of nodes of a certain functionality, determined, at complete conversion of functionalities, by the composition of the monomer mixture. To verify this formula, researchers have had to use specific techniques to drive the reaction to completion.[147,148] However, these attempts, appear to be unnecessary, since in the case of uncompleted reaction or the nonstoichiometric composition of initial monomer mixture (i.e., when some functionalities are reduntant), it suffices to substitute in the Becker formula fractions of the nodes of various kinds for the determinate coefficients. Theoretical calculation of these fractions or their measurement is an important problem in predicting the physicochemical properties of polymers.

One of the most efficient methods for the experimental measurement of the probabilities of different subgraphs is the nuclear magnetic resonance (NMR) method, which is based on the dependence of chemical shifts of atoms (NMR frequency change) on their local environment. For linear polymers the NMR technique enables us to determine the probabilities of sequences of up to five to seven monomeric units.[149,150] Significant progress has been achieved in the application of high-resolution NMR techniques to branched polymers.[151,152] At low resolution, to each kind of unit there corresponds a single peak in the NMR spectrum. Increasing the resolution may affect the splitting of signals, depending on the local environment of a node in the direction of the connecting edges (Fig. 2.2), with the intensity of signal being,

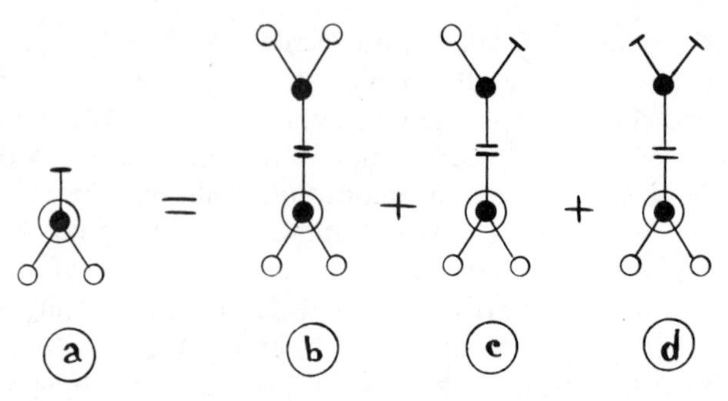

Figure 2.2. Resonating atom in monad ($U_{1,1}$) and three of its possible vicinities. Correspondingly, with improvement of resolution, NMR signal split in three peaks.

in general, proportional to the number of resonating atoms within a given environment. Thus, relative areas of NMR signals are equal to the probabilities that an arbitrarily taken unit is the root of the corresponding rooted graph. The resonating atom may belong not only to the unit but also to the reacted or unreacted functionality; hence, in this approach[29,30] a pendant vertex or even a link can be chosen as a subgraph root (Fig. 2.3). The probabilities of rooted subgraphs, differing only by the choice of the root, are interrelated and can be expressed via the concentration of molecular fragments represented by common nonrooted subgraph. To avoid sorting out all subgraph elements to serve as a root,[29,30] it is much more convenient to use nonrooted k-ads introduced in the preceding. Their relative fraction will be specified by $c(U_{k,q})$, which is the ratio of the concentration of the respective molecular fragment to the total concentration of monomeric units. Experimentally measurable fractions of rooted subgraphs are obtained by reducing $c(U_{k,q})$ to a single resonating nucleus and multiplying by the number of ways to choose a required root. For example, if a nucleus is found only in an unreacted functionality, then to a single nucleus there correspond $c(U_{k,q})/f(1-p)$ k-ads ($U_{k,q}$). The rooted subgraph shown in Fig. 2.3 a is obtained by selecting any of two functionalities in the unit of the first kind of the sequence $U_{2,1}$ as a root, and thus, the relative intensity of the signal for the functionality with the environment shown in this figure is $2c(U_{2,1})/f(1-p)$.

Various authors[29,30,141–143] have constructed a hierarchic sequence of subgraphs following their diameter and not the number of nodes. In fact, when a resonating atom is located in the centre of a unit, all connecting edges are equivalent and are expected to affect identically the chemical shift of the fragments arranged at identical distances in the direction of any connecting edge of the root. However, an asymmetric arrangement of a

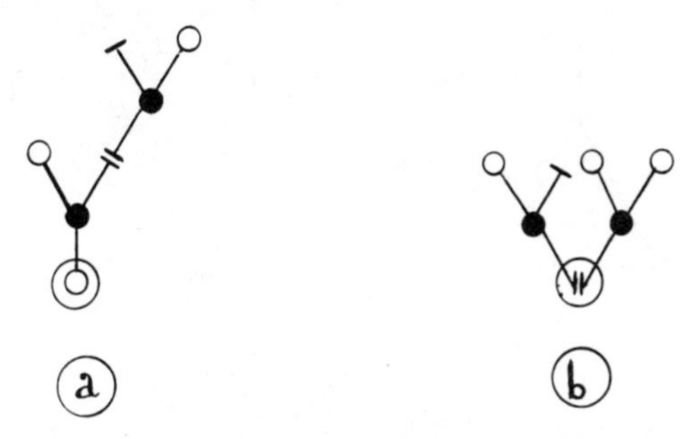

Figure 2.3. (*a*) Unreacted functional group or (*b*) bond must be taken as graph root if they contain resonating atom.

nucleus cannot be ruled out. For example, the chemical shift for the nitrogen atom in the urea monomer unit will naturally be more sensitive to environment in the direction of nearest connecting edges (Fig. 2.4). This justifies inclusion of all possible *k*-ads in the total list.

B. Subgraph Stoichiometry

The amount of various *k*-ads cannot be arbitrary. In fact, a unit of one of three kinds may be found in the direction of the connecting edge of the monad $(U_{1,1})$, so that the probability of the rooted monad in Fig. 2.2*a* is formed from the probabilities of three rooted dyads in Figs. 2*b*–*d*. Recalling that there are two ways to choose the root in deriving the subgraph in Fig. 2.2*b* from the dyad $(U_{2,0})$, we arrive at the equality $\lambda_1 \equiv c(U_{1,1}) = 2c(U_{2,0}) + c(U_{2,1}) + c(U_{2,2})$.

To derive the relationship between probabilities of various sequences, let us examine in an arbitrary molecular graph a certain $(k-1)$-ad $(U_{k-1,r})$ containing $L(k-1,r)$ connecting edges. If $(U_{k-1,r})$ is not identical to the complete graph, that is, $L(k-1,r) \neq 0$, then it is a subgraph of $L(k-1,r)$ various *k*-ads. Each of the latter contains, in addition to the $k-1$ nodes of the sequence $(U_{k-1,r})$, another node of the molecular graph with which $(U_{k-1,r})$ is connected by its external connecting edge. Consequently, the

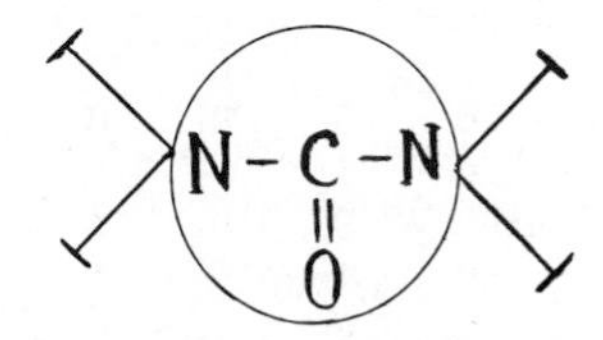

Figure 2.4. Example of unit with nonsymmetric disposition of resonating atoms.

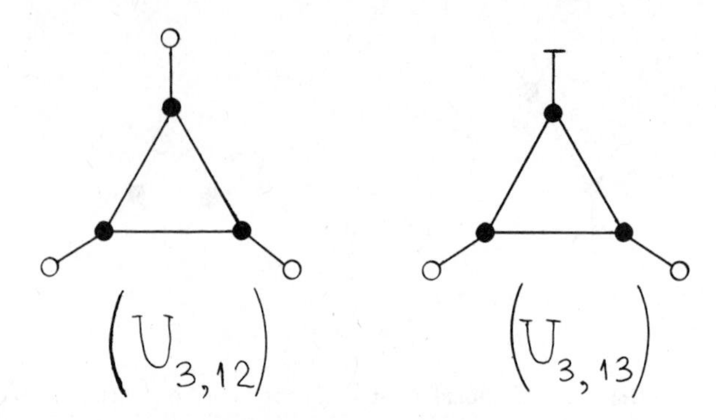

Figure 2.5. Example of cyclic triads.

number $c(U_{k-1,r})$ of such $(k-1)$-ads per monomer unit is one $L(k-1,r)$th fraction of the number of various k-ads if each of these is counted as many times as there are subgraphs of the type $U_{k-1,r}$:

$$L(k-1,r)c(U_{k-1,r}) = \sum_q \kappa(k-1,r;k,q)c(U_{k,q}). \tag{2.2}$$

Here the coefficient of topological stoichiometry $\kappa(k-1,r;k,q)$ denotes the number of subgraphs $(U_{k-1,r})$ in $(U_{k,q})$, that is, the number of ways in which one of the nodes of $(U_{k,q})$ can be removed by breaking the proper bond so that the remaining $k-1$ nodes make up a sequence $(U_{k-1,r})$. Thus, by removing either extreme node of the triad $(U_{3,0})$ (Fig. 2.1), we obtain a dyad $(U_{2,1})$ and hence $\kappa(2,1;3,0)=2$. In a similar way, from the triad $(U_{3,1})$ we have the dyads $(U_{2,1})$ and $(U_{2,3})$ (see Table II). The coefficient $\kappa(k-1,r;k,q)$ in the list of graph-theoretic definitions by Essam and Fisher[13] are referred to as the weak lattice constant.

TABLE II Values of $\kappa(2,r;3,q)$ for Dyads and Triads of Figure 2.1

	q											
r	0	1	2	3	4	5	6	7	8	9	10	11
0	0	0	0	0	0	0	0	0	0	0	0	0
1	2	1	0	1	0	0	0	0	0	0	0	0
2	0	0	2	0	0	1	1	0	0	0	0	0
3	0	1	0	0	2	0	0	1	0	0	0	0
4	0	0	0	1	0	1	0	1	2	2	1	0
5	0	0	0	0	0	0	1	0	0	0	1	2

It is noteworthy that the numbers $c(U_{k,q})$ of sequences with the same k and different q are, in general, interrelated in that between them some linear relations may exist that are obtained by considering the topological stoichiometry of the sequences $(U_{k,q})$. Each contains some of the items on the r.h.s. of Eq. (2.2) and is derived by generalization of (2.2) in the following way.

Let $L(k-1,r)$ external connecting edges of the sequence $(U_{k-1,r})$ generate $\mathrm{A}(k-1,r)$ equivalence classes[2] with $\sigma_\alpha(k-1,r)$ elements in the αth class. In Fig. 2.1 the equivalence classes are shown as Roman numbers. A certain sequence $(U_{k-1,r})$ is a subgraph of $\sigma_\alpha(k-1,r)$ various k-ads, each obtained when its $k-1$ nodes are supplemented with another molecular unit connected with $(U_{k-1,r})$ by its external edge from the equivalence class α. Note that such a k-ad $(U_{k,q})$ cannot be obtained by adding a supplementary unit to a connecting edge from a different equivalence class. Consequently, all k-ads $(U_{k,q})$ with nonzero values of the coefficient $\kappa(k-1,r;k,q)$ are split into $\mathrm{A}(k-1,r)$ nonintersecting classes. The number $c(U_{k-1,r})$ is $\sigma_\alpha(k-1,r)$ times smaller than the total number of k-ads from the αth class counted with due regard for the multiplicity of inclusion of $(U_{k-1,r})$ into $(U_{k,q})$:

$$c(U_{k-1,r}) = \sum_{q}{}_{\alpha} \kappa(k-1,r;k,q)c(U_{k,q})/\sigma_\alpha(k-1,r), \tag{2.3}$$

where the subscript α on the summation symbol denotes the summation of $(U_{k,q})$ only from the αth equivalence class. Thus, each sequence with $\mathrm{A}(k-1,r)$ equivalence classes of external connecting edges is associated with $\mathrm{A}(k-1,r)-1$ relations between the numbers $c(U_{k,q})$. A single such relation for the triads shown in Fig. 2.1 is (see Table II)

$$\begin{aligned} c(U_{2,4}) &= c(U_{3,3}) + c(U_{3,7}) + 2c(U_{3,9}) \\ &= \tfrac{1}{2}[c(U_{3,5}) + 2c(U_{3,8}) + c(U_{3,10})] \end{aligned} \tag{2.4}$$

since only in $(U_{2,4})$ do the connecting edges form more than one equivalence class. For 32 tetrads of three-functional nodes containing no cycles, we have seven such relations because each of the triads $(U_{3,5})$, $(U_{3,6})$, $(U_{3,7})$, $(U_{3,8})$, and $(U_{3,11})$ contains two equivalence classes of connecting edges and $(U_{3,10})$ contains three such classes.

In a similar way the relations between the numbers $c(U_{1,q})$ of monads for branched copolymers can be deduced. To this effect it is necessary to define an 0-ad consisting of a single edge without nodes and then to work out equations[6] similar to (2.3). As a consequence, we arrive at as many linear relations between the numbers $c(U_{1,q})$ as there are different types of bonds connecting pairwise different units.

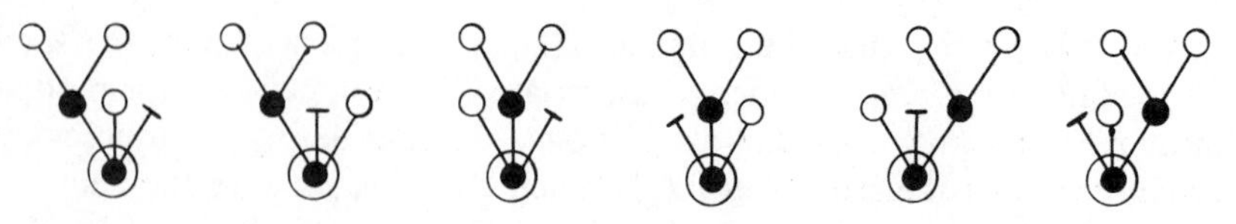

Figure 2.6. Different ordered dyads corresponding to nonrooted dyad $(U_{2,1})$ presented in Fig. 2.1.

The relations thus obtained are of a topological nature and are independent of polymer formation conditions. These can be used in treating the NMR spectroscopic data while assigning signals, testifying measurement accuracy, and so on, similar to their analogues for linear copolymers.[138]

In essence, this approach can be generalized to take into account cycle-containing subgraphs. In this case to the number $c(U_{k,q})$ calculated by formula (2.2) or (2.3) should be added the number of subgraphs contained in the cyclic k-ads. Thus, the dyad $(U_{2,3})$ is contained three times in the triad $(U_{3,12})$ and once in $(U_{3,13})$ (Fig. 2.6). Hence, to the right-hand side (r.h.s.) of Eq. (2.3), accounting for $c(U_{2,3})$, should be added in item $3c(U_{3,12}) + c(U_{3,13})$. Still, calculation of the amount of cycle-containing subgraphs of arbitrary sizes is no longer a straightforward matter (cf. Section I.F), and thus in this section we confine ourselves to treelike k-ads.

C. Probability Measure on Subgraphs

The relations of the preceding section allow us, with the amount of k-ads available, to calculate the probabilities of subgraphs of smaller size. However, the inverse problem to describe the configurational statistics of the polymer starting from concentrations of small molecular fragments known from experiment, is usually of greater interest. A constructive algorithm is required to calculate probabilities of k-ads of arbitrary size k. In principle, an approach suggested elsewhere[29,30] can be used for this purpose. However, a recurrent application of the procedure for constructing random graphs, described in Section I.D, is quite cumbersome. In this case it is much more profitable to make use of the methods of the theory of branching processes, where the set of realizations can be regarded as a random set of molecular graphs. Then, it is only necessary to relate the probabilities of rooted ordered subgraphs, arising in branching processes, to the number of nonrooted k-ads.

In models I and II, which neglect cycle formation, the set of molecular graphs represent the molecular forest with the probability of a certain tree $(\mathbf{l}, q)$ coinciding with the fraction of the respective molecules in the polymeric pattern. The vector $\mathbf{l}$ characterizes either the number of vertices of different color, which depict monomeric units of different types in the copolymer, or units of different kinds in the homopolymer. Let us associate with each nonrooted tree of a molecular forest a set of rooted trees obtained by selecting all its nodes, one after another, as a root (cf. Fig. 1.5). The set of rooted trees thus obtained forms a clone.[153] From the construction it follows that the probability of finding a certain rooted tree is equal to that of choosing the node associated with the root in a random choice among all nodes of the molecular forest, that is, directly related to the weight MSD $f_W(\mathbf{l}, q)$. If the transformation of the automorphism group of the molecular graph transforms a root into $\sigma_\beta(\mathbf{l}, q)$ other nodes (the subscript denotes equivalence classes of nodes of the graph), then by selecting any of those as a root, we obtain the same rooted tree (the trees b and d as well as f, g, and h in Fig. 1.5), the probability being $f_W(\mathbf{l}, q)\sigma_\beta(\mathbf{l}, q)/l$.

Let us now associate each rooted tree of the clone with all different ordered rooted trees (Fig. 1.8) by permuting its vertices in different ways. Different ordered trees obtained from the same rooted tree will be regarded, by definition, as equally probable, and their total probability should equal that of the original rooted tree. Thus, the probability of each $D_\beta(\mathbf{l}, q)$ ordered trees with a root from the equivalence class β is

$$P\{\mathbf{l}, q\}_\beta = f_W(\mathbf{l}, q)\sigma_\beta(\mathbf{l}, q)/lD_\beta(\mathbf{l}, q), \tag{2.5}$$

with $D_\beta(\mathbf{l}, q)$ obtained from the sum of the r.h.s. of Eq. (1.21) by retaining the items associated with the selection of the node of degree f_β from the equivalence class β as a root:

$$D_\beta(\mathbf{l}, q) = \frac{f_\beta \sigma_\beta(\mathbf{l}, q) D(\mathbf{l}, q)}{\sum_\nu l_\nu f_\nu}, \tag{2.6}$$

$$P\{\mathbf{l}, q\}_\beta f_\beta = \frac{f_W(\mathbf{l}, q)\sum_\nu l_\nu f_\nu}{lD(\mathbf{l}, q)}. \tag{2.7}$$

In a similar way, from the set of nonrooted k-ads we can turn to the rooted ordered sequences (Fig. 2.6). Selecting all nodes of any of $(U_{k,q})$ one after another to serve as a root, we obtain the number $D(k, q)$ of various ordered sequences $\{U_{k,q}\}_r$. The relation (2.6) remains valid as well for subgraphs; it is merely necessary to replace the argument $(\mathbf{l}, q)$, referring to the qth isomer of the $\mathbf{l}$-mer, by (k, q).

Let us now consider $D(k, q)$ families of ordered trees, rth of which include all the trees starting from $\{U_{k,q}\}_r$, that is, having $\{U_{k,q}\}_r$ as a subgraph (Fig. 2.7). The probability $P\{U_{k,q}\}_r$ of the ordered k-ad $\{U_{k,q}\}_r$ is the total

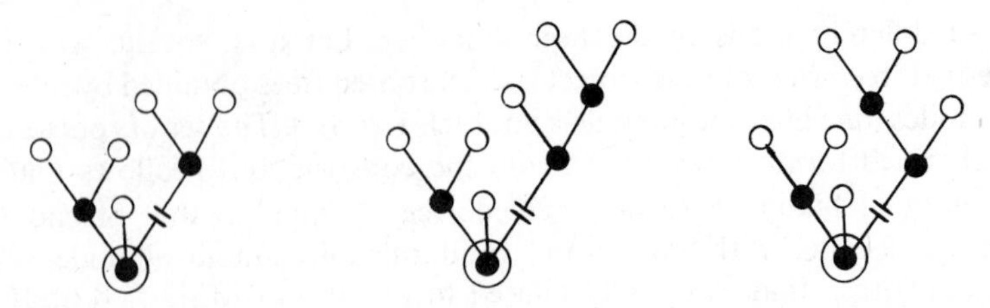

Figure 2.7. Several trees that belong to family generated by first ordered dyad in Fig. 2.6.

probability of trees from the rth family. Note that the sum of probabilities of all families is not unity, since, first, these do not include trees with the number of nodes less than k and, second, some tree from the ordered clone may be a member of several such families, namely of as many families as there are different k-ads $(U_{k,q})$ containing its root. Thus, the tree in Fig. 2.8a is a member of all three families generated by the dyad $(U_{2,2})$ since its root

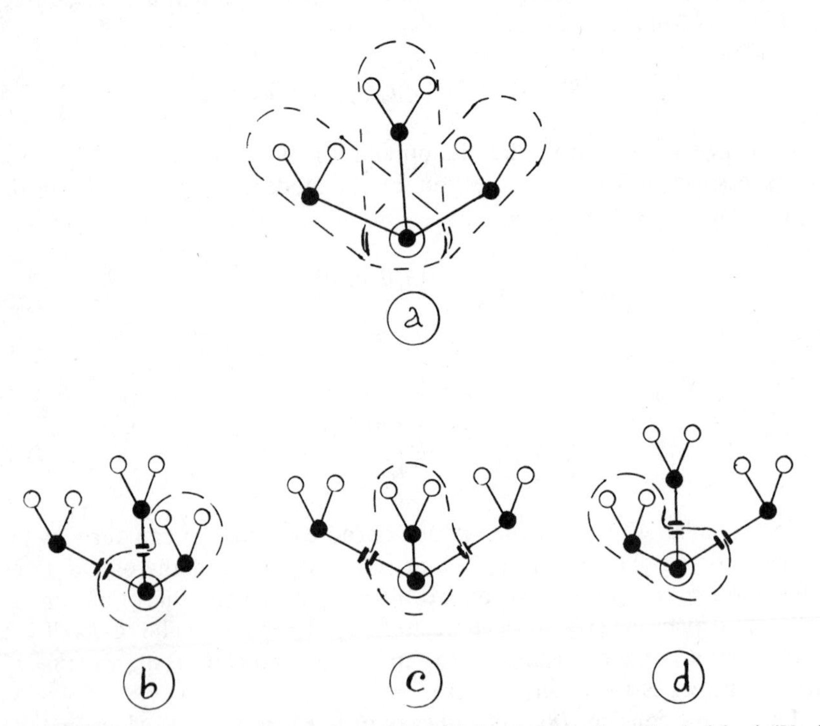

Figure 2.8. (a) Example of tree having its root simultaneously in three identical dyads $(U_{2,2})$. Tree present in three families that originate from different ordered dyads [(b–d)], circled by dashed line.

enters three such dyads. Consequently, the total probability of trees of all families is equal to the total concentration of monomeric units contained in the k-ads $(U_{k,q})$, with each counted as many times as there are k-ads $(U_{k,q})$ containing it. Observing that the number $c(U_{k,q})$ k-ads $(U_{k,q})$ is one kth fraction of the total number of units entering the k-ads, with due regard for the multiplicity of inclusion, we arrive at

$$kc(U_{k,q}) = \sum_r P\{U_{k,q}\}_r. \tag{2.8}$$

In each family the ordered trees, corresponding to some $(\mathbf{l}, q)$-isomer, are found in identical amounts, since regardless of the choice of a root in some k-ad and the order of arrangement of its vertices, the remaining vertices of the $(\mathbf{l}, q)$-mer can be permuted in just the same number of ways. Since the product of the probability of any ordered tree by the degree of its root is constant, according to Eq. (2.7), the same holds true for the probabilities of ordered k-ads. Combining in the sum (2.8) $D_\beta(k, q)$ equal items $P\{U_{k,q}\}_r = P\{U_{k,q}\}_\beta$ corresponding to the root from the βth equivalence class of nodes of $(U_{k,q})$ and summing over all β's with due regard for the relation (2.5), that is, replacing (l, q) by (k, q), we obtain a generalization of Eq. (2.7):

$$\begin{aligned} kc(U_{k,q}) &= \sum_\beta D_\beta(k, q) P\{U_{k,q}\}_\beta \\ &= \frac{kP\{U_{k,q}\}_\beta f_\beta D(k, q)}{\sum_\beta \sigma_\beta(k, q) f_\beta} = \frac{kP\{U_{k,q}\}_\beta D_\beta(k, q)}{\sigma_\beta(k, q)}. \end{aligned} \tag{2.9}$$

This expression relates unambiguously the numbers $c(U_{k,q})$ of nonrooted k-ads to the probabilities $P\{U_{k,q}\}$ of the associated ordered sequences. The quantities $D(k, q)$, $D_\beta(k, q)$, $\sigma_\beta(k, q)$, and f_β are the topological characteristics of the k-ad $(U_{k,q})$, whereas the factor $P\{U_{k,q}\}_\beta$ (identical for all equivalence classes with equal degrees of root) represents the individuality of the configurational structure of the set of macromolecules of the particular polymeric pattern. For sequences with small k the numbers $D(k, q)$ and $D_\beta(k, q)$ can be easily determined by ordinary sorting out, whereas in the case of long sequences one can take advantage of the results of ref. 154, which describes the structure of the automorphism group for a tree and actually gives an algorithm for calculating its order.

The relations listed here are valid for an arbitrary number of vertex types and are independent of the method of representation of the molecular graph. In the particular case of f-functional monomers, for the trees with pendant vertices corresponding to functionalities, the degree of all nodes is identical ($f_\beta = f$ for all β's). Hence, the probabilities of all ordered k-ads is independent of the choice of a root, so that

$$kc(U_{k,q}) = D(k,q)P\{U_{k,q}\}. \tag{2.10}$$

The only currently available constructive algorithm to build the probability measure on trees is that induced by branching processes. The realizations of branching processes make up a set of random ordered trees, a statistical forest.[153] In Section I, for some models of polymer formation, the probabilities of various realizations of the branching process were shown to coincide with the weight fractions of the molecules represented by these realizations, that is, the statistical forest is identical to the clone of ordered rooted molecular graphs. In other circumstances the probability measure on the statistical forest can be used as a certain approximation to the description of clone trees distribution.[26] The probability parameters of the branching process represent the fractions of various subgraphs of small size, and thus they provide direct opportunity to express the probabilities $P\{U_{k,q}\}_r$ and, via formula (2.9), the numbers $c(U_{k,q})$ for arbitrary k-ads.

Let us consider as an example the branching process corresponding to model I (Section I.C). The only parameter of this model is the conversion, which is equal to the fraction of reacted functionalities. The conversion can be expressed via the probabilities of 0-ads consisting of a single edge (Section II.B), since any edge (with cuts) connecting two black nodes appears when two functionalities are converted to a link. In terms of the theory of branching processes, such an edge is associated with generating a black individual with the probability p, whereas a pendant vertex corresponds to the appearance of a white descendant with the probability $1-p$. A k-ad involving k' connecting edges represents the history of a family[21] with offspring of $n = k - 1 + k'$ black individuals and the remaining $m = (f-2)k + 2 - k'$ white individuals (Fig. 2.9). The probability of this event is $p^n(1-p)^m$, and from Eq. (2.10), we obtain

$$kc(U_{k,q}) = D(k,q)p^n(1-p)^m. \tag{2.11}$$

In calculating the numbers $D(k,q)$, let us restrict ourselves with a simplest case of linear subgraphs, that is, the sequences $(S(i_1),\ldots,S(i_k))$ of monomeric units of the kinds $i_1, i_2, \ldots, i_k$, respectively. For instance, the triad in Fig. 2.9 in this notation would read $(U_{3,8}) = (S(2), S(3), S(2))$. Taking the extreme units $S(i_1)$ to serve as a root, its $f - i_1$ pendant vertices can be arranged in $C_f^{i_1}$ ways among i_1 edges with bars. One of these edges, namely that internal, is extended to the node $S(i_2)$, whereas all others remain as external edges of the k-ad. Since this internal edge can be chosen in i_1 ways, the number of different transpositions of branches of the root is $iC_f^{i_1} = fC_{f-1}^{i_1-1}$. Similarly,

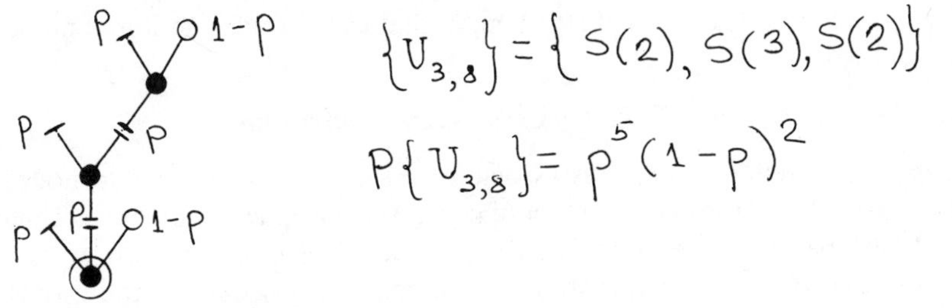

$$\{U_{3,8}\} = \{S(2), S(3), S(2)\}$$

$$P\{U_{3,8}\} = p^5(1-p)^2$$

Figure 2.9. Ordered triad $\{U_{3,8}\}$ represented as realization of branching process and factorization of its probability in probabilities that particles generate branching process yielding descendants of given color.

for other nodes $S(i_n)$ $(1 < n < k)$ there are $(i_n - 1)C_{f-1}^{i_n - 1}$ transpositions, whereas for the latter there are $C_{f-1}^{i_k - 1}$, since it bears none of the internal edges. Multiplying all these quantities yields the number $D_1(k, q)$ of ordered linear sequences of units of the kinds $S(i_1), S(i_2), \ldots, S(i_k)$ with the root $S(i_1)$. The total number of ordered k-ads U_k is obtained from $D_1(k, q)$ by using formula (2.6), where $(\mathbf{l}, q)$ is replaced by (k, q) and account is taken of equal functionality of all nodes:

$$D(k, q) = kf C_{f-1}^{i_1 - 1} \prod_{n=2}^{k-1} [(i_n - 1)C_{f-1}^{i_n - 1}] C_{f-1}^{i_k - 1} / \sigma_1(k, q), \tag{2.12}$$

where $\sigma_1(k, q) = 2$ for the symmetric k-ads coinciding with their inversion, $\{S(i_1), S(i_2), \ldots, S(i_k)\} = \{S(i_k), S(i_{k-1}), \ldots, S(i_1)\}$, and $\sigma_1(k, q) = 1$ for asymmetric sequences. The numbers $c(U_{k,q})$ are given by formulas (2.11) and (2.12) with due regard for

$$n = \sum_{j=1}^{k} i_j - k + 1, \qquad m = fk - \sum_{j=1}^{k} i_j. \tag{2.13}$$

In order to find the total number $c(U_{k,\text{lin}})$ of linear sequences of k units, the numbers $c(U_{k,q})$ should be summed up over all q's, that is, over all possible sets $\{i_1, \ldots, i_k\}$. As a consequence, for $k \geqslant 2$, one has the simple formula

$$c(U_{k,\text{lin}}) = \tfrac{1}{2} f p [p(f-1)]^{k-2} = \tfrac{1}{2} N_{k-1}. \tag{2.14}$$

The value of $c(U_{k,\text{lin}})$ is thus half the average number N_{k-1} of individuals in the $(k-1)$th generation of the branching process. The factor $\frac{1}{2}$ in the formula (2.14) accounts for the fact that in turning from the molecular forest to the clone, both ends of any linear graph are selected to serve as a root.

The number of sequences of unit length, that is, the total fraction of monads, is evidently 1.

D. Trails and Molecular Conformations

Each linear sequence ($U_{k,q}$) is associated with a trail between two nodes of the molecular graph. The total set of its trails determines many physicochemical parameters of the molecule, which depend on its conformation, on the position $\mathbf{r}_i$ of nodes of this molecule in space. For example, the radius of gyration, hydrodynamic radius, and intensity of light scattering for a particular (l, q)-mer of a certain configuration can be represented in the form[65,155–164]

$$S_{l,q} = \frac{1}{l^2} \sum_{i,j=1}^{l} S(\mathbf{r}_i - \mathbf{r}_j). \tag{2.15}$$

The choice of a function $S(\mathbf{r})$ is defined by the property considered [e.g., $S(\mathbf{r}) = |\mathbf{r}|^2/2$ for the radius of gyration squared].

The specificity of treelike molecules consists in the fact that the distribution of the distance between two units is defined merely by the number of units between these units, that is, by the length of the trail between the vertices of the molecular graph. Thus, by averaging formula (2.15) over all conformations of an (l, q)-isomer, one obtains

$$\langle S_{l,q} \rangle = \frac{2}{l^2} \sum_{n=0}^{\infty} N_n(l, q) \langle S(n) \rangle, \tag{2.16}$$

where $N_n(l, q)$ is the number of trails of length n in the molecular graph of the (l, q)-isomer, whereas $\langle S(n) \rangle$ is the average value of the property S for the pair of units separated on the graph by a trail of n edges and $n-1$ nodes. Next, this formula should be MSD averaged with a weight defined by the method of measurement of the property S. For example, light-scattering experiments yield the Z-average values $\langle\langle S \rangle\rangle_Z$:

$$\begin{aligned} \langle\langle S \rangle\rangle_Z &\equiv \sum_{l,q} \frac{l^2 c(l, q)}{\sum l^2 c(l, q)} \langle S_{l,q} \rangle \\ &= \frac{2}{P_W} \sum_{n=0}^{\infty} c(U_{n+1,\mathrm{lin}}) \langle S(n) \rangle \\ &= \frac{1}{P_W} \sum_{n=0}^{\infty} N_n \langle S(n) \rangle. \end{aligned} \tag{2.17}$$

The last of these equations follows from (2.14), whereas the next to the last permits us, due to the formula

$$c(U_{n+1,\mathrm{lin}}) = \sum_{l,q} c(l,q) N_n(l,q), \tag{2.18}$$

to replace averaging over the molecular distribution by averaging over the probability measure N_n/P_W, defined on the set of trails regardless of whether they belong to any molecule. This distribution of trails (2.14) by their lengths can be calculated using the theory of branching processes. The same methods allow us, from the total number of N_n, to single out specific contributions of trails belonging to the molecular graphs of l-mers. To this effect, one should make use of a counter s that counts the number of individuals in the realization of the branching process. A realization chosen at random consists of l individuals with the probability $f_W(l)$, whereas there are, on the average, N_{nl} trails of length n outgoing from any root of the l-mer. Thus, in the nth generation of the branching process, there are, on the average, $f_W(l)N_{nl}$ descendants belonging to the l-mers. The g.f. of these numbers,

$$N_n(s) = \sum_l N_{nl} f_W(l) s^l, \tag{2.19}$$

can be calculated by standard procedures[155–157].

Comparing (2.14) and (2.18) allows to infer the equality

$$N_{nl} = \frac{\sum_q \left[N_n(l,q) \frac{2}{l} \right] c(l,q)}{\sum_q c(l,q)}. \tag{2.20}$$

The square brackets contain, as can be easily verified, the average number of trails outgoing from a single node of the (l,q)-isomer. In turning from the nonrooted trails to those rooted by making use of formula (2.20), we can express[2,156] the weight-average and number-average characteristics S via the generating function $N_n(S)$[159,2]:

$$\begin{aligned} \langle\langle S \rangle\rangle_W &\equiv \sum_{l,q} f_W(l,q) \langle S_{l,q} \rangle \\ &= \sum_n \langle S(n) \rangle \sum_l c(l) N_{nl} \\ &= \sum_n \langle S(n) \rangle \int_0^1 N_n(s) \frac{ds}{s}, \end{aligned} \tag{2.21}$$

$$
\begin{aligned}
\langle\langle S \rangle\rangle_N &\equiv \sum_{l,q} f_N(l,q)\langle S_{l,q}\rangle \\
&= P_N \sum_n \langle S(n)\rangle \sum_l c(l) N_{nl}/l \\
&= \sum_n \langle S(n)\rangle \int_0^1 \frac{ds}{s} \int_0^s N_n(s')\frac{ds'}{s'}.
\end{aligned}
\tag{2.22}
$$

The form of the g.f. (2.19) depends on the type of branching process used, which, in turn, is defined by the adopted model of polymer formation. For example,[2,157] within the framework of the Flory model,

$$
N_n(s) = f p(u(s))^2 \left[\frac{(f-1)pu(s)}{1-p+pu(s)} \right]^{n-1}, \tag{2.23}
$$

where the function $u(s)$ is given by formula (1.12).

It is noteworthy that the treelike form of the molecule has been essentially used in deriving Eq. (2.16) from Eq. (2.15). Consideration of cyclization requires the development of new approaches discussed in the following section.

III. MOLECULES AS GRAPHS WITH COORDINATES AND CONSIDERATION OF CYCLIZATION

A. Graphs Embedded in Three-Dimensional Space

In the description of molecules by graphs in the preceding sections account was taken merely of their topology, and the spatial arrangement of units and groups was completely neglected. However, some physicochemical properties of the polymeric systems, such as those systems discussed in Section I.A, depend not only on the configuration of macromolecules but also on their conformations. The consideration of the mutual arrangement of molecular fragments in space is of particular importance in the theory, taking into account the reaction of intermolecular cyclization. The results discussed here are of general character, but for the sake of simplicity these will be exemplified with f-functional polycondensation. In this way the formulas can be simplified to the greatest extent while retaining all the principal peculiarities of the theoretical approaches under consideration.

In order to fix an arrangement of the graph in space, it is sufficient to supply each vertex with a label, a coordinate of the molecular fragment depicted by this vertex (Fig. 3.1). Each chemical bond in the molecule can either be broken to give two functional groups at the site or appear as a result of the chemical reaction between these functionalities. Hence, the arrangement of bonds between the reacted functionalities should also be

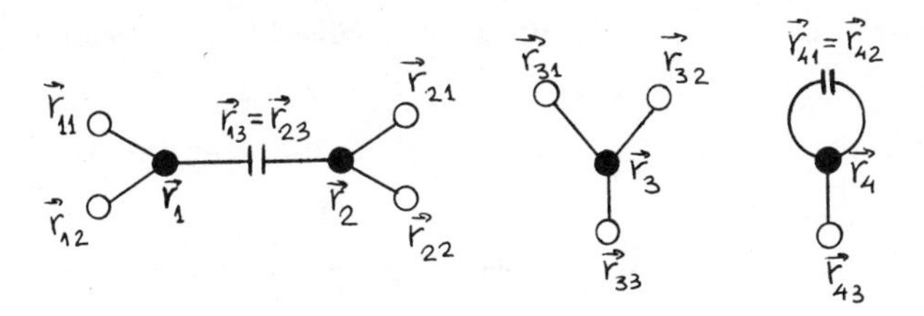

Figure 3.1. Graph $\mathscr{G}_4\{\mathbf{r}\}$ for system consisting of dimer and two monomers, one of them containing a cycle. Labels at graph elements indicate coordinates of corresponding fragments of molecule.

specified by marking them with bars. Any pair of bars belonging to one bond has identical labels corresponding to its coordinates. In this method of labelling, the graph of the system consisting of N monomeric units requires exactly $(f+1)N$ labels. Besides labeling the elements of the graph, we also number them. Thus, the units will be given the numbers from 1 to N, and the functionalities (including those reacted) for each unit independently will be numbered $1, 2, \ldots f$. The coordinate of the ith unit is denoted as $\mathbf{r}_i$, and its jth group–as $\mathbf{r}_{ij}$. The numbered graph (not necessarily connected) with N nodes is denoted $\mathscr{G}_N$ and the labeled numbered graph is denoted $\mathscr{G}_N\{\mathbf{r}\}$, where $\{\mathbf{r}\}$ corresponds to the set of coordinates of all units and functionalities.

Thus, to each arrangement of macromolecules in the system there corresponds some labeled graph. The construction of the probability measure on the set of labeled graphs is evidently equivalent to fixing the probability of any state of the polymeric system and, in the case of thermodynamic equilibrium, is described by the Gibbs distribution. In order to deduce this distribution, it is necessary to set down some model assumptions on the nature of macromolecular interaction in the reaction system.

The model considered, model III, assumes no interaction between the molecular fragments except for the chemical (reversible) reaction between functionalities. The reactivity of the functionalities is assumed to be indentical and constant, as in model I; however, the formation of cyclic structures is now permitted. The energy of such a system in the absence of external fields is the product of the energy of a single bond F_0 and their number N_b. The coordinates of some functionalities (e.g., $\mathbf{r}_{31}$ and $\mathbf{r}_{32}$ in Fig. 3.1) may be identical even if there is no chemical bond between them. The arrangement of functionalities of one unit in space is, in general, interrelated (cf., e.g., the rigid monomers in Fig. 1.17*b*). However, for simplicity, in what follows we assume each of these functionalities to be connected with the monomer by a flexible linear chain of root-mean-square length a, the distribution of endpoint distances of which $\mathbf{r}_i - \mathbf{r}_{ij}$ is described by the function $\lambda(\mathbf{r}_i - \mathbf{r}_{ij})$.

B. Gibbs Distribution and Probability Measure on Graphs in Space

The standard expression for the probability of some state of a system with variable number N_1, $N_2, N_n, \ldots$ of particles of different type is[165,166]

$$\mathscr{P} = \exp\left\{\left(\Omega + \sum_{\nu} \mu_\nu N_\nu - E\right)\Big/ T\right\}, \tag{3.1}$$

where Ω is the thermodynamic potential of the system determined from the probability normalization conditions, μ_ν is the chemical potential of a particle of the νth type, E is the energy of a given state, and T is absolute temperature.

Within the model considered, the particles in the statistical ensemble are not regarded as the polymeric molecules but as monomeric units, unreacted functionalities, and chemical bonds. With any particle in the partition function of the large canonical ensemble there is associated a factor, namely, its activity (u_u, z_f, z_b), which involves the chemical potential of the particle and thermal wavelength arising on integration of the distribution function over the particles' momenta. The factor $\exp(-F_0/T)$ associated with the energy contribution of each bond is incorporated into the activity z_b of the latter. The number of different particles is characterized by the vector $\mathbf{N} = \{N_u, N_f, N_b\}$. In some cases we will omit the subscript of its first component. If the first component is fixed, the remaining two are not independent since the three components are interrelated via the simple stoichiometric relation

$$f N_u = N_f + 2N_b. \tag{3.2}$$

Hence, actually, the system is not a three-component but a two-component one.

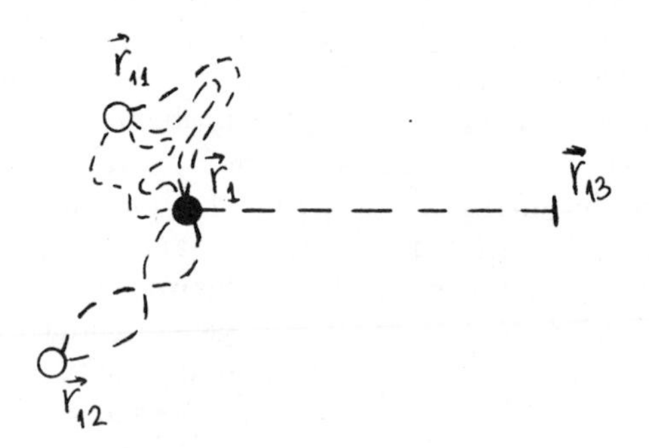

Figure 3.2. Dependence of number of possible conformations of chain on distance between its ends. This number is proportional to distribution of $\lambda(\mathbf{r}_i - \mathbf{r}_{ij})$.

In case of starlike monomers the method of fixing the graph in space unites several states of the system differing by the configurations of linear chains that connect units functionalities (both reacted and unreacted) into one state (Fig. 3.2). To account for the multiplicity of degeneration of the state with the setup in space of the units and functionalities to each chain connecting the monomeric unit (i) to the adjacent functionality (i, j), there should be associated in the statistical weight of this state a factor $c(\mathbf{r}_i - \mathbf{r}_{ij})$ that is equal to the number of conformations of the chain. It should be emphasized that the assumption of equal probability of all conformations is essentially employed in this case, since in the model considered the energy of the system is independent of the conformations. The function $c(\mathbf{r}_i - \mathbf{r}_{ij})$ coincides (up to the renormalization factor c) with the distribution $\lambda(\mathbf{r}_i - \mathbf{r}_{ij})$ introduced in the preceding:

$$\lambda(\mathbf{r}_i - \mathbf{r}_{ij}) = c(\mathbf{r}_i - \mathbf{r}_{ij}) \Big/ \int c(\mathbf{r}_i - \mathbf{r}_{ij})\, d\mathbf{r}_{ij} = c(\mathbf{r}_i - \mathbf{r}_{ij})/c. \tag{3.3}$$

If the system is exposed to external fields $H_u(\mathbf{r})$, $H_f(\mathbf{r})$, $H_b(\mathbf{r})$ affecting the particles, then, according to (3.1), with each particle with the coordinate $\mathbf{r}$ there should be associated a factor $\exp(-H_v(\mathbf{r})/T) \equiv \exp h_v(\mathbf{r})$. The subscript v assumes the value u, f, or b depending on the type of particle, and the ratio of the scalar fields h_v to the absolute temperature T taken with opposite sign will in what follows be considered as the components of the vector field $\mathbf{h} = \{h_u, h_f, h_b\}$. The probability of an arbitrary graph $\mathscr{G}_N\{\mathbf{r}\}$, taking into account the preceding, is defined by

$$\begin{aligned}\mathscr{P}(\mathscr{G}_N\{\mathbf{r}\}) = e^{\Omega/T}(f!M)^N L^{N_b} \prod_{i=1}^{N} \left\{ \exp h_u(\mathbf{r}_i) \prod_{j=1}^{f} \lambda(\mathbf{r}_i - \mathbf{r}_{ij}) \right\} \\ \times \prod_{(i,j)} \exp h_f(\mathbf{r}_{ij}) \prod \exp h_b(\mathbf{r}_k),\end{aligned} \tag{3.4}$$

where the next to the last product involves only unreacted functionality, and the last product involves all N_b bonds. According to the number of independent components, the system is characterized by two thermodynamic parameters M and L related to activities Z_v the chemical bond strength and the value c (3.3) via the simple relationships

$$f!M \equiv z_u (z_f)^f c^f, \qquad L \equiv z_b/(z_f)^2. \tag{3.5}$$

In deriving expressions (3.4), (3.5) for the statistical weight, use was made of the stoichiometric equality (3.2). From this definition it follows that M coincides with the activity of the monomer since the functionalities are statistically independent and, hence, their activities are multiplied. The factor $f!$ accounts for the symmetric (equivalent) position of these functionalities in the monomeric molecule. In the absence of external fields the parameter L, according to its definition, coincides with the equilibrium constant for the elementary reaction of chemical bond formation.

The thermodynamic Ω potential, which is a functional of the vector field $\{\mathbf{h}\}$, is defined by the probability normalization condition (3.4). To determine $\Omega\{\mathbf{h}\}$, the r.h.s. of Eq. (3.4) should be integrated over the coordinates of all particles and divided by $N!(f!)^N$ to account for the identical nature of all monomeric units and their functionalities. The expression obtained should further be summed over all numbered graphs $\mathscr{G}_N$ with N black nodes and then over the number N of the latter. Equating the final result to unity, one can arrive at the formula

$$\exp\left(\frac{-\Omega\{\mathbf{h}\}}{T}\right) = \sum_{N=0}^{\infty} \frac{1}{N!} M^N \sum_{\mathscr{G}_N} L^{N_b} I(\mathscr{G}_N | \{\mathbf{h}\}) \tag{3.6}$$

$$I(\mathscr{G}_N|\{\mathbf{h}\}) = \int \cdots \int \prod_{i=1}^{N} \left\{ [\exp h_u(\mathbf{r}_i) d\mathbf{r}_i] \prod_{j=1}^{f} \left[\exp h_f(\mathbf{r}_{ij}) \lambda(\mathbf{r}_i - \mathbf{r}_{ij}) d\mathbf{r}_{ij} \right] \right\}$$
$$\times \prod \{ \delta(\mathbf{r}_{ij} - \mathbf{r}_{pq}) \exp [h_b(\mathbf{r}_{ij}) - h_f(\mathbf{r}_{ij}) - h_f(\mathbf{r}_{pq})] \}, \tag{3.7}$$

where each expression in the braces in the last product corresponds to the chemical bond formed as a result of the chemical reaction between the functionalities (i,j) and (p,q) of the graph $\mathscr{G}_N$. In contrast to Eq. (3.4), the product of factors $\exp h_f(\mathbf{r}_{ij})$ is extended in Eq. (3.7) to include the reacted functionalities, but to compensate for the superfluous factors thus arising with each bond, there is associated the factor $\exp(-h_f(\mathbf{r}_{ij}) - h_f(\mathbf{r}_{pq}))$. The identity of lables $\mathbf{r}_{ij}$ and $\mathbf{r}_{pq}$ of a pair of corresponding to the common chemical bond is accounted for in Eq. (3.7) by using the Dirac's δ functions, thus allowing integration over all $(f+1)N$ coordinates $\mathbf{r}_i$ and $\mathbf{r}_{ij}$.

To determine the values of the thermodynamic parameters, the Ω potential of the system should be differentiated with respect to the chemical potentials μ_v of its components or with respect to their activities z_v[166]:

$$\bar{N}_v = -\frac{\partial \Omega}{\partial \mu_v} = -\frac{z_v}{T}\frac{\partial \Omega}{\partial z_v}. \tag{3.8}$$

Taking into account the dependence (3.5) of the parameters M and L on the activities, the parameters can be expressed through the mean densities of units, bonds, and functionalities:

$$-\frac{M}{T}\frac{\partial\Omega}{\partial M}=\bar{N}_u=\bar{\rho}_u V, \qquad -\frac{L}{T}\frac{\partial\Omega}{\partial L}=\bar{N}_b=\bar{\rho}_b V,$$

$$-f\frac{M}{T}\frac{\partial\Omega}{\partial M}+2\frac{L}{T}\frac{\partial\Omega}{\partial L}=\bar{N}_f=\bar{\rho}_f V, \tag{3.9}$$

where V is the volume of the system. The values of these three densities are interrelated because of the stoichiometric condition (3.2) by the linear equation, and so any two of them are independent. It is most convenient to use as the determining parameters, together with $\bar{\rho}_u$, the dimensionless conversion p of functionalities,

$$p=2\bar{\rho}_b/f\bar{\rho}_u=1-\bar{\rho}_f/f\bar{\rho}_u. \tag{3.10}$$

Having determined the thermodynamic parameters M and L in this way, the principal problem of constructing the Gibbs probability measure may be regarded as being completely solved. The remaining part of Section III is concerned with the description of the approaches that allow us, through appropriate averaging procedures over this measure, to calculate various thermodynamic and structural characteristics of the polymeric system.

C. Transition to Connected Graphs

The summation in expression (3.6) for the thermodynamic Ω potential is performed over all possible graphs, each depicting a certain state of the polymeric system. This sum can be reduced to that over the connected components of such graphs associated with individual molecules by using the combinatorial theorem,[167] which has been employed to derive the first Mayer theorem in the theory of nonideal gases.[168,169]

Let the quantity $J(\mathcal{G}_N)$, which depends on the topology and depends not on the way of numbering, the graph $\mathcal{G}_N$, be represented as the product of identical quantities $J(\mathcal{G}_l)$ over all connected components $\mathcal{G}_l$ of the graph $\mathcal{G}_N$. Then the following relationship is valid:

$$\sum_N \frac{1}{N!}\sum_{\mathcal{G}_N} J(\mathcal{G}_N)=\exp\left\{\sum_l \frac{1}{l!}\sum_{\mathcal{G}_l} J(\mathcal{G}_l)\right\}, \tag{3.11}$$

where the internal summation on the r.h.s. is performed only over the connected numbered graphs.

The integrand in (3.7) and, hence, the whole integral, is actually factorized into the product of the form specified above, as well as the multipliers in (3.6).

Thus, the Ω potential

$$-\frac{\Omega\{\mathbf{h}\}}{T}=\sum_{l}\frac{1}{l!}M^{l}\sum_{\mathscr{G}_l}L^{b}I(\mathscr{G}_l|\{\mathbf{h}\}) \tag{3.12}$$

is represented as a sum of contributions of individual molecules with l units and b bonds. To each molecule (l, q) in the sum (3.12) there will correspond as many items as there are ways to build the molecule of l monomers. The number $W(l, q)$ is expressed by using formula (1.5) via the order of the automorphism group $\mathscr{S}(l, q)$ of the graph associated with such an (l, q)-mer. The value of the integral $I(\mathscr{G}_l|\{\mathbf{h}\}) \equiv I((l, q)|\{\mathbf{h}\})$, as it observed in the preceding, is independent of the method of numbering the graph $\mathscr{G}_l$. Hence, relationships (3.12) and (1.5) allow us to represent the Ω potential as a sum over the nonnumbered molecular graphs of individual isomers (l, q), which do not interact physically in model III. Thus, the system considered represents an ideal gas where the components are (l, q)-mers. The quantity Ω for such a multicomponent gas is known[166,170] to be the sum of the number of molecules of each of its components. Hence, using formulas (3.12) and (1.5) the volume-mean equilibrium concentration of isomers can be derived:

$$\bar{c}((l, q)|\{\mathbf{h}\})=\frac{(f!M)^{l}L^{b}}{V\mathscr{S}(l, q)}I((l, q)|\{\mathbf{h}\}). \tag{3.13}$$

This expression allows us to realize the physical meaning of the parameter M. For $h_u(\mathbf{r}) \equiv h_f(r) \equiv h_b(\mathbf{r}) \equiv 0$ the integral $I((1)|\{\mathbf{0}\})$ corresponding to the monomer is equal to the volume of the system V, since, according to (3.3), the function $\lambda(\mathbf{r}_i - \mathbf{r}_{ij})$ is normalized to unity, and the number of integrations in $I((1)|\{\mathbf{0}\})$ exceeds the number of functions λ by 1. Since $\mathscr{S}(1) = f!$, M coincides with the concentration of monomers in the absence of external fields and, hence, can be measured experimentally. The parameter L, by (3.5) in the case of the chemical equilibrium between formation and the breaking of bonds, coincides with the equilibrium constant k for the elementary condensation reaction, provided the latter proceeds without liberation of any by-product. In the opposite case $L = k/z$, where z is the concentration of this by-product. However, the system may exist in a state of partial, rather than complete, equilibrium. For example, if the rate of interchain exchange reactions is greater than that of the reaction of chain growth, the former can "equilibrate" the molecular distribution to the state of thermodynamic equilibrium according to molecular energies.[171] In this case the value of the conversion p of functionalities may be in general nonequilibrium. For such systems the parameter L is no longer identical to the equilibrium constant and should be expressed, for example, via the conversion p and $\bar{\rho}_u$.

D. Correlation Functions and Their Generating Functionals

The probability density to find simultaneously k particular particles in given points of space is called the k-point correlation function.[165,172] For example, the joint probability $\hat{\theta}^{u\cdots u}(\mathbf{r}_1,\ldots,\mathbf{r}_k)$ that the units $1,2,\ldots,k$ have the coordinates $\mathbf{r}_1,\mathbf{r}_2,\ldots,\mathbf{r}_k$ is obtained by integrating the Gibbs distribution over the coordinates of all groups and the remaining units for all possible ways of binding of functionalities (including those adjacent to the units involved). For the calculation of various experimentally measured characteristics of the polymeric system, the most significant are the one-point $\hat{\theta}^u(\mathbf{r}_1)$ and two-point $\hat{\theta}^{uu}(\mathbf{r}_1,\mathbf{r}_2)$ correlation functions (correlators). In a similar way these can be defined the correlation functions for positions in the space of functionalities $(\hat{\theta}^f,\hat{\theta}^{ff})$ and bonds $(\hat{\theta}^b,\hat{\theta}^{bb})$ as well as the joint correlator $\hat{\theta}^{\nu\mu}$, with indices ν and μ being either u, f, or b depending on the type of particle.

Another equivalent method to define the correlation function is based on the consideration of the microscopic density[172] of the particles of the system, for example, the units in a given point of space r,

$$\rho_u^m(\mathbf{r}) = \sum_{i=1}^{N} \delta(\mathbf{r}-\mathbf{r}_i). \tag{3.14}$$

The mean value of this random quantity, which, by definition, is equal to $\theta^u(\mathbf{r})$, is obtained by averaging the former over the Gibbs probability measure (3.4). The similar averaging of the product $\rho_u^m(\mathbf{r}^{(1)})\rho_u^m(\mathbf{r}^{(2)})$ of microscopic densities in points $\mathbf{r}^{(1)}$ and $\mathbf{r}^{(2)}$ yields, to within the additive term $\hat{\theta}^u(\mathbf{r}^{(1)})\,\delta(\mathbf{r}^{(1)}-\mathbf{r}^{(2)})$, the value of the correlator $\hat{\theta}^{uu}$ for the pair of units in points $\mathbf{r}_1=\mathbf{r}^{(1)}$ and $\mathbf{r}_2=\mathbf{r}^{(2)}$. Therefore, the correlator $\hat{\theta}^{uu}$ for $\mathbf{r}_1\neq\mathbf{r}_2$ is identical everywhere to the correlator of the microscopic density of units θ^{uu}. The same result holds for the correlation function of any order

$$\theta^{\nu\cdots\mu}(\mathbf{r}^{(1)},\ldots,\mathbf{r}^{(k)}) \equiv \langle \rho_\nu^m(\mathbf{r}^{(1)})\cdots\rho_\mu^m(\mathbf{r}^{(k)})\rangle. \tag{3.15}$$

In what follows the subscript will refer to the number of a monomer, whereas the superscript will refer to the point in space. Averaging (3.15) is equivalent to summing the probabilities of k-rooted graphs, where the roots are labeled by the vectors $\mathbf{r}^{(1)},\ldots,\mathbf{r}^{(k)}$ while the remaining labels can be arbitrary.

The correlation functions are the generalization for the spatial case of the notion of moments $m_k=\langle\xi^k\rangle$ of a random quantity ξ. They can be obtained by differentiating the generating function (g.f.) of moments. In a similar way, the correlators can be obtained by differentiating the generating functional (G.F.) of correlators,[173] the arguments of which are arbitrary functions $\{\mathbf{h}(\mathbf{r})\}$.

The main property of the functional differentiation that we need is that of taking the derivative of an integral functional

$$\frac{\delta}{\delta h(\mathbf{r})} \int \xi(\mathbf{r}') h(\mathbf{r}')\, d\mathbf{r}' = \xi(\mathbf{r}). \tag{3.16}$$

In particular, putting $\xi(\mathbf{r}') = \delta(\mathbf{r}' - \mathbf{r}'')$ we arrive at the formula $\delta h(\mathbf{r}'')/\delta h(\mathbf{r}) = \delta(\mathbf{r}'' - \mathbf{r})$. The majority of the properties of ordinary differential calculus holds also for functional differentiation. For instance, the differentiation of an exponential function is reduced to multiplying it by the derivative of its argument, and the derivative of a product is the sum of terms in which each cofactor is differentiated in sequence.

Keeping these rules in mind, it is easily verified that the differentiation of the integral $I(\mathscr{G}_N | \{\mathbf{h}\})$ [Eq. (3.7)] or the Ω potential [Eq. (3.6)] with respect to $h_u(\mathbf{r})$ leads to multiplying the integrand by the microscopic density of units (3.14) since

$$\begin{aligned} &\frac{\delta}{\delta h_u(\mathbf{r})} \prod_{i=1}^{N} \exp h_u(\mathbf{r}_i) \\ &= \frac{\delta}{\delta h_u(\mathbf{r})} \exp \int \rho_u^m(\mathbf{r}') h_u(\mathbf{r}')\, d\mathbf{r}' = \rho_u^m(\mathbf{r}) \exp \int \rho_u^m(\mathbf{r}') h_u(\mathbf{r}')\, d\mathbf{r}' \\ &= \rho_u^m(\mathbf{r}) \prod_{i=1}^{N} \exp h_u(\mathbf{r}_i). \end{aligned} \tag{3.17}$$

Similarly, the differentiation of (3.6) with respect to $h_f(\mathbf{r})$ or $h_b(\mathbf{r})$ yields the factors $\rho_f^m(\mathbf{r})$ or $\rho_b^m(\mathbf{r})$, respectively. Having observed that the multiplication of the integrand in (3.6) by some quantity is equivalent, to within a factor of $\exp(\Omega\{\mathbf{h}\}/T)$, to averaging this quantity over the measure (3.4), we arrive at the remarkable formula

$$\theta^{\nu \cdots \mu}(\mathbf{r}^{(1)}, \ldots, \mathbf{r}^{(k)}) = \exp\left(\frac{\Omega\{\mathbf{h}\}}{T}\right) \frac{\delta}{\delta h_\nu(\mathbf{r}^{(1)})} \cdots \frac{\delta}{\delta h_\mu(\mathbf{r}^{(k)})} \exp\left(-\frac{\Omega\{\mathbf{h}\}}{T}\right) \tag{3.18}$$

From Eq. (3.18) it follows that the G.F. of correlators can be expressed via the Ω potential of the system:

$$\Phi\{\mathbf{s}\} = \exp[(\Omega\{\mathbf{h}\} - \Omega\{\mathbf{h} + \mathbf{h}^\nu\})/T], \tag{3.19}$$

where the virtual vector field $\mathbf{h}^\nu$ has the components $h_\nu^\nu \equiv \ln s_\nu$. Instead of taking derivatives with respect to the real fields $h_\nu(\mathbf{r})$, it is much convenient to differentiate with respect to the virtual fields $h_\nu^{\text{v}} \equiv \ln s_\nu(\mathbf{r})$, which in final formulas should be set equal to zero [i.e., $s_\nu(\mathbf{r}) = 1$]. Therefore, we shall write the vector $\{\mathbf{s}\}$ as the argument of the G.F. Φ, in contrast to the Ω potential. Since, according to Eq. (3.19), the virtual field appears as an additive item to the real field, the function $s_\nu(\mathbf{r})$ are multiplied throughout by $\exp h_\nu(\mathbf{r})$. For convenience, such a combination is denoted as $\tilde{s}_\nu(\mathbf{r})$:

$$\tilde{s}_\nu(\mathbf{r}) \equiv s_\nu(\mathbf{r}) \exp h_\nu(\mathbf{r}). \tag{3.20}$$

In the absence of external fields, $\tilde{s}_\nu(\mathbf{r})$ is naturally identical to $s_\nu(\mathbf{r})$.

The second variational derivatives of the Ω potential itself yield the values of the so-called irreducible correlators

$$\frac{\delta^2\Omega\{\mathbf{h}\}}{\delta h_\nu(\mathbf{r}_1)\delta h_\mu(\mathbf{r}_2)} = \theta^{\nu\mu}(\mathbf{r}_1, \mathbf{r}_2) - \theta^\nu(\mathbf{r}_1)\theta^\mu(\mathbf{r}_2), \tag{3.21}$$

which, in the absence of correlations in the position of particles, vanish.

It can be easily verified that the correlators $\hat{\theta}$ are obtained by differentiating the g.f. $\Phi\{\mathbf{s}\}$ (3.19) not with respect to the virtual fields $\ln s_\nu(\mathbf{r})$ but with respect to $s_\nu(\mathbf{r})$. Hence, while θ is the generalization of the moments m_k of a random quantity, $\hat{\theta}$ generalizes the factorial moments.

The formal scheme developed in the foregoing can be applied to the statistical system consisting of an individual molecule of the (l, q)-mer. Its conformational Gibbs probability measure is described by the integral (3.7) with $\mathscr{G}_N = \mathscr{G}(l, q)$. The variational differentiation of the latter gives rise to a factor, the microscopic density $\rho_{l,q}^{m\nu}$, which naturally involves the contribution of the only isomer considered rather than of all molecules of the system, as is the case when differentiating $I(\mathscr{G}_N|\{\mathbf{h}\})$. Thus, the ratios of the derivatives of $I((l,q)|\{\mathbf{h}\})$ to this integral yield, similar to (3.18), the correlators of an individual (l, q)-mer:

$$\chi_{l,q}^{\nu\cdots\mu}(\mathbf{r}^{(1)}, \ldots, \mathbf{r}^{(k)}) = \langle \rho_{l,q}^{m\nu}(\mathbf{r}^{(1)}) \cdots \rho_{l,q}^{m\mu}(\mathbf{r}^{(k)}) \rangle \tag{3.22}$$

so that the G.F. of these correlators is

$$\Psi_{l,q}\{\mathbf{s}\} = \frac{I((l,q)|\{\mathbf{h} + \mathbf{h}^{\text{v}}\})}{I((l,q)|\{\mathbf{h}\})}. \tag{3.23}$$

Now, taking into account Eqs. (3.13), (3.19), and (3.23), the analogue of the first theorem of Mayer [eq. (3.12)] can be rewritten in the form

$$\ln \Phi\{\mathbf{s}\} - \frac{\Omega\{\mathbf{h}\}}{T}$$
$$= -\frac{\Omega\{\mathbf{h}+\mathbf{h}^{\nu}\}}{T} = \sum_{l,q} \frac{(f!M)^l L^b}{\mathcal{S}(l,q)} I((l,q)|\{\mathbf{h}+\mathbf{h}^{\nu}\})$$
$$= V\sum_{l,q} \bar{c}((l,q)|\{\mathbf{h}\})\Psi_{l,q}\{\mathbf{s}\} \equiv \Psi\{\mathbf{s}\}. \tag{3.24}$$

Differentiation of the functional $\Psi\{\mathbf{s}\}$ with respect to the functions $\ln s_\nu(\mathbf{r})$ results in averaging over the MSD of density correlators of individual molecules (3.22).

With such an averaging procedure the contribution of molecules with l units can be extracted by using the "counter" $s_u \equiv s$, as was done in Eqs. (1.12) and (1.14). For example, the most significant among those correlators of molecular density are

$$\chi^{\nu}(\mathbf{r}; s_u) \equiv \sum_{l} \chi_l^{\nu}(\mathbf{r}) s_u^l = \sum_{l,q} V\bar{c}((l,q)|\{\mathbf{h}\}\chi_{l,q}^{\nu}(\mathbf{r}) s_u^l,$$
$$\chi^{\nu\mu}(\mathbf{r},\mathbf{r}'; s_u) \equiv \sum_{l} \chi_l^{\nu\mu}(\mathbf{r},\mathbf{r}') s_u^l = \sum_{l,q} V\bar{c}((l,q)|\{\mathbf{h}\})\chi_{l,q}^{\nu\mu}(\mathbf{r},\mathbf{r}') s_u^l. \tag{3.25}$$

For the spatially homogeneous system the correlator $\chi^{uu}(\mathbf{r},\mathbf{r}'; 1)$ is identical to the structural function $g(\mathbf{r}-\mathbf{r}')$ introduced elsewhere.[133] For such a system $\chi_{l,q}(\mathbf{r}-\mathbf{r}')$ is resembled to the "pair connectedness"[92] of the (l,q)-mer. The Fourier component of this correlator has the form

$$\tilde{\chi}_{l,q}^{uu}(\mathbf{q}) = \left\langle \sum_{j,k=1}^{l} \exp\left[i\mathbf{q}(\mathbf{r}_j - \mathbf{r}_k)\right] \right\rangle \tag{3.26}$$

and finds application in the theory accounting for volume interactions.

To obtain the correlators of individual (l,q)-isomers, it is necessary, after differentiation of the functional Ψ, to set the virtual field $\ln s_u(\mathbf{r})$ to be independent of coordinates. The counters of functionalities $(\nu = f)$ and bonds $(\nu = b)$, which count the numbers of the corresponding individuals in the molecules, are introduced in a similar way.

It should be emphasized that the simple logarithmic relationship between the G.F. of the correlators of total Φ and molecular Ψ densities holds merely in the absence of physical interactions between the molecular fragments adopted in the model considered. In this case the correlators are connected via the equations

$$\chi^{\nu}(\mathbf{r}) = \theta^{\nu}(\mathbf{r}), \qquad \chi^{\nu\mu}(\mathbf{r}', \mathbf{r}'') = \theta^{\nu\mu}(\mathbf{r}', \mathbf{r}'') - \theta^{\nu}(\mathbf{r}')\theta^{\mu}(\mathbf{r}''), \tag{3.27}$$

that is, the correlators $\chi^{\nu\mu}$ of individual molecules coincide with the irreducible correlators (3.21). There are no simple connections between the correlators [e.g., (3.27)] in the presence of volume interactions (see Section IV).

In practice, of all moments m_k of the distribution of a random quantity ξ, use is usually made of the first two, m_1 and m_2, since they determine the majority of characteristics of this distribution. Similarly, the first two correlators are sufficient, as a rule, for the calculation of many physico-chemical properties of the polymeric system. If the system is spatially homogeneous, the one-point correlation functions $\hat{\theta}(\mathbf{r}) = \theta(\mathbf{r}) = \chi^{\nu}(\mathbf{r}) \equiv \chi^{\nu}$ are independent of the coordinate $\mathbf{r}$. They allow us to calculate those characteristics of molecules determined solely by their topology (configuration) (e.g., the MWD). In fact, χ^{u} represents (independent of the coordinate $\mathbf{r}$) the density of units $\rho_u(\mathbf{r}) = \bar{\rho}_u$. The contribution $\bar{\rho}_{u,l} = \chi_l^u$ of molecules with l units can be extracted from the total density by introducing the counter; that is, if the one-point correlator is endowed with the counter s, it is proportional to the g.f. of the weight MWD:

$$\chi^{u}(s) = \sum_l \chi_l^u s^l = \sum_l \bar{\rho}_{u,l} s^l = \bar{\rho}_u G_W(s) = \chi^{u}(1) G_W(s). \tag{3.28}$$

To determine the characteristics of the system that depend on the conformations of macromolecules, the two-point correlators can be employed, which for the spatially homogeneous and isotropic system depend only on the distance $\mathbf{r} = \mathbf{r}' - \mathbf{r}''$ between the pair of particular points $\chi^{\nu\mu}(\mathbf{r}) = \chi^{\nu\mu}(\mathbf{r}', \mathbf{r}'')$. Similar to the one-point correlators, the two-point correlators consist of contributions of individual molecules (3.25). Each contribution resembles, in a sense, the function $\hat{\chi}_{l,q}(\mathbf{r})$ used formerly in the theory of polymers and defined as the mean overall conformation density of units of some isomer (l, q) at a distance $\mathbf{r}$ from its arbitrary chosen unit.[87] To establish $\hat{\chi}_{l,q}(\mathbf{r})$ it is necessary to average the microscopic density of units $\rho_{l,q}^{mu}(\mathbf{r} + \mathbf{r}')$ of this isomer under the condition that one of them is found at the point $\mathbf{r}'$. To extract the states of the system in which the ith unit is found at the point $\mathbf{r}'$, it is sufficient to add the factor $\delta(\mathbf{r}' - \mathbf{r}_i)$ to the Gibbs distribution, which describes the probabilities of the different conformations of the molecule. The summation of these δ functions in the subscript i with a weight of l^{-1}, which means that any unit is found at the point $\mathbf{r}'$ with equal probability, gives rise to the factor $\rho_{l,q}^{mu}(\mathbf{r}')/l$ in the function averaged. Hence, $\hat{\chi}_{l,q}(\mathbf{r})$ is connected with the correlator $\chi_{l,q}^{uu}(\mathbf{r})$ via the relation

$$\hat{\chi}_{l,q}(\mathbf{r}) = V\chi_{l,q}^{uu}(\mathbf{r})/l, \tag{3.29}$$

where the volume of the system arises from the normalization condition for the conditional Gibbs distribution for the macromolecules with one unit fixed.

An arbitrarily chosen unit of the system located at the point $\mathbf{r}'$ belongs to the (l, q)-isomer with the probability $f_W(l, q)$. Thus, after averaging, Eq. (3.29) gives the density of units located at the distance $\mathbf{r}'$ from an arbitrarily chosen unit and belonging to the same molecule:

$$\hat{\chi}(\mathbf{r}) = \sum_{l,q} f_W(l, q)\hat{\chi}_{l,q}(\mathbf{r}) = V\sum_{l,q} \frac{c(l,q)}{\bar{\rho}_u}\chi_{l,q}^{uu}(\mathbf{r}) = \frac{\chi^{uu}(\mathbf{r})}{\bar{\rho}_u} \tag{3.30}$$

Hence, the correlators $\hat{\chi}$ and χ^{uu} differ from one another only by a constant factor. In percolation theory $\hat{\chi}(\mathbf{r})$ is called the function of "pair connectedness," and the analogous averaging of the correlator $\hat{\chi}_{l,q}$ over the isomers of identical degree of polymerization l yields a function that coincides with the "density profile for the cluster of l nodes."[92]

The most significant distinction between the correlator of total density θ^{uu} on the one hand and $\hat{\chi}$, χ^{uu} on the other is that the latter, together with the correlation of positions of units in space, account for their connectedness. This enables us to calculate the physicochemical properties of the polymeric system, which are determined by the contributions of individual molecules, provided the contribution of any isomer is determined by the set of its conformations. The calculation of such properties requires three steps of averaging. The first step involves averaging over the set of conformations of each (l, q)-mer, the second step is over all isomers q, and the final step is over the MWD with the appropriate weight.

To express the mean characteristics via the correlator, it is convenient to rewrite formula (2.15) in the form

$$S_{l,q} = \frac{1}{l^2}\iint S(\mathbf{r})\rho_{l,q}^{mu}(\mathbf{r}')\rho_{l,q}^{mu}(\mathbf{r}' + \mathbf{r})\, d\mathbf{r}\, d\mathbf{r}'. \tag{3.31}$$

Averaging both sides of this expression over the sets of conformations of the (l, q)-mer gives rise to the correlator $\chi_{l,q}^{uu}(\mathbf{r})$ in the integrand:

$$\langle S_{l,q}\rangle = \frac{V}{l^2}\int S(\mathbf{r})\chi_{l,q}^{uu}(\mathbf{r})\, d\mathbf{r}. \tag{3.32}$$

Further averaging can be performed in a most simple way in the calculation of the Z-mean characteristics

$$\langle\!\langle S \rangle\!\rangle_Z = \int S(\mathbf{r}) V\left[\sum_{l,q} c(l,q) \chi_{l,q}^{uu}(\mathbf{r}) \Big/ \sum_{l,q} l^2 c(l,q) \right] d\mathbf{r}$$

$$= \int S(\mathbf{r}) \chi^{uu}(\mathbf{r})\, d\mathbf{r} / \bar{\rho}_u P_W. \tag{3.33}$$

In particular, for $S(\mathbf{r}) = 1$ the preceding formula gives the expression for the weight-average degree of polymerization:

$$P_W = \int \chi^{uu}(\mathbf{r})\, d\mathbf{r} / \bar{\rho}_u. \tag{3.34}$$

In the calculation of the weight-mean and number-mean values of S, the contribution of the l-mers in the averaging procedure (3.31), which can be extracted by using the counters, must be taken with the factors l^{-1} and l^{-2}, respectively. This can be achieved, as in formulas (2.21) and (2.22), by the action of the integral operator, transforming $\chi^{uu} \equiv \chi_Z^{uu}$ into χ_W^{uu} and χ_W^{uu} into χ_N^{uu}, on the correlator $\chi^{uu}(\mathbf{r}, s)$:

$$\chi_W^{uu}(\mathbf{r}) = \int_0^1 \chi^{uu}(\mathbf{r}, s) \frac{ds}{s}, \qquad \chi_N^{uu}(\mathbf{r}) = \int_0^1 \frac{ds}{s} \int_0^s \chi^{uu}(\mathbf{r}, s') \frac{ds'}{s'}. \tag{3.35}$$

The integration of the correlators modified in this way, together with the function $S(\mathbf{r})$, yields the required characteristics [cf. (2.21) and (2.22)]:

$$\langle\!\langle S \rangle\!\rangle_W = \frac{\int S(\mathbf{r}) \chi_W^{uu}(\mathbf{r})\, d\mathbf{r}}{\bar{\rho}_u}, \qquad \langle\!\langle S \rangle\!\rangle_N = \frac{P_N \int S(\mathbf{r}) \chi_N^{uu}(\mathbf{r})\, d\mathbf{r}}{\bar{\rho}_u}. \tag{3.36}$$

E. Equations for Generating Functionals

It was already observed in Section I that the statistical momenta of the MWD can be obtained most simply from equations of the form of (1.12) for the g.f. of this distribution through term-by-term differentiation at the point $s = 1$. In this section we generalize ideology and derive the respective equations (now, however, of the functional, not the algebraic, form) for the G.F. of the correlators of individual molecules. A term-by-term functional differentiation of this equation allows us to derive in a simple manner the equations for arbitrary correlators and to obtain their solutions.

To derive the equations for the G.F. of correlators in model III we make use of the so-called contraction of graphs; the meaning of this operation will be explained later.

Any edge of a graph either belongs to a cycle or it does not. In the latter case the edge is called a bridge.[174] When all bridges are eliminated, the graph is split into the connected components, which are called leaves (Fig. 3.3*b*). When each leaf is contracted into a vertex, the vertices will be connected only by the bridges, that is, this contracted graph will contain no cyclic edges. Such a tree, with vertices depicting the leaves, is called the leaf composition[174] of the initial graph (Fig. 3.3*c*). This construction can be formally viewed as the graph of the molecule consisting of units of quasi-monomers, which differ from each other in the functionality as well as in the number of "true" monomeric units contained in them and their distribution in space. To such quasi-monomers there correspond the graphs (Fig. 3.3*d*) obtained from the free cycles (Fig. 3.3*b*) by adding white vertices corresponding to unreacted functionalities.

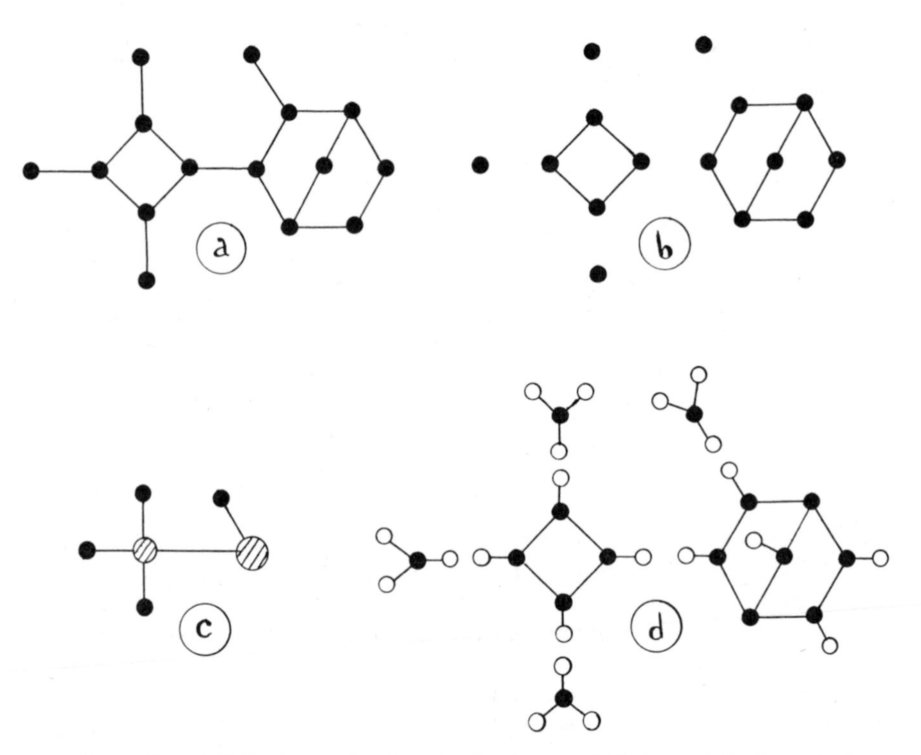

Figure 3.3. (*a*) Molecular graph with no functional groups, (*b*) its leaves, (*c*) leaf composition, and (*d*) constituent quasimonomers.

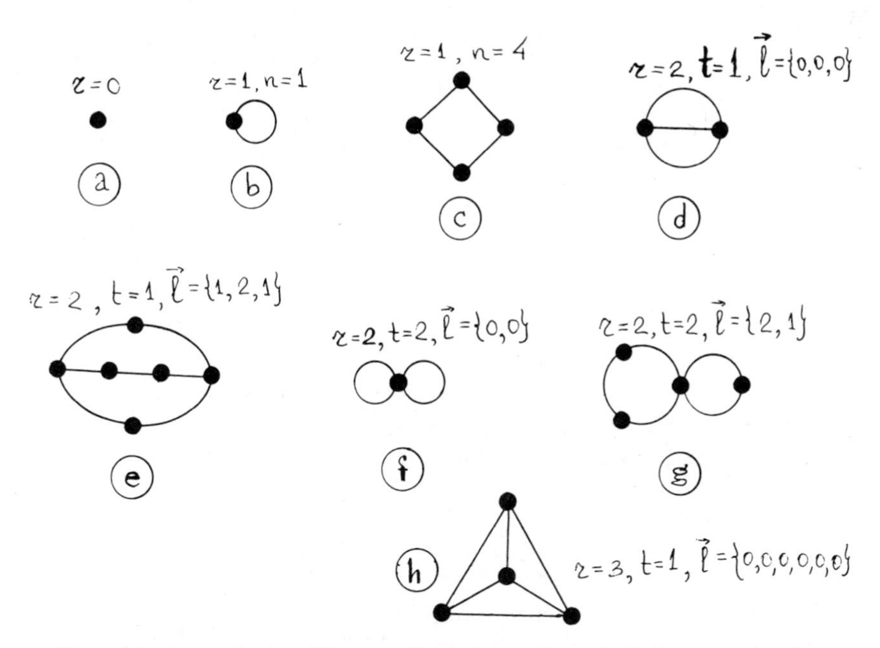

Figure 3.4. Leaves having different cyclic ranks r and topological structures t. Homeomorphic graph pairs are (b) and (c), (d) and (e), and (f) and (g). Graphs (b), (d), (f), and (h) are elementary representatives of their topological classes.

One of the most important characteristics of a leaf is its cyclic rank, that is, the minimal number of edges to be eliminated in order to obtain the tree (Fig. 3.4). The degenerated "cycle" of the rank $r = 0$ consists of the only monomeric unit. The simplest cycles ($r = 1$) differ only in the number of units involved. More complex cycles ($r \geqslant 2$) may have different topology, which will be specified by the superscript t, for each r separately. By definition, two graphs have identical topology if they are homeomorphic[13,47]; that is, they can be obtained from one another by the sequential substitution of each of the edges by the linear chain with all vertices of degree 2 (Fig. 3.5). An elementary representation of the topology is the graph having none of such vertices. By substituting its edges with linear chains of different lengths l_α, all graphs of a given topology can be obtained. Thus, the cycle $(r, t, \mathbf{l})$ is completely defined by its cyclic rank, the number of topology, and the vector $\mathbf{l} = \{l_\alpha\}$ of lengths of its edges.

The position of the tree vertex in model I was specified by the only label $\mathbf{r}_i$. But to describe the cycle embedded in space, it is necessary to specify the coordinates $\mathbf{r}_i$ of all its vertices. Since the position of the functionalities is also of interest, we supply the graph of the cycle consisting of black vertices

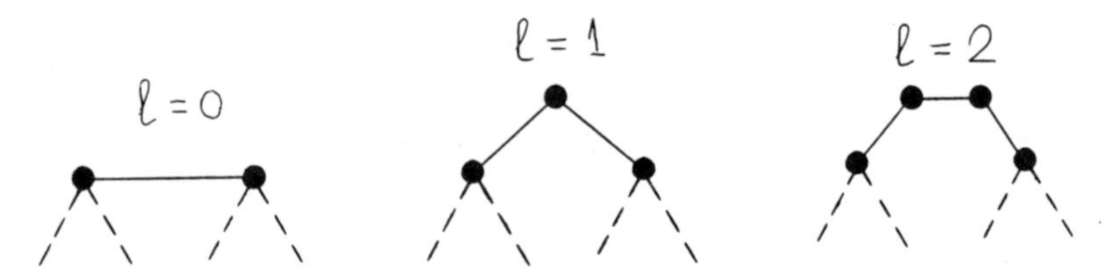

Figure 3.5. Substitution of graph edge for chain of l vertices, which does not change graph topology.

(Figs. 3.4–3.5) with white terminal vertices in such a way that the degree of each node is f (Figs. 3.3d and 3.6). The functionality of the cycle, that is, the number of its white vertices, is defined unambiguously by its rank r and the number n of its black vertices,

$$f_{r\mathbf{l}} \equiv f_{rn} = (f-2)n - 2(r-1), \tag{3.37}$$

since $2(n-r+1)$ groups have contributed to the cyclic linkages. The functional groups of the cycles are called *effective* groups since they define the effective functionality of quasi-monomers, which make up the leaf composition. The order of the automorphism group of the free-cycle graph with the effective groups is denoted $\mathscr{S}(r\mathbf{l})$ (Fig. 3.6). This quantity is defined in a standard way.[20] For instance, the automorphisms of the six-membered free cycle of rank $r=1$ in Fig. 3.6 are the six rotations, two reflections, and independent permutations of the $f-2$ groups of each of the six units.

Since the probabilities of different molecules in the equilibrium system do not depend on the way they are prepared, the creation of the macromolecular ensemble can be considered as the two-step process of their formation. In the first step the quasi-monomers are formed, which in the course of the second step yield the polymer with the equilibrium molecular-structural distribution, due to reactions between effective groups without the formation of a new cycle. The calculation of this distribution is thus reduced to consideration of the ideal polycondensation of quasi-monomers, which requires the knowledge of their initial distribution. If the g.f. of this distribution is known, the g.f. for the MSD can be obtained in a standard way by using the theory of branching processes. In what follows these ideas will be generalized for the case, when, together with the structure of polymeric molecules, the arrangement of fragments in space is also taken into account, so that the g.f. will be replaced by the generating functionals.

The significant role is played by the G.F. of correlators of individual molecules on the set of quasi-monomers, which is obtained by summing

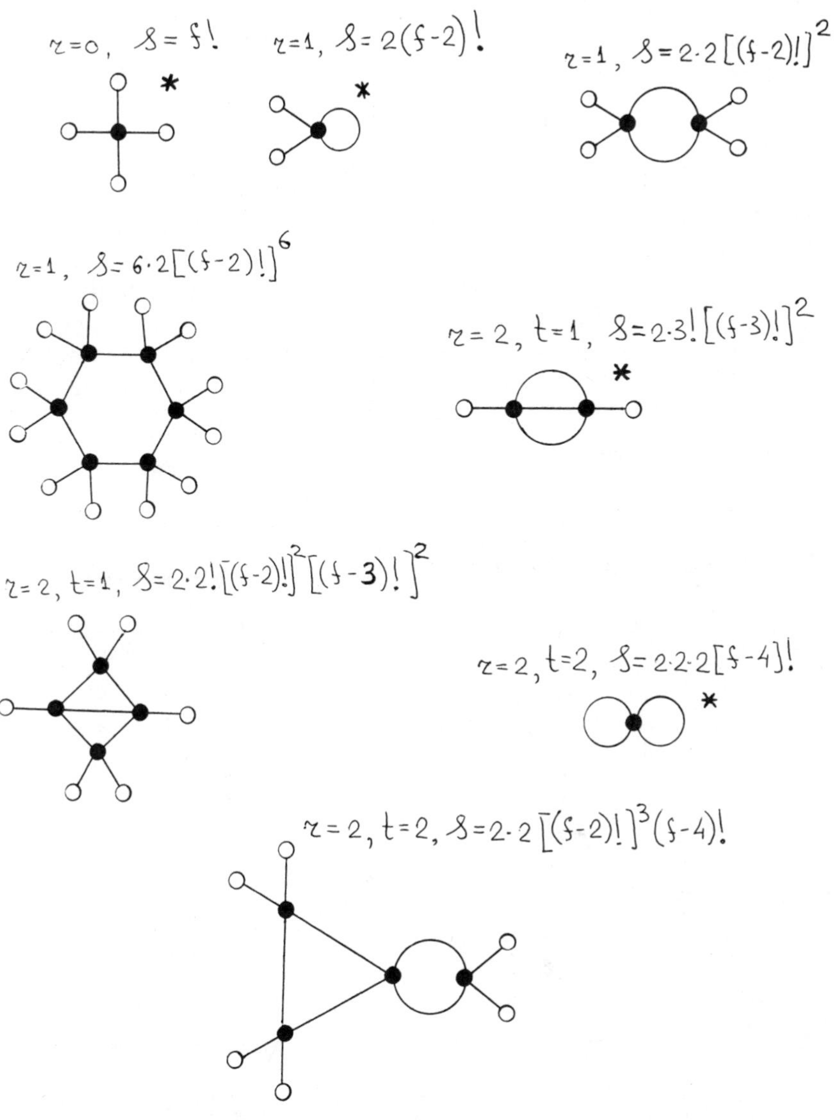

Figure 3.6. Representations of simplest topologies with shown effective groups for the example of a four-functional ($f = 4$) monomer and orders of automorphism groups. Elementary representatives of their topological classes indicated with asterisks.

(3.24) over the free cycles ($rt\mathbf{l}$):

$$\Gamma\{\mathbf{s}\} = \sum_{rt\mathbf{l}} \Gamma_{rt\mathbf{l}}\{\mathbf{s}\} = \sum_{rt\mathbf{l}} [(f!M)^n L^{n+r-1}/\mathcal{S}(rt\mathbf{l})] \times \int \cdots \int \prod_{i=1}^{n} \{d\mathbf{r}_i \tilde{s}_u(\mathbf{r}_i)\} \prod \{d\mathbf{r}_{ij} \tilde{s}_f(\mathbf{r}_{ij})\} \lambda_{rt\mathbf{l}}(\{\mathbf{r}\}, \{s_b\}). \tag{3.38}$$

The first product here involves all n units, whereas the second involves f_{rn} effective groups of the cycle. The functional $\lambda_{rt\mathbf{l}}(\{\mathbf{r}\}, \{s_b\})$ accounts for the loss of conformational entropy accompanying the cyclization, and for $s_b \equiv 1$ it is equal, to within the normalization factor, to the probability density for the distribution of units and functionalities of the quasi-mononer in space. This functional is obtained by integrating the same integrand as in $I(\mathcal{G}_{rt\mathbf{l}}|\{\mathbf{h}\})$ for the vanishing virtual fields $h_u^v = h_f^v = 0$. Still, in contrast to Eq. (3.7), the

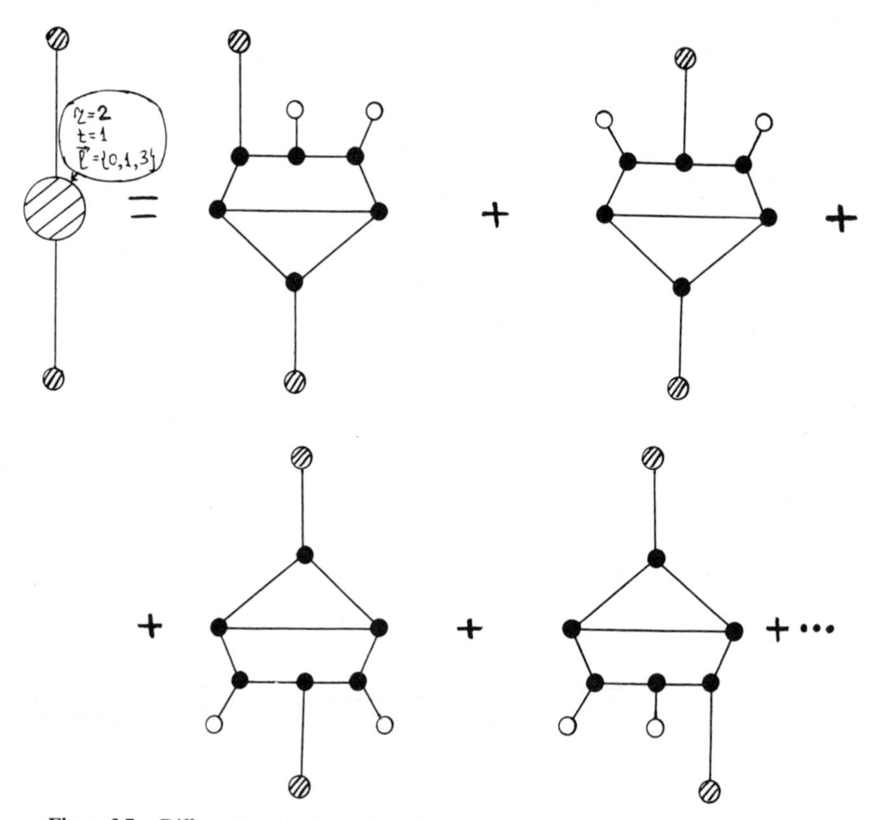

Figure 3.7. Different topological orientations of a cycle, contracted to same vertex in leaf composition.

integration is performed only over the coordinates of the reacted groups. The functional derivatives

$$\Gamma^{\nu}(\mathbf{r}, \{\mathbf{s}\}) = \frac{\delta\Gamma\{\mathbf{s}\}}{\delta s_{\nu}(\mathbf{r})}, \qquad \nu = u, f, b, \tag{3.39}$$

which involve the contributions of quasi-monomers only, substitute for the expansion series, arising in the calculation of correlators, each term of which corresponds to a single isomer.

In Section I.D it was establish that, by using the formulas connecting the number of ordered trees with the order of their automorphism groups, the summation over all molecules can be reduced to the sum over the types of monomeric units [Eq. (1.22)], in our case over the cycle types ($rt\mathbf{l}$). This procedure resembles the transition from the connected diagrams to their irreducible components in Mayer's second theorem in the theory of non-ideal gases.[168,169] In fact, in the variational differentiation of the G.F. [formula (3.24)] all nodes of the molecular graph are taken successively as the root located at the point $\mathbf{r}$, whereas the coordinates of the remaining vertices are integrated out. Each of the nodes belongs to some cycle ($rt\mathbf{l}$), whose image becomes the root of the leaf composition. Since the vertex of a given type may be the image of different contracted cycles, which are connected with their neighbors in a topologically different way, (Fig. 3.7), one leaf composition represents, in general, different isomers, each entering the sum (3.24) with its own factor $1/\mathcal{S}(l, q)$ (cf. Section I.E). The summation of these inverse orders of the automorphic groups can be accomplished by using the invariance principle (Section I.E):

$$\sum \frac{1}{\mathcal{S}(l,q)} = \frac{1}{\mathcal{S}(m,q')} \prod_{rt\mathbf{l}} \left[\frac{f_{rn}!}{\mathcal{S}(rt\mathbf{l})} \right]^{m_{rt\mathbf{l}}}. \tag{3.40}$$

Here the summation is performed over all molecular graphs, which are represented by the same leaf composition having $m_{rt\mathbf{l}}$ vertices of the type ($rt\mathbf{l}$). The order of the automorphism group $\mathcal{S}(m, q')$ of this leaf composition is expressed by formula (1.21) via the number of rooted ordered trees associated with the composition. Thus, the problem of constructing the spatial measure for the cyclic molecules is reduced to the enumeration of trees with vertices of various types.

In differentiating $\Psi\{\mathbf{s}\}$ with respect to $\ln s_{\nu}(\mathbf{r})$, the trees are enumerated with the root being, depending on the index ν, either black nodes ($\nu = u$) or white pendant vertices ($\nu = f$) or a pair of bars, that is, bonds ($\nu = b$). To each such tree representing the leaf composition of the molecular graph there can be associated exactly one item, which arises in the expansion of the corresponding derivative of the functional Γ, where the argument is replaced by $\hat{\mathbf{s}} = \{s_u, s_f(1 + L\Psi^f \tilde{s}_b/(\tilde{s}_f)^2), s_b\}$:

$$\Psi^\nu(\mathbf{r}, \{\mathbf{s}\}) \equiv \frac{\delta\Psi\{\mathbf{s}\}}{\delta \ln s_\nu(\mathbf{r})} = s_\nu(\mathbf{r})\Gamma^\nu(\mathbf{r}, \{\hat{\mathbf{s}}\}), \qquad \nu = u, f, \tag{3.41}$$

$$\Psi^b(\mathbf{r}, \{\mathbf{s}\}) = \frac{\delta\Psi\{\mathbf{s}\}}{\delta \ln s_b(\mathbf{r})} = s_b(\mathbf{r})\left[\Gamma^b(\mathbf{r}, \{\hat{\mathbf{s}}\}) + \frac{L}{2}\exp(h_b(\mathbf{r}) - 2h_f(\mathbf{r}))(\Gamma^f(\mathbf{r}, \{\hat{\mathbf{s}}\}))^2\right].$$

If we set all $s_\nu(\mathbf{r}) \equiv 1$ in these formulas, relations (3.41) transform into the system of integral equations for the density distributions $\rho_\nu(\mathbf{r}) = \Psi^\nu(\mathbf{r}, \{\mathbf{1}\})$ of units, functionalities, and bonds for given external fields $h_\nu(\mathbf{r})$.

The functional Ψ can be recovered via its derivatives (3.41):

$$\Psi\{\mathbf{s}\} = \Gamma\{\hat{\mathbf{s}}\} - \frac{L}{2}\int \frac{\tilde{s}_b(\mathbf{r})}{(\tilde{s}_f(\mathbf{r}))^2}(\Psi^f(\mathbf{r}, \{\mathbf{s}\}))^2\, d\mathbf{r}. \tag{3.42}$$

It is noteworthy that if Ψ^f in (3.42) is regarded as an independent variable, then Eq. (3.41) for this function coincides with the extremum condition for the functional Ψ (3.42).

Subsequent differentiation of formulas (3.41) yields equations for correlation functions of any order. In case of spatially homogeneous fields the equations are greatly simplified. Thus, for example, the second derivatives $\Psi^{\nu\mu}$ of the functional Ψ are functions of the difference of the arguments, $\mathbf{r}' - \mathbf{r}'' = \mathbf{r}$. The Fourier transform reduces the integral equations to the algebraic system. As a consequence, the following solution can be obtained for the Fourier transform of the correlator of the molecular density of units:

$$\begin{aligned} \tilde{\Psi}^{uu}(\mathbf{q}, \{\mathbf{s}\}) &\equiv \int \frac{\delta^2\Psi\{\mathbf{s}\}}{\delta \ln s_u(\mathbf{r}')\delta \ln s_u(\mathbf{r}'')} e^{-i\mathbf{q}\mathbf{r}}\, d\mathbf{r} = \Psi^u(\{\mathbf{s}\}) \\ &\quad + \tilde{\Gamma}^{uu}(\mathbf{q}, \{\hat{\mathbf{s}}\}) + \frac{(\tilde{\Gamma}^{uf}(\mathbf{q}, \{\hat{\mathbf{s}}\}))^2}{L^{-1} - \tilde{\Gamma}^{ff}(\mathbf{q}, \{\hat{\mathbf{s}}\})}, \end{aligned} \tag{3.43}$$

where the double superscripts denote the Fourier transforms of the corresponding second derivatives of the functional Γ.

F. Case Study: Model I

The formulas for the generating functionals acquire the most simple form in the theory of Flory. The only admissible type of "cycle" is the trivial one: for $r = 0$ it corresponds to the usual monomeric unit. The functional Γ in this case consists of the only term:

$$\Gamma\{\mathbf{s}\} = M \int d\mathbf{r}_1 \tilde{s}_u(\mathbf{r}_1) \prod_{j=1}^{f} \left[\int \lambda(\mathbf{r}_1 - \mathbf{r}_{1j}) \tilde{s}_f(\mathbf{r}_{1j})\, d\mathbf{r}_{1j} \right]. \tag{3.44}$$

Its derivatives

$$\Gamma^u(\mathbf{r}, \{\mathbf{s}\}) = M \exp h_u(\mathbf{r}) \prod_{j=1}^{f} \left[\int \lambda(\mathbf{r} - \mathbf{r}_{1j}) \tilde{s}_f(\mathbf{r}_{1j})\, d\mathbf{r}_{1j} \right],$$

$$\Gamma^f(\mathbf{r}, \{\mathbf{s}\}) = f M \exp h_f(\mathbf{r}) \int d\mathbf{r}_1 \tilde{s}_u(\mathbf{r}_1) \lambda(\mathbf{r} - \mathbf{r}_1) \prod_{j=2}^{f} \left[\int \lambda(\mathbf{r}_1 - \mathbf{r}_{1j}) \tilde{s}_f(\mathbf{r}_{1j})\, d\mathbf{r}_{1j} \right], \tag{3.45}$$

define the functionals Ψ^u, Ψ^f; the ratios of these functionals to the densities ρ_u and ρ_f will be denoted G_W and $\mathscr{U}$. The equations for the latter quantities according to Eqs. (3.41) and (3.45) are of the form

$$G_W(\mathbf{r}, \{\mathbf{s}\}) = \frac{\Psi^u(\mathbf{r}, \{\mathbf{s}\})}{\rho_u(\mathbf{r})} = \frac{\tilde{s}_u(\mathbf{r}) M}{\rho_u(\mathbf{r})} \times \left\{ \int d\mathbf{r}' \lambda(\mathbf{r} - \mathbf{r}') [\tilde{s}_f(\mathbf{r}') + \tilde{s}_b(\mathbf{r}') L \rho_f(\mathbf{r}') \exp(-h_f(\mathbf{r}')) \mathscr{U}(\mathbf{r}')] \right\}^f, \tag{3.46}$$

$$\mathscr{U}(\mathbf{r}', \{\mathbf{s}\} = \frac{\Psi^f(\mathbf{r}', \{\mathbf{s}\})}{s_f(\mathbf{r}') \rho_f(\mathbf{r}')} = \frac{M f \exp h_f(\mathbf{r}')}{\rho_f(\mathbf{r}')} \int d\mathbf{r}'' \tilde{s}_u(\mathbf{r}'') \lambda(\mathbf{r}' - \mathbf{r}'') \times \left\{ \int d\mathbf{r}''' \lambda(\mathbf{r}'' - \mathbf{r}''') [\tilde{s}_f(\mathbf{r}''') + \tilde{s}_b(\mathbf{r}''') L \rho_f(\mathbf{r}''') \mathscr{U}(\mathbf{r}''', \{\mathbf{s}\}) / \exp h_f(\mathbf{r}''')] \right\}^{f-1}. \tag{3.47}$$

The physical meaning of these expressions can be elucidated most simply in the absence of external fields. In order to determine the parameters in the derivatives Ψ^v, let $h_v^v(\mathbf{r}) \equiv h_v \equiv 0$. In this case the functions s_v, G_W, and $\mathscr{U}$ are independent of coordinates and become equal to unity, so that Eqs. (3.46) and (3.47) afford the relation of the parameters M and L with constant throughout the space densities of units and groups:

$$1 = M(1 + L\bar{\rho}_f)^f / \bar{\rho}_u, \quad 1 = M f (1 + L\bar{\rho}_f)^{f-1} / \bar{\rho}_f. \tag{3.48}$$

Using the conversion of groups p instead of the density of groups transforms relations (3.48) to the form

$$M = \bar{\rho}_u(1-p)^f, \qquad L = p/f\,\bar{\rho}_u(1-p)^2. \tag{3.49}$$

Inserting these relations into formulas (3.46) and (3.47), we arrive at the quite elegant equations[210]

$$G_W(\mathbf{r}, \{\mathbf{s}\}) = s_u(\mathbf{r})\left\{\int \lambda(\mathbf{r}-\mathbf{r}')[(1-p)s_f(\mathbf{r}') + p s_b(\mathbf{r}')\mathcal{U}(\mathbf{r}', \{\mathbf{s}\})]\,d\mathbf{r}'\right\}^f, \tag{3.50}$$

$$\mathcal{U}(\mathbf{r}', \{s\}) = \int d\mathbf{r}''\lambda(\mathbf{r}'-\mathbf{r}'')s_u(\mathbf{r}'') \times \left\{\int \lambda(\mathbf{r}''-\mathbf{r}''')[(1-p)s_f(\mathbf{r}''') + p s_b(\mathbf{r}''')\mathcal{U}(\mathbf{r}''', \{\mathbf{s}\})]\,d\mathbf{r}'''\right\}^{f-1}, \tag{3.51}$$

which have attractive probabilistic meaning. They describe the history of the branching random process, where individuals migrate in space and are labeled with coordinates. These processes are called "general" branching processes.[21] Just as the information about the usual branching process is contained in the generating functions describing the probabilities for each individual to produce some litter, so the general branching process is characterized by the generating functionals. These G.F.'s represent, in addition to the number of descendants for each individual, the probabilities for the descendant to be located at some point in space. The generating functional of such pointwise distributions[21] have obvious meaning.

An individual of the zeroth generation corresponding to the monomeric unit at the point $\mathbf{r}$ produces f descendants. Each individual, regardless of the others, is found at point $\mathbf{r}'$ with the probability density $\lambda(\mathbf{r}-\mathbf{r}')$ (Fig. 3.8*a*, *b*). This migration of individuals is the principal distinction between

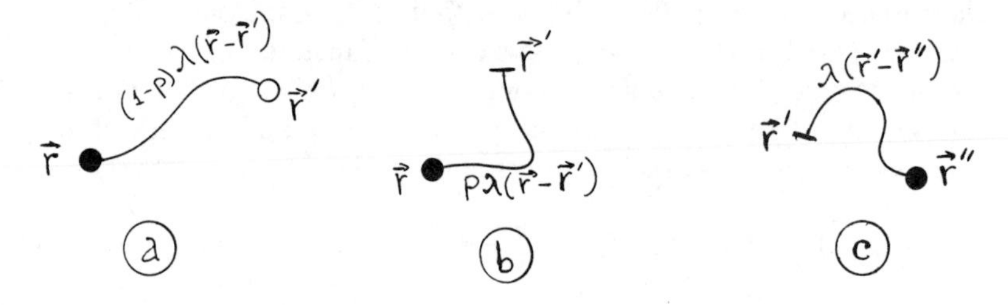

Figure 3.8. Displacements of particles in branching process.

the general process and those used formerly (see Section I). A descendant located at a point may be associated either with an unreacted group or with a bond. The probabilities of these two events, just as in the traditional branching process [Eq. (1.12)], are $1-p$ and p, respectively. Therefore, the G.F. for the point distribution of the litter of the zeroth-generation individual is of the form

$$F_0^u(\mathbf{r},\{\mathbf{s}\})=\left\{\int\lambda(\mathbf{r}-\mathbf{r}')[(1-p)s_f(\mathbf{r}')+ps_b(\mathbf{r}')]\,d\mathbf{r}'\right\}^f. \tag{3.52}$$

The descendants representing the unreacted groups are incapable of further reproduction, and development of the branching process in this direction terminates. The bond generates one individual, a monomeric unit, located with the probability density $\lambda(\mathbf{r}'-\mathbf{r}'')$ at the point $\mathbf{r}''$ (Fig. 3.8*c*). Hence, the G.F. for the litter of the bond is

$$F^b(\mathbf{r}',\{\mathbf{s}\})=\int\lambda(\mathbf{r}'-\mathbf{r}'')s_u(\mathbf{r}'')\,d\mathbf{r}''. \tag{3.53}$$

The process of reproduction is further repeated recurrently with a distinction that the unit of any nonzeroth generation produces $f-1$ descendants, so that its G.F. F^u differs from Eq. (3.52) by the occurrence of $f-1$, not f, factors. According to the principles of pointwise distributions of the general branching process,[21,175] the G.F. for the point distribution of all individuals of some realization of the branching process is represented by expressions (3.50) and (3.51).

When the coordinates of the unreacted groups and bonds can be discarded, the counters $s_f(\mathbf{r})$, $s_b(\mathbf{r})$ may be set equal to unity. The substitution

$$U(\mathbf{r}'',\{s\})=s(\mathbf{r}'')\left\{1-p+p\int\lambda(\mathbf{r}''-\mathbf{r}''')\mathscr{U}(\mathbf{r}''',\{s\})\,d\mathbf{r}'''\right\}^{f-1} \tag{3.54}$$

simplifies the former formulas:

$$G_W(\mathbf{r},\{s\})=s(\mathbf{r})\left\{1-p+p\int\lambda_{uu}(\mathbf{r}-\mathbf{r}'')U(\mathbf{r}'',\{s\})\,d\mathbf{r}''\right\}^f, \tag{3.55}$$

$$U(\mathbf{r}'',\{s\})=s(\mathbf{r}'')\left\{1-p+p\int\lambda_{uu}(\mathbf{r}''-\mathbf{r}''')U(\mathbf{r}''',\{s\})\,d\mathbf{r}'''\right\}^{f-1}. \tag{3.56}$$

Here account is taken of the reproduction of individuals (units only), where

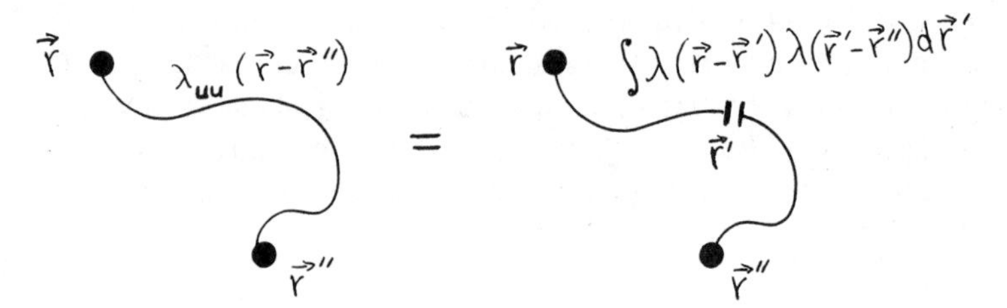

Figure 3.9. Displacements of particles corresponding to monomer units.

the distribution of descendants with respect to the position in space of their fathers is described by the distribution function for the distance between the adjacent units of the chain, which is the convolution of two functions $\lambda_{uf} \equiv \lambda$ (Fig. 3.9):

$$\lambda_{uu}(\mathbf{r} - \mathbf{r}'') = \int \lambda(\mathbf{r} - \mathbf{r}')\lambda(\mathbf{r}' - \mathbf{r}'')\, d\mathbf{r}'. \tag{3.57}$$

The generating functional for the point distribution of the litter of one individual is, in this case,

$$F(\mathbf{r}, \{s\}) = \left\{1 - p + p \int \lambda_{uu}(\mathbf{r} - \mathbf{r}'')s_u(\mathbf{r}'')\, d\mathbf{r}''\right\}^f \tag{3.58}$$

for the zeroth-generation individuals, or a similar expression with the power factor $f - 1$ for other generations.

The branching processes introduced in this section differ from those traditionally used[2,15] with respect to the information about the arrangement of individuals involved. When this information is of no interest, the counters $s_v(\mathbf{r})$ should be set constant, $s_v(\mathbf{r}) = s_v$. The functional $U(\mathbf{r}', \{s\})$ in this case transforms into the function $u(s)$. Since the integral of λ_{uu} is unity, formulas (3.55) and (3.56) become the familiar Eqs. (1.12).

The general branching process in space enables us to calculate the correlators by differentiating formulas (3.50) and (3.51) or (3.55) and (3.56) and solving the integral equations thus arising. For example, the Fourier transform of the correlator χ^{uu} is represented in the form

$$\begin{aligned}\tilde{\chi}^{uu}(\mathbf{q}, s) &\equiv \int \chi^{uu}(\mathbf{r}, s)\exp(-i\mathbf{q}\mathbf{r})\, d\mathbf{r} \\ &= \bar{\rho}_u G_W(s)\left[1 + \frac{f p u(s)\tilde{\lambda}^2(\mathbf{q})}{1 - p + p u(s) - (f-1) p u(s)\tilde{\lambda}^2(\mathbf{q})}\right],\end{aligned} \tag{3.59}$$

where $\tilde{\lambda}(\mathbf{q})$ is the Fourier transform of $\lambda(\mathbf{r})$, and the latter quantity squared, $\tilde{\lambda}^2(\mathbf{q}) = \tilde{\lambda}_{uu}(\mathbf{q})$, is, according to Eq. (3.57), the Fourier transform of the function $\lambda_{uu}(\mathbf{r})$. The same result can be obtained from the general formula (3.43) by inserting into it the corresponding derivatives of the functional Γ (3.44).

The contribution of the l-mers $\tilde{\chi}_l^{uu}$ into the correlator χ^{uu} can be found via expansion of (3.59) into the power series in s. Making use of the mean number of trails of length n, outgoing from a single unit in the treelike l-mer, N_{nl}, the latter contribution can be expressed as

$$\tilde{\chi}_l^{uu}(\mathbf{q}) = \bar{\rho}_u f_W(l) \sum_{n=0}^{l} (\tilde{\lambda}_{uu}(\mathbf{q}))^n N_{nl}. \tag{3.60}$$

Having observed that $(\tilde{\lambda}_{uu}(\mathbf{q}))^n$ is the Fourier transform for the distribution $W_n(\mathbf{r})$ of distances between the ends of the linear chain of $(n+1)$ units, the correlator χ_l^{uu} can be expressed in the coordinate representation:

$$\chi_l^{uu}(\mathbf{r}) = \bar{\rho}_u f_W(l) D_l(\mathbf{r}), \qquad D_l(\mathbf{r}) = \sum_{n=0}^{l} W_n(\mathbf{r}) N_{nl}. \tag{3.61}$$

The integral of the function $D_l(\mathbf{r})$ is the number of units l, and hence,

$$\tilde{\chi}^{uu}|_{\mathbf{q}=\mathbf{0}} \equiv \int \chi^{uu}(\mathbf{r})\, d\mathbf{r} = \int \sum_l \chi_l^{uu}(\mathbf{r})\, d\mathbf{r} = \bar{\rho}_u P_W. \tag{3.62}$$

According to Eq. (3.61), the function $D_l(\mathbf{r})$ has the following meaning. Take at random a black vertex of an arbitrary graph of the l-mer as the root and mark it with the label $\mathbf{r}' = \mathbf{0}$ and then establish the density of the number of other units of the same molecule at a distance $\mathbf{r}$ from the fixed unit, and average it over all ways to choose the root. This definition is identical to that of the two-point correlator $\hat{\chi}_l$ and of the percolation function of "pair connectedness"[164] introduced in the foregoing. The Fourier transform of the correlator $\hat{\chi}$, the Debye function, can be measured in numerous scattering experiments (e.g., in the scattering of light, X rays, neutrons). Since, according to Eq. (3.30), the correlator χ^{uu} differs from $\hat{\chi}$ by merely a factor, the normalized Fourier transform of this correlator is identical to the scattering factor $P_Z(\theta) = \tilde{\chi}^{uu}(\mathbf{q})/\tilde{\chi}^{uu}(\mathbf{0})$. The scattering angle θ in this case is connected with the modulus of the vector $\mathbf{q}$ and the length λ of the scattering wave via the relation[164] $|\mathbf{q}| = 4\pi\lambda^{-1} \sin\theta/2$.

Direct calculations by formula (3.60) can be carried out by using a method[2] that employs the generating function (2.19) for the numbers N_{nl}. In fact, comparison of Eq. (3.60) and (2.19) shows that the G.F. for the correlators $\tilde{\chi}_l^{uu}(\mathbf{q})$ is represented by the function $N_n(s)$ with the renormalized argument

$s \to s\tilde{\lambda}_{uu}(\mathbf{q})$. Hence, the methods described in Section II.D can be employed for the calculation of correlators and the physico chemical properties defined by them.

Now let us return to the general case of arbitrary fields $h_v(\mathbf{r})$ acting on particles. Here the denominators in Eq. (3.46) and (3.47) are determined from the condition that they should coincide with the numerators $\rho_v = \Psi_v$ when the counters in the numerators are set equal to unity, $s_v(\mathbf{r}) \equiv 1$. For example, the equation for the density of functionalities is of the form

$$\rho_f(\mathbf{r}) = \Psi^f(\mathbf{r}, \{1\}) = \exp[h_f(\mathbf{r})] Mf \int d\mathbf{r}' \exp h_u(\mathbf{r}') \lambda(\mathbf{r} - \mathbf{r}') \{Z(\mathbf{r}')\}^{f-1}, \tag{3.63}$$

$$Z(\mathbf{r}') = \int \lambda(\mathbf{r}' - \mathbf{r}'') \exp h_f(\mathbf{r}'') [1 + K(\mathbf{r}'') \rho_f(\mathbf{r}'')] \, d\mathbf{r}'', \tag{3.64}$$

$$K(\mathbf{r}'') = L\exp[h_b(\mathbf{r}'') - 2h_f(\mathbf{r}'')]. \tag{3.65}$$

This density serves as a parameter in the equations for the G.F. of the branching process, which are obtained from Eqs. (3.46) and (3.47) by inserting $\rho_f(\mathbf{r})$ and $\rho_u(\mathbf{r})$:

$$G_W(\mathbf{r}, \{\mathbf{s}\}) = s_u(\mathbf{r}) \Xi^f(\mathbf{r}, \{\mathbf{s}\}), \tag{3.66}$$

$$\Xi(\mathbf{r}, \{\mathbf{s}\}) = \int [\Lambda_f(\mathbf{r}, \mathbf{r}') s_f(\mathbf{r}') + \Lambda_b(\mathbf{r}, \mathbf{r}') s_b(\mathbf{r}') \mathscr{U}(\mathbf{r}', \{\mathbf{s}\})] \, d\mathbf{r}', \tag{3.67}$$

$$\mathscr{U}(\mathbf{r}', \{\mathbf{s}\}) = \int \Lambda(\mathbf{r}', \mathbf{r}'') s_u(\mathbf{r}'') \Xi^{f-1}(\mathbf{r}'', \{\mathbf{s}\}) \, d\mathbf{r}'', \tag{3.68}$$

$$\Lambda_f(\mathbf{r}, \mathbf{r}') = \lambda(\mathbf{r} - \mathbf{r}') \exp h_f(\mathbf{r}') / Z(\mathbf{r}), \qquad \Lambda_b(\mathbf{r}, \mathbf{r}') = \Lambda_f(\mathbf{r}, \mathbf{r}') K(\mathbf{r}') \rho_f(\mathbf{r}'), \tag{3.69}$$

$$\Lambda(\mathbf{r}', \mathbf{r}'') = \lambda(\mathbf{r}' - \mathbf{r}'') \exp h_u(\mathbf{r}'') Z^{f-1}(\mathbf{r}'') \Big/ \int \lambda(\mathbf{r}' - \mathbf{r}'') \exp h_u(\mathbf{r}'') Z^{f-1}(\mathbf{r}'') \, d\mathbf{r}''. \tag{3.70}$$

These equations can be interpreted[210] on the basis of general principles.[21,175] The individuals of the branching process do not differ in any way from those whose reproduction is described by formulas (3.50) and (3.51). However, their migration in space is affected by external fields. Additionally, the functional group enters the reaction with a probability that depends on its position

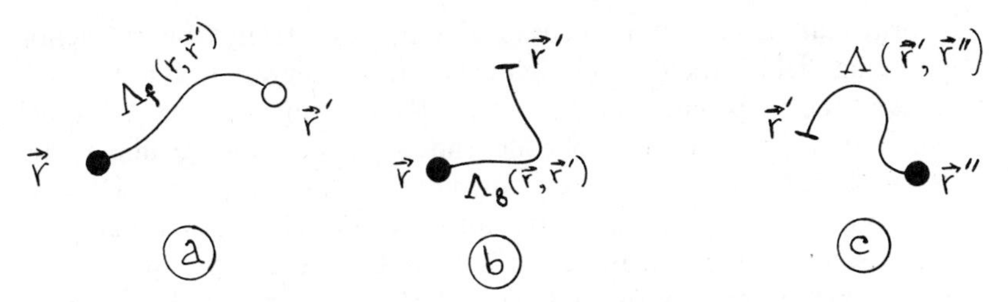

Figure 3.10. Displacements of particles in branching process taking into account external fields acting on molecular fragments.

since the equilibrium constant $K(\mathbf{r}'')$ depends on the coordinate $\mathbf{r}''$. Expression (3.65) for $K(\mathbf{r}'')$ accounts not only for the chemical part F_0 (entering $L = z_b/z_f^2$) of the change in free energy in the course of the elementary act of the reaction but also for the physical component $h_b(\mathbf{r}'') - 2h_f(\mathbf{r}'')$, connected with the difference in the potential energy of the pair of groups before and after the reaction.

The individual of the zeroth generation, which corresponds to the monomeric unit located at the point $\mathbf{r}$, produces f independent functional groups, and hence, the G.F. (3.66) for such an individual is the product of f identical factors (3.67). Each describes the fate of one of the descendants, which is the functional group, either reacted or not. Depending on the latter alternative, the probability density (3.69) of finding one such descendant at the point $\mathbf{r}'$ is $\Lambda_b(\mathbf{r}, \mathbf{r}')$ or $\Lambda_f(\mathbf{r}, \mathbf{r}')$ (Fig. 3.10). As opposed to the spatially homogeneous case, the distribution function $\lambda(\mathbf{r} - \mathbf{r}')$ for the distance between the group and the unit is multiplied by the Boltzmann factor $\exp h_f(\mathbf{r}')$, which accounts for the difference in the potential energy of the group in the external field. The functional group at the point $\mathbf{r}'$ may be found in the two states: either to form the chemical bond or to remain unreacted. The ratio of the probabilities of these two states is expressed by the law of acting masses,

$$2\rho_b(\mathbf{r}')/\rho_f(\mathbf{r}') = K(\mathbf{r}')\rho_f(\mathbf{r}'), \tag{3.71}$$

which is reflected in the difference of the transition probabilities $\Lambda_b(\mathbf{r}, \mathbf{r}')$ and $\Lambda_f(\mathbf{r}, \mathbf{r}')$. The normalization conditions for these probabilities determine the denominator $Z(\mathbf{r})$ in Eqs. (3.69), which has the meaning of the partition function for the functional group, under the condition that the adjacent unit is fixed at the point $\mathbf{r}$.

Sorting out all possible locations $\mathbf{r}'$ of the functional group in space and taking into account the probabilities of its states, we arrive at the functional Ξ (3.67), which, thus, describes the fate of one group of the unit located at the point $\mathbf{r}$.

The principal distinction in the branching process arising from the action of the external field is the interrelation between the migration of individuals in space and their reproduction. As a result, the functions Λ_f and Λ_b cannot be factorized into the product of independent factors $(1-p)\lambda$ and $p\lambda$ (cf. Figs. 3.8 and 3.10). In addition, the probability densities $\Lambda_f(\mathbf{r},\mathbf{r}')$ and $\Lambda_b(\mathbf{r},\mathbf{r}')$ depend not only on the translation vector $\mathbf{r}-\mathbf{r}'$, but also on its origin $\mathbf{r}$.

The unreacted group does not contribute to the further reproduction, and each bond produces exactly one unit. Next, the history is repeated recurrently, so that units and bonds alternate regularly. This circumstance allows us to confine ourselves to the construction of the G.F. $\mathscr{U}$ [Eq. (3.68)] for the litter of the bond only and accounts for the positions of units via the counters $s_u(\mathbf{r})$. If this bond is found at the point $\mathbf{r}'$, then the probability $\Lambda(\mathbf{r}',\mathbf{r}'')$ of finding the next unit at $\mathbf{r}''$ is defined not only by the field $h_u(\mathbf{r}'')$ affecting the unit directly at this point but also by the indirect effect of the field h_f acting on its functional groups. The fact that units are linked by the linear chains with the groups leads, when the groups migrate, to the appearance of an additional force of entropy origin, where the potential $(f-1)\ln Z(\mathbf{r}'')$ accounts for the additive contributions of each of $f-1$ virtually migrating functional groups. The fate of these groups is described, as in the preceding generation, by the functional Ξ. This enables us to express the G.F. of a single bond $\mathscr{U}$ via Ξ and thus to complete the system of the two equations (3.67) and (3.68) for these functionals.

When we are not interested in the location of the descendants of individuals, the counters $s_v(\mathbf{r})$ can be set equal to constants s_v. In the absence of external fields we arrive at the equations for the MWD of all molecules of the system. However, under the action of fields h_v the MWD will be formed depending on the coordinate $\mathbf{r}$ of the point. In this case it is senseless to speak about the concentration of some isomer (l,q) at a given point, which characterizes the number MWD. In return, the concentration of its units $\rho^u_{l,q}(\mathbf{r})$ of the isomer has clear meaning and defines the weight MWD, the generating function of which is

$$G_W(\mathbf{r},s) = s\Xi^f(\mathbf{r},s), \tag{3.72}$$

$$\Xi(\mathbf{r},s) = \int \Lambda_f(\mathbf{r},\mathbf{r}')\,d\mathbf{r}' + \int \Lambda_b(\mathbf{r},\mathbf{r}')\mathscr{U}(\mathbf{r}',s)\,d\mathbf{r}', \tag{3.73}$$

$$\mathscr{U}(\mathbf{r}',s) = s\int \Lambda(\mathbf{r}',\mathbf{r}'')\Xi^{f-1}(\mathbf{r}'',s)\,d\mathbf{r}''. \tag{3.74}$$

Differentiation of this function with respect to s at the point $s=1$ allows us to determine the mean degree of polymerization of the molecule, to which there belongs the random unit with the coordinate **r**:

$$P_W(\mathbf{r}) = \left.\frac{dG_W(\mathbf{r}, s)}{ds}\right|_{s=1} = f\Xi'(\mathbf{r}) + 1, \tag{3.75}$$

where the derivative $\Xi'(\mathbf{r})$ of the function $\Xi(\mathbf{r}, s)$ with respect to s at $s = 1$ satisfies the integral equation

$$\Xi'(\mathbf{r}) = \int \Lambda_b(\mathbf{r}, \mathbf{r}')\, d\mathbf{r}' + (f-1) \int\left[\int \Lambda_b(\mathbf{r}, \mathbf{r}')\Lambda(\mathbf{r}', \mathbf{r}'')\, d\mathbf{r}'\right]\Xi'(\mathbf{r}'')\, d\mathbf{r}''. \tag{3.76}$$

We close consideration of the treelike molecules and turn our attention to the description of polymers with cyclic fragments.

G. One-Point Correlators and MWD

In the spatially homogeneous case the functions $s_v(\mathbf{r})$ in Eqs. (3.38), (3.42), and (3.43), are independent of the coordinates. The densities $\rho_v(\mathbf{r})$ also are constant in the whole volume, and the functionals Ψ, Γ reduce to the ordinary functions ψ, γ:

$$\psi(\mathbf{s}) = \gamma(s_u, s_f + L\psi^f(\mathbf{s})/s_f, s_b) - LV(\psi^f)^2 s_b/2s_f^2, \tag{3.77}$$

$$\gamma(\mathbf{s}) = \sum_{rt\mathbf{l}} \gamma_{rt\mathbf{l}}(\mathbf{s}) = \sum_{rt\mathbf{l}} \frac{(f!M)^n L^{n+r-1}}{\mathscr{S}(rt\mathbf{l})} (s_u)^n (s_f)^{frn} \int \cdots \int \lambda_{rt\mathbf{l}}(\{\mathbf{r}\}, s_b) \prod d\mathbf{r}_{ij}. \tag{3.78}$$

Taking the usual (not functional) derivatives of γ, we obtain, in place of the correlators of the microscopic density of individual molecules, the correlators of the number of individuals, which, in view of summation (3.24), with the weight $c(l, q)$, determine the MWD of the system. When taking the functional derivative, the dimension of the spatial domain of integration decreases by 1. In the case of usual differentiation this is equivalent to relating the derivatives $\gamma^v = \partial\gamma/\partial s_v$ of the function γ to the unit volume. Thus, Eqs. (3.41) assume the form

$$\psi^v = \frac{s_v}{V}\gamma^v(\hat{s}), \qquad v = u, f,$$

$$\psi^b = s_b\left[\frac{\gamma^b(\hat{s})}{V} + \left(\frac{L\gamma^f(\hat{s})/v)^2}{2}\right)\right], \quad \hat{s} = \{s_u, s_f + L\psi^b/s_f, s_b\}. \tag{3.79}$$

If the arguments s_v of these functions are set equal to unity, we obtain the connection between the densities $\rho_v = \psi^v(1)$ and the parameters M, L. The contributions γ_{rtl} to the function γ at $s_v = 1$ coincide with the concentration of the corresponding cycles. The expression for these concentrations for the case of three-functional monomers was established by Erukhimovich[62] by using alternative (field-theory) mathematical tools. It should be noted that in the approach of Ref. 179 information on the distribution of cycles in molecules was lacking. In order to retain this information, the arguments s_v must remain unequal to unity.

According to Eq. (3.24), the function ψ represents the generating function for the number MWD, and the derivative (3.79) (the one-point correlator) represents the generating function for the weight MWD [cf. (3.28)]. In a similar way the MWD can be obtained, which accounts for the number $m_{rt\mathbf{l}}$ of cycles of each type in molecules. To this end the contribution of cycles of the type ($rt\mathbf{l}$) in Eqs. (3.77)–(3.79) should be multiplied by the "counter of cycles" $s_{rt\mathbf{l}}$. For example, consider the weight distribution $f_W(\mathbf{m})$ of units, which enter the molecules with the "vector of cycles" $\mathbf{m} = \{m_{rt\mathbf{l}}\}$. Its generating function is of the form

$$G_W(\mathbf{s}) \equiv \sum_{\mathbf{m}} f_W(\mathbf{m}) \prod_{rt\mathbf{l}} (s^n s_{rt\mathbf{l}})^{m_{rt\mathbf{l}}} = \sum_{rt\mathbf{l}} s^n s_{rt\mathbf{l}} d_{rt\mathbf{l}} (\xi(\mathbf{s}))^{f_{rt\mathbf{l}}} \tag{3.80}$$

$$\xi(\mathbf{s}) = 1 - \alpha + \alpha \sum_{rt\mathbf{l}} \beta_{rt\mathbf{l}} s^n s_{rt\mathbf{l}} (\xi(\mathbf{s}))^{f_{rt\mathbf{l}}-1}, \tag{3.81}$$

$$d_{rt\mathbf{l}} = \frac{\gamma^u_{rt\mathbf{l}}(1, 1/(1-\alpha), 1)}{V\rho_u}, \qquad \beta_{rt\mathbf{l}} = \frac{\gamma^f_{rt\mathbf{l}}(1, 1/(1-\alpha), 1)}{V\rho_f},$$

$$\alpha = \frac{L\rho_f}{1 + L\rho_f}. \tag{3.82}$$

The sum of $d_{rt\mathbf{l}}$ is equal to 1 in view of the relationship $\rho_u = \psi^u(\mathbf{1}) = \gamma^u(\hat{\mathbf{1}})/V$, and the same holds for the sum of $\beta_{rt\mathbf{l}}$.

Formulas (3.80) and (3.81) were established for the first time in ref. 63 and were interpreted in terms of the theory of branching processes. With a probability of $d_{rt\mathbf{l}}$ [which has the meaning of a fraction of units entering the cycles ($rt\mathbf{l}$)], the branching process starts from the individual ($rt\mathbf{l}$), which, being an individual of the zeroth generation, always produces exactly f_{rn} individuals, in accordance with the effective functionality (3.37) of the cycle. In all other generations this cycle yields $f_{rn} - 1$ descendants. These descendants, turn out to be "white" (i.e.., the unreacted groups) with a probability $1 - \alpha$, and individuals of the type ($rt\mathbf{l}$) with a probability $\alpha\beta_{rt\mathbf{l}}$; α refers to the probability that an arbitrarily taken effective group of the cycle will enter the reaction, that is, α is the conversion of the effective groups. The realizations of this process are

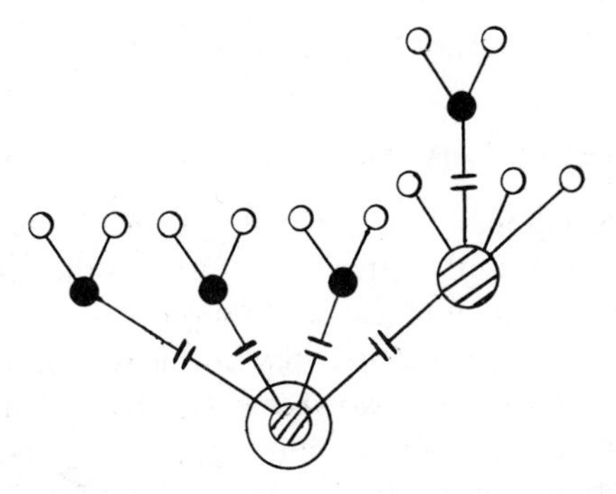

Figure 3.11. Realization of branching process with cycles, which corresponds to graph of Fig. 3.3 and originates from simple cycle in the zeroth generation.

represented by the leaf compositions of molecular graphs, which show explicitly both unreacted and reacted effective groups (Fig. 3.11).

For the calculation of the correlators $\chi^{\nu\cdots\mu}$, it is necessary to take into account the mutual arrangement of units and groups in space. This can be achieved if we consider, instead of the branching process just described, the process involving the individuals bearing the information about the coordinates of units and groups of the cycles (cf. Section III.F). Formulas (3.80) and (3.81) and (3.50) and (3.51) are generalized in this case to

$$G_W(\mathbf{r},\{\mathbf{s}\}) = \sum_{rtl} d_{rtl} s^n s_{rtl} \int \prod_{i=1}^{n} (d\mathbf{r}_i s_u(\mathbf{r}_i)) \left\{ \frac{1}{n} \sum_{i=1}^{n} \delta(\mathbf{r}-\mathbf{r}_i) \right.$$
$$\left. \times \frac{\lambda_{rtl}(\{\mathbf{r}\},\{s_b\})}{VK_{rtl}} \right\} \prod_{j=1}^{f_{rn}} (d\mathbf{r}'_j[(1-\alpha)s_f(\mathbf{r}'_j) + \alpha s_b(\mathbf{r}'_j)\mathscr{U}(\mathbf{r}'_j,\{\mathbf{s}\})]), \quad (3.83)$$

$$\mathscr{U}(\mathbf{r}',\{\mathbf{s}\}) = \sum_{rtl} \beta_{rtl} s^n s_{rtl} \int \prod_{i=1}^{n} d\mathbf{r}''_i s_u(\mathbf{r}''_i) \prod_{j=2}^{f_{rn}} (d\mathbf{r}'''_j [(1-\alpha)s_f(\mathbf{r}'''_j)$$
$$+ \alpha s_b(\mathbf{r}'''_j)\mathscr{U}(\mathbf{r}'''_j,\{\mathbf{s}\})]) \left\{ \frac{1}{f_{rn}} \sum_{j=1}^{f_{rn}} \delta(\mathbf{r}'-\mathbf{r}'''_j)\lambda_{rtl}(\{\mathbf{r}\},\{s_b\})/VK_{rtl} \right\}, \quad (3.84)$$

$$VK_{rtl} = \int \lambda_{rtl}(\{\mathbf{r}\},\{\mathbf{1}\}\, d\{\mathbf{r}\}. \quad (3.85)$$

The parameters d_{rtl}, β_{rtl} have the same meaning as previously, whereas the factors in curly braces represent the probability densities for the arrangement of units and groups of the cycle in given points, subject to the condition that one of the units (or groups) is found at the point $\mathbf{r}$ ($\mathbf{r}'$).

The sequential differentiation of Eqs. (3.79)–(3.81) yields correlators of any desired order.

H. Estimation of Contributions from Cycles of Different Topologies

The fundamental functional $\Gamma\{\mathbf{s}\}$ through which all characteristics of the system can be obtained is represented in Eq. (3.38) as the sum of contributions of all possible cycles (r, t). In practical applications it is usually highly desirable to fold the infinite series. As the first step in this direction one can perform a partial summation by one of two methods. The first method consists in calculating the total contribution of all cycles of a given topology (r, t) of any size. The second approach, in contrast, consists in summing the contributions of all graphs with a fixed number of black vertices. Both approaches imply the construction of the perturbation theory with the cut-off of small terms, but they correspond to different physical situations. The first method of summation, adopted in this section, appears to be reasonable when the system contains a small number of cycles, in particular, a complex structure with large r. In contrast, if there are many cycles, largely of small size n, the second method is preferable.

The homeomorphic graphs, although having identical topology (r, t), differ from each other only by the vector $\mathbf{l}$, that is, by the number of "bifunctional" vertices (units with two reacted groups) belonging to the αth edge (see Section III.E). Addition of each such vertex to the linear chain, which replaces the edge of the elementary representative (Figs. 3.5 and 3.12), increases the order of its automorphic group by $(f-2)!$ times due to permutations of the pendant vertices thus arising. The exponent n of the factor $(f!ML)$ in Eq. (3.38) increases by 1. In the function λ_{rtl}, instead of the product of one of the δ functions, $\delta(\mathbf{r}_{ij}-\mathbf{r}_{pq})$, corresponding to a bond, with the counter $s_b(\mathbf{r}_{ij})$ of this bond, there appears the functional of the inserted unit

$$
\begin{aligned}
&K(\mathbf{r}_{ij}-\mathbf{r}_{pq}, \{\mathbf{s}\}) \\
&\quad = \int \cdots \int s_u(\mathbf{r}_1)\, d\mathbf{r}_1 \prod_{k=3}^{f} [s_f(\mathbf{r}_{1k})\lambda(\mathbf{r}_1-\mathbf{r}_{1k})\, d\mathbf{r}_{1k}] \\
&\qquad \times \lambda(\mathbf{r}_{11}-\mathbf{r}_1)\lambda(\mathbf{r}_{12}-\mathbf{r}_1)\delta(\mathbf{r}_{11}-\mathbf{r}_{ij})\delta(\mathbf{r}_{12}-\mathbf{r}_{pq})\, d\mathbf{r}_{11}\, d\mathbf{r}_{12} s_b(\mathbf{r}_{ij}) s_b(\mathbf{r}_{pq}). \qquad (3.86)
\end{aligned}
$$

Each successive insertion of a unit into the linear chain again gives the factor $[f(f-1)ML]$ due to the increase in n and leads to similar replacement of one of the δ functions by $K(\mathbf{r}_{ij}-\mathbf{r}_{pq}, \{\mathbf{s}\})$. The summation of functionals Γ_{rtl} for cycles of identical topology is reduced in this way to summing a specific functional geometric progression, with each term representing the convolution of several functions $K(\mathbf{r}_{ij}-\mathbf{r}_{pq}, \{\mathbf{s}\})$. The total contribution Γ_{rt} of the

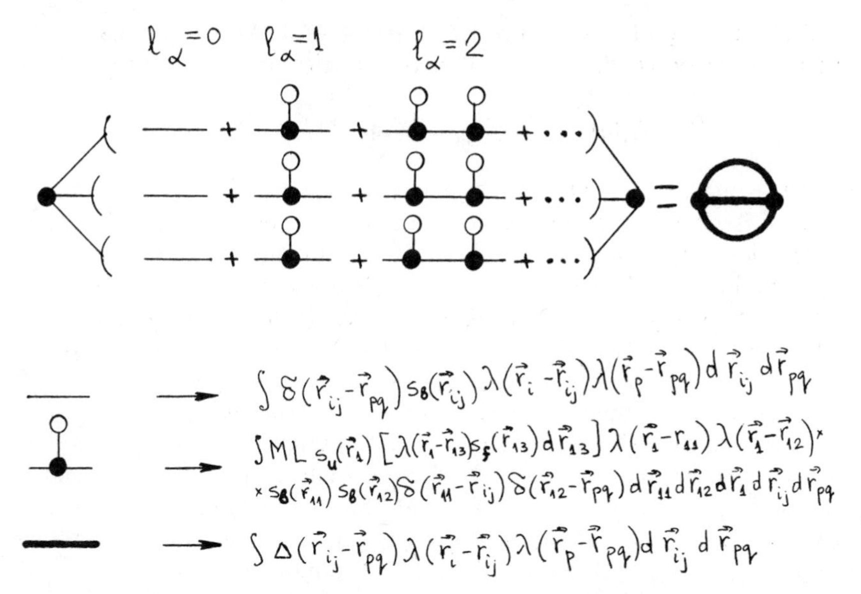

Figure 3.12. Substitution of bond function $\delta(\mathbf{r}_{ij}-\mathbf{r}_{pq})$ by the "boldface" bond function $\Delta(\mathbf{r}_{ij}-\mathbf{r}_{pq})$ as additional units are inserted into linear chain between branching units.

topology (rt) is obtained from the contribution Γ_{rto} of the elementary representative of this topology when each δ function $\delta(\mathbf{r}_{ij}-\mathbf{r}_{pq})$ in λ_{rto} is replaced by the sum of the functional geometric progression $\Delta(\mathbf{r}_{ij}-\mathbf{r}_{pq})$. The latter is given by its Fourier transform

$$\tilde{\Delta}(\mathbf{q}) = 1/[1-f(f-1)ML\tilde{K}(\mathbf{q},\{\mathbf{s}\})], \tag{3.87}$$

where $\tilde{K}(\mathbf{q})$ is the Fourier transform of the functional (3.86). With such replacement the function λ_{rto} is converted to the functional, depending on $\{\mathbf{s}\}$, that, in place of each edge of the elementary representative, "inserts" linear chains of arbitrary length. In the coordinate representation, to each edge of the elementary representative there corresponds the convolution

$$\begin{aligned}&\int \lambda(\mathbf{r}_i-\mathbf{r}_{ij})\Delta(\mathbf{r}_{ij}-\mathbf{r}_{pq})\lambda(\mathbf{r}_p-\mathbf{r}_{pq})\,d\mathbf{r}_{ij}\,d\mathbf{r}_{pq}\\ &=\frac{1}{(2\pi)^3}\int \tilde{\lambda}^2(\mathbf{q})\tilde{\Delta}(\mathbf{q})\exp\{i\mathbf{q}(\mathbf{r}_i-\mathbf{r}_p)\}\,d\mathbf{q}.\end{aligned} \tag{3.88}$$

The summation of simple cycles $(r=1)$ requires particular consideration, since the order $\mathscr{S}(1,n)=2n[(f-2)!]^n$ of the automorphism group of such a cycle, consisting of n units, involves a factor of n. As a consequence, the Fourier transform of the sum of the functional series for $r=1$ assumes the form

$$\tilde{\Delta}_1(\mathbf{q}) = -\ln[1 - ML\tilde{K}(\mathbf{q}, \{\mathbf{s}\}) f(f-1)]. \tag{3.89}$$

Thus, once the partial summation has been carried out, the functional Γ and its derivatives are represented by the sum of contributions of different topologies, (rt), the number of which is naturally infinite. Nevertheless, the contributions of different topologies are expected to be different, so that only a finite number of terms can be retained in the infinite series. For example, with growing density ρ_u the fraction of intermolecular reactions increases in comparison with the intramolecular ones. A similar effect will also cause an increase in the root-mean-square distance a between the functional groups of a single molecule. From this it follows that for small values of the dimensionless parameter $\varepsilon = 1/(\rho_u a^3)$ there will be a small number of cycles, and in the first approximation only the simple cycles $(r=1)$ can be taken into consideration. The next order of approximation should, in addition, take into account the cycles with $r=2, \ldots$. In a more rigorous manner this procedure is conducted in the next section.

To illustrate the construction of a sequence of increasingly accurate solutions, choose the coordinate-free spatially homogeneous case, when the functional Γ reduces to the function γ, the arguments of which are set equal to $s_u = s_b = 1$ and $s_f = 1 + L\rho_f = 1/(1-\alpha)$ in the final expressions. Equations (3.86) and (3.87), with constant $s_v(\mathbf{r}) \equiv s_v$, allow us to determine

$$\tilde{\Delta}(\mathbf{q}) = 1/[1 - f(f-1)ML\tilde{\lambda}^2(\mathbf{q}) s_u (s_f)^{f-2} s_b] \tag{3.90}$$

so that to each αth edge of the elementary representative, which is incident to the vertices $\mathbf{r}_i$ and $\mathbf{r}_p$, there corresponds, according to Eqs. (3.88) and (3.90), the factor

$$\Delta(\mathbf{r}_i - \mathbf{r}_p) = \frac{1}{(2\pi)^3} \int \frac{\exp[-i\mathbf{q}_\alpha(\mathbf{r}_i - \mathbf{r}_p)]\, d\mathbf{q}_\alpha}{(\tilde{\lambda}(\mathbf{q}_\alpha))^{-2} - f(f-1)MLs_u(s_f)^{f-2}s_b}. \tag{3.91}$$

If the elementary representative of the topology (r, t) has v vertices, then the preintegral factor, in accordance with Eq. (3.38), is $(f!ML)^v L^{r-1}$. In addition, the integrand gives rise to the factor $(s_u s_b)^v (s_b)^{r-1} (s_f)^{(f-2)v-2(r-1)}$, and the product of the cofactors (3.91) $(\alpha = 1, \ldots, u;\, u = v + r - 1)$, each being associated with the edge, is integrated over the coordinates of all v vertices, thus leading to integrals of the type

$$\int \exp\Bigl(-i\sum_\alpha b_{i\alpha}\mathbf{q}_\alpha \mathbf{r}_i\Bigr) d\mathbf{r}_i = (2\pi)^3 \delta\Bigl(\sum_\alpha b_{i\alpha}\mathbf{q}_\alpha\Bigr), \qquad b_{i\alpha} = 0, \pm 1. \tag{3.92}$$

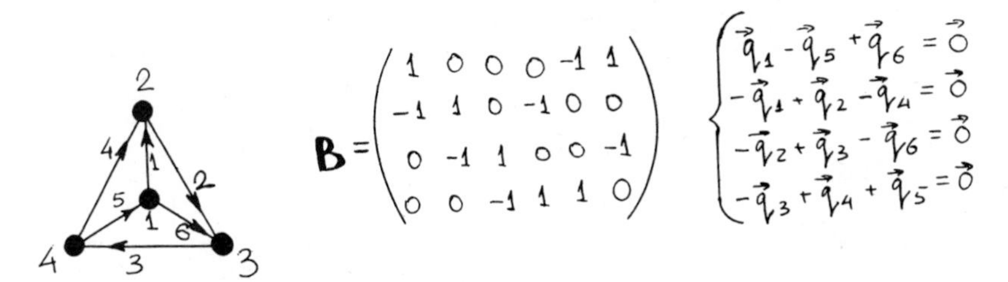

Figure 3.13. Orgraph of elementary topological representative, $(r, t) = (3, 1)$, its incidence matrix **B**, and "momentum conservation" equalities, which determine integration surface $S(r, t)$.

In order to relate the coefficients $b_{i\alpha}$ to the graph's topology, let us depict its elementary representative and fix in an arbitrary way the orientation of its edges (Fig. 3.13). Since the function $\Delta(\mathbf{r})$ is even, the difference $\mathbf{r}_i - \mathbf{r}_p$ in the exponent of Eq. (3.91) can be rewritten in such a way that the subscript i would denote the vertex from which the αth edge of the orgraph is outgoing; whereas p would denote the vertex into which the αth edge comes. Then the $b_{i\alpha}$ turn out to be exactly the elements of the incidence matrix **B** of the graph, whereas the argument of the δ function in Eq. (3.92) is the scalar product of the ith row of the matrix and the vector of momenta, $\mathbf{Q} = (\mathbf{q}_1, \mathbf{q}_2, \ldots, \mathbf{q}_n)$. One of the rows of the matrix **B** is a linear combination of the other rows, and hence, after performing $v - 1$ integrations, yielding the δ functions (3.92), the argument in the last integral will be null, and this integral will equal to the volume V. Thus, only r of u momenta will be independent, and the integration in the momentum space is carried out over the $3r$-dimensional surface $S(r, t)$, defined by the topology of the graph via its incidence matrix **B** by the matrix equation

$$\mathbf{B}\mathbf{Q}^T = \mathbf{0} \tag{3.93}$$

In ref. 63 the same surface was defined by

$$\mathbf{q}_\alpha - \sum_{i=v}^{u} c_{i\alpha}\mathbf{q}_i = \mathbf{0}, \qquad \alpha = 1, \ldots, v-1, \qquad c_{i\alpha} = 0, \pm 1, \tag{3.94}$$

via the elements of the reduced matrix of cycles, **C** (Fig. 3.14), which corresponds to a certain spanning tree of the graph[47,176] whose edges are numbered from 1 to $v - 1$ and whose chords are numbered from v to u. These equations follow from the fact that the sum of vectors $\mathbf{r}_i - \mathbf{r}_p$ over the edges of each cycle is null. This leads to the appearance in the product of integrals (3.91) of additional δ functions, the arguments of which are shown

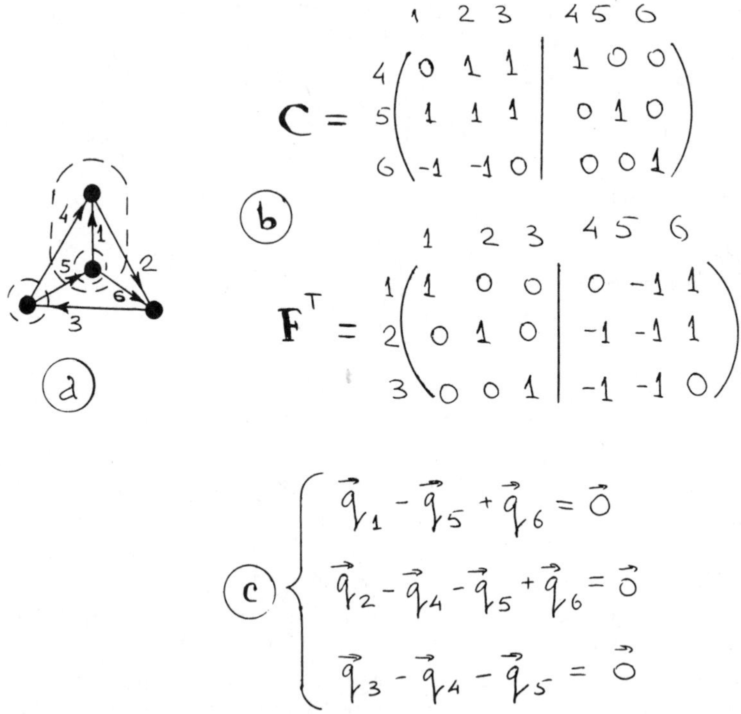

(a)

(b)

$$\mathbf{C} = \begin{array}{c} \\ 4 \\ 5 \\ 6 \end{array} \begin{pmatrix} 1 & 2 & 3 & | & 4 & 5 & 6 \\ 0 & 1 & 1 & | & 1 & 0 & 0 \\ 1 & 1 & 1 & | & 0 & 1 & 0 \\ -1 & -1 & 0 & | & 0 & 0 & 1 \end{pmatrix}$$

$$\mathbf{F}^T = \begin{array}{c} \\ 1 \\ 2 \\ 3 \end{array} \begin{pmatrix} 1 & 2 & 3 & | & 4 & 5 & 6 \\ 1 & 0 & 0 & | & 0 & -1 & 1 \\ 0 & 1 & 0 & | & -1 & -1 & 1 \\ 0 & 0 & 1 & | & -1 & -1 & 0 \end{pmatrix}$$

(c)

$$\begin{cases} \vec{q}_1 - \vec{q}_5 + \vec{q}_6 = \vec{0} \\ \vec{q}_2 - \vec{q}_4 - \vec{q}_5 + \vec{q}_6 = \vec{0} \\ \vec{q}_3 - \vec{q}_4 - \vec{q}_5 = \vec{0} \end{cases}$$

Figure 3.14. Dashed lines indicate cuts (cocycles) corresponding to spanning tree, which consists of edges 1, 2, 3 [(*a*)] and cycle and cocycle matrices **C** and $\mathbf{F}^T$ for spanning tree choice [(*b*)]. Equations determined by matrix **C** [(*c*)] are equivalent to those in Fig. 3.13.

in Eq. (3.94). To rewrite Eq. (3.94) in matrix form, discard the right part from the matrix **C**, the unit matrix , change the sign of the remaining matrix, and add the new unit matrix on top of it. Multiplying the vector **Q** by the matrix **F** obtained in this way, we arrive at equations (3.94):

$$\mathbf{C} = (\tilde{\mathbf{C}}\mathbf{E}); \qquad \mathbf{F} = \begin{pmatrix} \mathbf{E} \\ -\tilde{\mathbf{C}} \end{pmatrix}; \qquad \mathbf{QF} = \mathbf{0}. \tag{3.95}$$

Now it remains to observe that the transposed matrix **F** is identical to the reduced matrix of cocycles (cuts) of the graph, defined by the spanning tree chosen.[47,176] By permuting the rows and columns and replacing a row by its algebraic sum with the other row, such a matrix can be transformed[176,177] to the incidence matrix **B**. Therefore, Eqs. (3.93) and (3.94) and (3.95) define the same surface of integration $S(rt)$.

We are now in a position to write down the contribution $\gamma_{r,t}$ of all diagrams associated with the topology (r, t):

$$\gamma_{r,t}(\mathbf{s}) = V\frac{[f!MLs_u(s_f)^{f-2}s_b]^v[s_bL(s_f)^{-2}]^{r-1}}{\mathscr{S}(r,t,\mathbf{0})(2\pi)^r}\int\cdots\int_{S(rt)}$$

$$\times \prod_{\alpha=1}^{u} \frac{d\mathbf{q}_\alpha}{(\tilde{\lambda}(\mathbf{q}_\alpha))^{-2} - f(f-1)MLs_us_b(s_f)^{f-2}} \tag{3.96}$$

Differentiation of this formula with respect to s_v allows us to find the contributions $\gamma_{r,t}^v$ to the derivatives of the function γ.

Contribution from the simple cycles has the somewhat different form

$$\gamma_1(\mathbf{s}) = V\frac{-1}{2(2\pi)^3}\int \ln\{1 - [\tilde{\lambda}(\mathbf{q})]^2 f(f-1)MLs_us_b(s_f)^{f-2}\}\, d\mathbf{q}, \tag{3.97}$$

$$\gamma_1^u(\mathbf{s}) = V\frac{1}{2(2\pi)^3}\int \frac{f(f-1)MLs_b(s_f)^{f-2}\, d\mathbf{q}}{(\tilde{\lambda}(\mathbf{q}))^{-2} - f(f-1)MLs_us_b(s_f)^{f-2}},$$

$$\gamma_1^f(\mathbf{s}) = \frac{(f-2)\gamma_1^u s_u}{s_f}. \tag{3.98}$$

To estimate the contributions, it is convenient to turn to dimensionless variables and functions. First, we observe that the integral (3.96) has the dimensionality of the rth power of concentration, and since the characteristic scale of changes of the function λ is a, the transformation of the integral to dimensionless form is accomplished by multiplying it by a^{3r}. The contributions $\gamma_{rt}, \gamma_{rt}^v$ are calculated for the argument values $s_u = s_b = 1$, $s_f = 1/(1-\alpha)$, and in what follows these arguments will be omitted. To estimate the integral by the value of its denominator, define the parameter

$$\tau = 1 - f(f-1)ML/(1-\alpha)^{f-2}. \tag{3.99}$$

Finally, make L dimensionless by using the density of monomeric units

$$v = L\rho_u(1-\alpha)^2. \tag{3.100}$$

The combination $\rho_u a^3$ arising in this procedure will be used as an important characteristic of the system

$$\varepsilon = 1/(\rho_u a^3). \tag{3.101}$$

With these specifications contributions (3.96)–(3.98) can be rewritten in the form

$$\begin{aligned}
\gamma_{rt} &= V\rho_u \varepsilon^r v^{r-1} H_{rt}^{(0)}(\tau), \qquad r \geqslant 2,\\
H_{rt}^{(0)}(\tau) &= \frac{[(f-2)!(1-\tau)]^v}{\mathcal{S}(r,t,\mathbf{0})} \frac{a^3}{(2\pi)^3} \int \cdots \int_{S(r,t)} \prod_\alpha \frac{d\mathbf{q}_\alpha}{(\tilde{\lambda}(\mathbf{q}_\alpha))^{-2} - 1 + \tau},\\
\gamma_{rt}^u &= V\rho_u \varepsilon^r v^{r-1} H_{rt}^{(1)}(\tau), \qquad r \geqslant 1;\\
\gamma_{rt}^f &= (1-\alpha)[(f-2)\gamma_{rt}^u - 2(r-1)\gamma_{rt}],\\
H_{rt}^{(i)} &= (\tau - 1)\frac{dH_{rt}^{(i-1)}(\tau)}{d\tau}, \qquad r \geqslant 2,\\
H_1^{(1)}(\tau) &= \frac{1-\tau}{2}\left(\frac{a}{2\pi}\right)^3 \int \frac{d\mathbf{q}}{(\tilde{\lambda}(\mathbf{q}))^{-2} - 1 + \tau}.
\end{aligned} \tag{3.102}$$

The contribution of the acyclic units is, by Eq. (3.38),

$$\gamma_0(\mathbf{s}) = Vf!Ms_u(s_f)^f, \tag{3.103}$$

whence it follows, for $s_u = 1$, $s_f = 1/(1-\alpha)$,

$$\begin{aligned}
\gamma_0 = \gamma_0^u &= V\rho_u(1-\tau)/f(f-1)v,\\
\gamma_0^f &= V\rho_u(1-\tau)(1-\alpha)/(f-1)v.
\end{aligned} \tag{3.104}$$

The parameter ε, together with the conversion p, is a characteristic parameter of the system, defining the values of τ, v, and α, which can be found from Eqs. (3.79), (3.82), and (3.10), which assume the form

$$v = \frac{1-\tau}{f(f-1)} + \sum_{r \geqslant 1, t} \varepsilon^r v^r H_{rt}^{(1)}(\tau), \tag{3.105}$$

$$\alpha = \frac{2(1-\tau)}{f(f-1)} + (f-2)v - 2\sum_{r \geqslant 1, t} (r-1)\varepsilon^r v^r H_{rt}^{(0)}(\tau), \tag{3.106}$$

$$p = 1 - \frac{\alpha(1-\alpha)}{fv}. \tag{3.107}$$

The expansions in powers of the parameter ε on the r.h.s. are typical of perturbation theory.

For small values, $\varepsilon \ll 1$, the main contribution to the infinite series (3.105) and (3.106) is offered by the first terms, provided the coefficients $H_{rt}^{(i)}$ remain bounded. However, Eq. (3.102) shows that the coefficients increase with decreasing τ. For $\tau = 0$ many of them grow infinitely. The order of divergence of integrals in Eq. (3.102) for $\tau \to 0$ depends on the topology (rt). If the elementary representative $(rt\mathbf{0})$ involves points of conjunction,[47] then the associated integral is factorized into the integrals for blocks, since the latter split the set of edges and simple cycles into nonintersecting subsets. Thus, it is sufficient to estimate the divergence of the integrals without the points of conjunction. At large $q = |\mathbf{q}|$ the function $(\tilde{\lambda}(\mathbf{q}))^{-2}$ grows as $\exp(a^2q^2/3)$, so for the upper limit all the integrals do converge. For small q, $(\tilde{\lambda}(\mathbf{q}))^{-2} \simeq 1 + a^2q^2/3 + O(q^4)$, and for such $\mathbf{q}$ the denominator in the integral $H_{rt}^{(0)}$ involves the product of u factors, $(a^2q_\alpha^2/3 + \tau)$, each associated with an edge of the graph. The integration here is carried out over r independent momenta q_α. Hence, the difference between the exponent of momenta in the denominator of the integrand and the dimension of the integration domain is

$$\omega(rt) = 2(v + r - 1) - 3r = 2v - r - 2. \tag{3.108}$$

Since the scale of momenta changes is characterized by the parameter $\tau \sim a^2q^2$ only, the dimensional considerations define unambiguously the asymptotic behavior of the contribution $H_{rt}^{(0)}$ for $\tau \to 0$ as

$$H_{rt}^{(0)}(\tau) \sim \tau^{-\omega(rt)/2}. \tag{3.109}$$

At $\omega(rt) = 0$ the integral diverges logarithmically, whereas for negative $\omega(rt)$ it remains finite at $\tau = 0$.

Differentiation of the function $H_{rt}^{(0)}$ increases the number of factors in the denominator of the integral by 1, so that

$$H_{rt}^{(i)}(\tau) \sim \tau^{-(\omega(rt) + 2i)/2}, \qquad \omega(rt) > -2i. \tag{3.110}$$

Thus, according to (3.108)–(3.110), among the topologies of identical cyclic rank the main contribution at small τ offer those ("main") where the elementary representatives have the largest number of vertices v. It can be easily verified that elementary representatives with all nodes of degree 3 (Fig. 3.15) comply with the condition. The characteristic $\omega(rt)$ of such cycles does not depend on the topology t and equals $\omega(r) = 3r - 6$. If the functionality $f = 3$, all possible topologies are main ones, and none contain the cut points. In fact, one of the edges incident to the cut point of degree 3 must represent the bridge, whereas the cycles-leaves contain no bridges

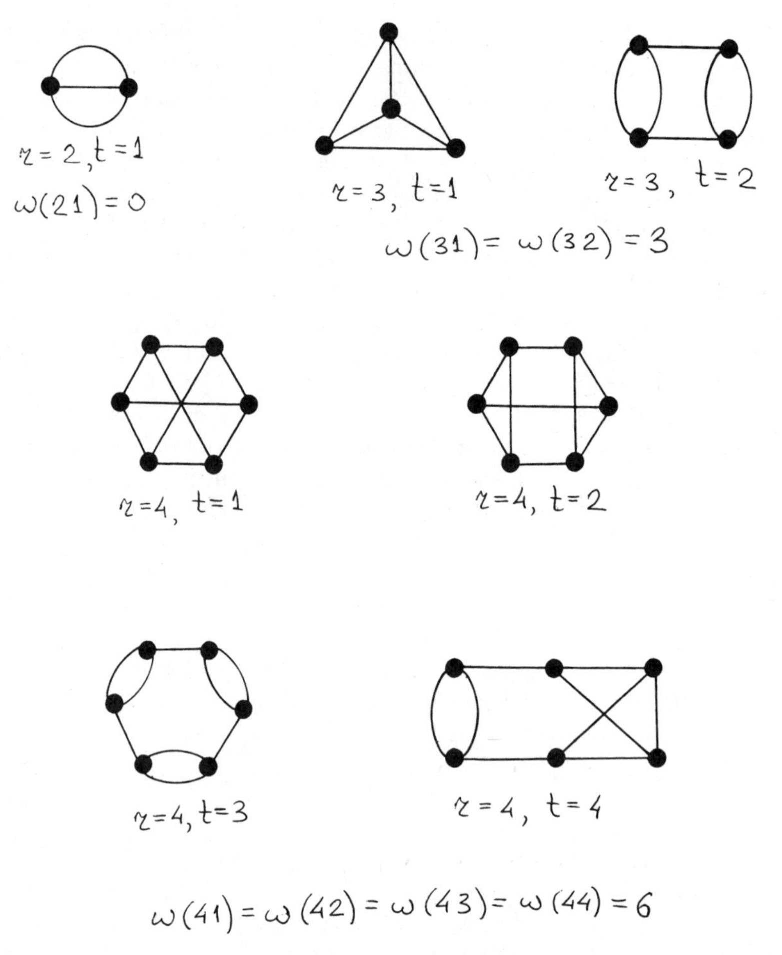

Figure 3.15. "Main" topologies providing dominating contribution to function $H_r^{(i)}$. Functional groups not depicted.

by definition. Elementary representatives of topology contain, by definition (Section III.E), no vertices of degree 2 as well.

For large functionalities $f \geqslant 4$ there occur also secondary topologies together with the main ones. If the former topologies contain no loops, the order of their divergence can be determined, as before, from Eqs. (3.108)–(3.110). The contribution of a loop is determined by the integral $H_1^{(1)}(\tau)$ [Eq. (3.102)], which is finite at $\tau = 0$. Hence, the divergence of integrals for such topologies can be found be "erasing" all loops and calculating the divergence of the remaining graph via Eq. (3.108).

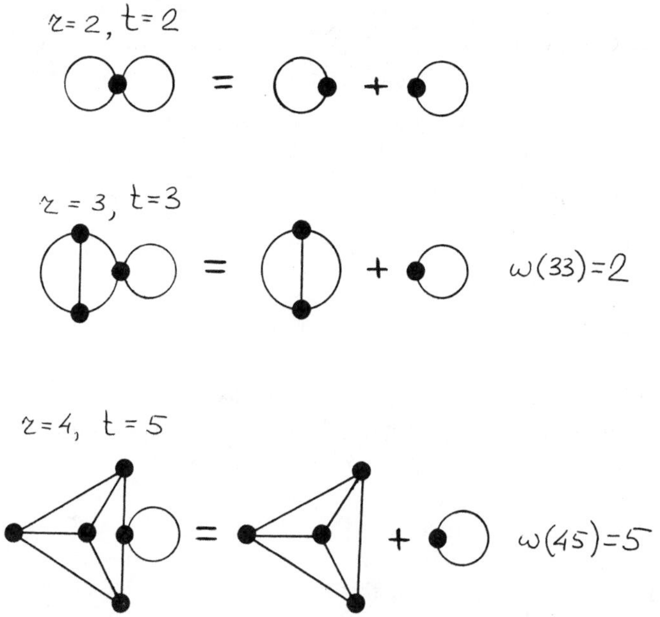

$r=2,\ t=2$

$r=3,\ t=3$ $\quad \omega(33)=2$

$r=4,\ t=5$ $\quad \omega(45)=5$

Figure 3.16. Topologies responsible for second-order contribution to function $H_r^{(i)}$ and the method of their generation.

The contribution of the topologies with loops is asymptotically larger than that of the remaining secondary graphs. In fact, the secondary diagram without loops contains at least one vertex less than the main one, so that $\omega(rt) \leqslant \omega(r) - 2$. At the same time, by "inserting" a loop in the middle of an edge of any main diagram of cyclic rank $r-1>1$, we obtain the topology with the divergence $\omega(rt) = \omega(r) - 1$ (Fig. 3.16).

The "eight"-graph with $r=2$ and $t=2$ (Figs. 3.4 and 3.16) is a peculiar one, and its contribution $H_{2,2}^{(0)}(\tau)$ is finite and proportional to the loop contribution squared.

To summarize, the results of this section can be formulated as follows: the coefficients of the series (3.105) and (3.106) behave asymptotically as

$$H_r^{(i)}(\tau) = \sum_t H_{rt}^{(i)}(\tau) = \tau^{-3r/2+3-i}(H_r^{i1} + H_r^{i2}\tau^{1/2} + O(\tau)), \qquad \tau \ll 1, \quad (3.111)$$

where H_r^{ij} are the sums of integrals of the type

$$\frac{1}{(2\pi)^r}\int \cdots \int_{S(rt)} \prod_\alpha \left[d\mathbf{q}_\alpha \left(\frac{1}{q_\alpha^2+1} \right)^{\nu_\alpha} \right], \qquad \nu_\alpha = 1, 2, \ldots . \quad (3.112)$$

This general formula fails to describe only the functions $H_2^{(0)}(\tau)$, which diverges logarithmically, $H_1^{(0)}(\tau)$, which enters none of the formulas, as well as $H_1^{(1)}(\tau)$, the expansion of which starts from the free term.

I. Perturbation Theory

Now we turn to the solution of Eqs. (3.105)–(3.107) by using the expansions in powers of the small parameter $\varepsilon \ll 1$. In the zeroth order of the perturbation theory we set $\varepsilon = 0$. In this case

$$v = v_0 = \frac{1-\tau}{f(f-1)}, \qquad \alpha = \alpha_0 = \frac{1-\tau}{f-1} = p. \tag{3.113}$$

This solution corresponds to the model I of ideal polycondensation without cyclization. Hence, it follows that the decrease in τ is equivalent to the increase in the conversion, and $\tau = 0$ corresponds to the ideal gelation point $p^* = 1/(f-1)$.

In the next order of perturbation theory Eqs. (3.105)–(3.107) should retain the terms, the order of which in ε does not surpass ε^1. In the region $\tau \sim 1$ it is sufficient to take into account the extra contribution of simple cycles:

$$\begin{aligned} v &= \frac{1-\tau}{f(f-1)} + \varepsilon v_0 H_1^{(1)}(\tau) = \frac{1-\tau}{f(f-1)}[1 + \varepsilon H_1^{(1)}(\tau)], \\ \alpha &= \frac{1-\tau}{f-1}\left[1 + \varepsilon \frac{f-2}{f} H_1^{(1)}(\tau)\right], \\ p &= \frac{1-\tau}{f-1} + \varepsilon \frac{3(f-2)-(f-4)\tau}{f(f-1)} H_1^{(1)}(\tau). \end{aligned} \tag{3.114}$$

The first two equations yield the parametric dependence of v and α on p and ε through the parameter τ, which is determined from the third equation.

In the region $\tau \sim 1$ the neglected terms are small, but with decreasing τ they grow infinitely. However, at $\tau \sim \varepsilon^{2/3}$ gelation comes, and the formulas should be renormalized. Prior to the gelation point $\tau \gtrsim \varepsilon^{2/3}$, and the terms neglected are of the order

$$\begin{aligned} &\varepsilon^r H_r^{(0)}(\tau) \lesssim (\varepsilon\tau^{-3/2})^r \tau^3 \ll \varepsilon, \\ &\varepsilon^r H_r^{(1)}(\tau) \lesssim (\varepsilon\tau^{-3/2})^r \tau^2 \ll \varepsilon, \qquad r \geqslant 2. \end{aligned} \tag{3.115}$$

Thus, solution (3.114) of Eqs. (3.105)–(3.107) is asymptotically accurate to within the terms of order ε and uniform in τ within the region $\varepsilon^{2/3} \lesssim \tau \lesssim \varepsilon^0$.

In the first order in the whole region up to the gelation point, any formulas containing γ and γ^{v} may retain merely the contributions of degenerated ($r = 0$) and simple ($r = 1$) cycles, since

$$\gamma_r \sim \varepsilon^r H_r^{(0)}(\tau) \ll \varepsilon, \qquad \gamma_r^{v} \sim \varepsilon^r H_r^{(1)}(\tau) \ll \varepsilon. \tag{3.116}$$

The function γ and its derivatives determine, for example, the concentration of cycles and the fraction of units in them. However, the expression for the weight-mean degree of polymerization will involve the second derivatives of γ, for which the order of contributions of cycles of all ranks $r \geqslant 1$ in the proximity of the gelation point, $\tau \sim \varepsilon^{2/3}$, is identical:

$$\gamma_r^{uu} \sim \gamma_r^{uf} \sim \gamma_r^{ff} \sim \varepsilon^r H_r^{(2)}(\tau) \sim (\varepsilon\tau^{-3/2})^r \tau \sim \varepsilon^{2/3}. \tag{3.117}$$

Nevertheless, it is possible to build up the solution uniform in τ. To this end, it is necessary to find the solution separately for $\tau \sim 1$ where ε is small and for $\tau \sim \varepsilon^{2/3}$ where τ is small and then to "join" the solutions together.[178] Such a procedure is also required for the solution of Eqs. (3.105)–(3.107) in second-order and higher order approximations.

Now we exemplify the procedure of joining solutions with the calculation of the weight-mean degree of polymerization in the first-order approximation:

$$P_W = 1 + \frac{\gamma^{uu}}{\rho_u V} + \frac{L\rho_u[\gamma^{uu}/\rho_u V]^2}{1 - L\rho_u \gamma^{ff}/V}. \tag{3.118}$$

In deriving this formula, use was made of relations (3.34) and (3.43). Since, at the gelation point, $P_W \to \infty$, it is more convenient to deal with the inverse value, $1/P_W$. It is quite easy to write the solution $(1/P_W)_0$ in the region $\tau \sim \varepsilon^0$, for which it is merely sufficient to take into account in γ the contribution of degenerated and simple cycles. In the proximity of the gelation point ($\tau \sim \varepsilon^{2/3}$), the construction of the solution $(1/P_W)_{2/3}$ requires the terms γ_r for all possible r, but since τ is small, it is sufficient to retain in the expansions (3.111) the first two terms, H_r^{i1} and H_r^{i2}, which are determined only by main topologies (Fig. 3.15) and one-loop graphs (Fig. 3.16). These two solutions can be "joint" since the limit of the first solution (as the implicit function of two variables p, ε) for $\tau \to 0$ is equal to the limit of the second for $\tau \to \varepsilon^0$, $\chi \equiv \tau\varepsilon^{-2/3} \to \infty$:

$$\lim_{\tau \to 0} (1/P_W)_0 = \lim_{\chi \to \infty} (1/P_W)_{2/3} = (1/P_W)_{\text{common}}. \tag{3.119}$$

This property allows us to find the solution that is uniform in the whole region:

$$\frac{1}{P_W} = \left(\frac{1}{P_W}\right)_0 + \left(\frac{1}{P_W}\right)_{2/3} - \left(\frac{1}{P_W}\right)_{\text{common}}$$

$$= \frac{1-(f-1)p}{1+p} - \frac{\varepsilon(f-2)^2(1-\tau)}{f(f-\tau)}[H_1^{(2)}(\tau) - H_1^{(1)}(\tau)]$$

$$- \frac{\varepsilon(f-2)(f-3)(1-\tau)}{f(f-\tau)} H_1^{(1)}(\tau)$$

$$- \frac{\varepsilon\tau(f-1)}{f(f-\tau)}[f^2 - 4 - (f-4)\tau][H_1^{(2)}(\tau) - H_1^{(1)}(\tau)]$$

$$+ (f-6)H_1^{(1)}(\tau) - \frac{f-1}{f}\varepsilon^{2/3} \sum_{r \geqslant 2} \frac{(f-2)^2 H_r^{21}}{[f(f-1)]^r} \chi^{-3r/2+1}$$

$$- \frac{f-1}{f}\varepsilon \sum_{r \geqslant 2} \frac{(f-2)^2 H_r^{22}}{[f(f-1)]^r} \chi^{-3r/2+3/2} + o(\varepsilon), \qquad \chi \equiv \tau\varepsilon^{-2/3}. \tag{3.120}$$

In the region $\tau \sim 1$ the first term, representing formula (1.13), is the main one, and the remaining terms are corrections of order ε. With growing conversion, the first term, $\sim \tau$, decreases, whereas its correction increases as $\varepsilon H_1^{(2)}(\tau) \sim \varepsilon\tau^{-1/2}$, $\tau \to 0$. At $\tau \sim \varepsilon\tau^{-1/2}$, that is, for $\varepsilon \sim \tau^{2/3}$, both these are of the same order. At this point, for $\chi \sim 1$, the contribution of the last two terms in Eq. (3.120) become essential. Outside this narrow neighborhood of the gelation point, $\chi \gg 1$, and as before, in first-order approximation, account may be taken only of degenerated and simple cycles.

The uniform τ solutions for other characteristics of the system, including the correlators, can be constructed in a similar way. In constructing the higher (kth-) order corrections, it is sufficient to take into account the contributions of cycles of rank $r \leqslant k$ everywhere, with the exception of the gelation point.

J. Statistics of Cyclic Fragments and Possible Generalizations of Model

As was noted in the foregoing, the molecular graphs can be represented as trees (the leaf compositions), with the majority of vertices for $\varepsilon \ll 1$ depicting the monomeric units, whereas the remaining vertices represent cyclic fragments connected by linear chains. Now we consider the asymptotic behavior of the distribution of these fragments in the proximity of the gelation point, regardless of to which molecules they belong. As was shown in Section III.H, among the cycles of identical rank, the predominant are those that, in neither unit, there enter more than three cyclic bonds. Together with this problem of revealing the main topologies among the cyclic fragments with given r, another problem may be posed, namely, to find the mean number of units in one cycle of definite topology.

To solve this problem, we carry out the summation of contributions $\gamma_{rt\mathbf{l}}$ (which the concentrations of cycles), not over all $\mathbf{l}$, but merely over those that contain exactly $l = n - v$ units with two reacted groups between the branching units. The contribution of cycles of the topology (rt) and size n, according to Eqs. (3.78) and (3.99)–(3.101), the definition of $\lambda_{rt\mathbf{l}}$, and taking into account the difference in $[(f-2)!]^l$ times between the orders of the automorphism groups $\mathscr{S}(rt\mathbf{l})$ and $\mathscr{S}(rt\mathbf{0})$, is

$$\gamma_{rtn} = \frac{\varepsilon^r v^{r-1}[(f-2)!]^v(1-\tau)^n}{\mathscr{S}(rt\mathbf{0})} \sum_{l=n-v}^{\infty} \int \cdots \int \prod_{\alpha=1}^{u} [\tilde{\lambda}(\mathbf{q}_\alpha)]^{2(l_\alpha+1)} d\mathbf{q}_\alpha. \quad (3.121)$$

We look for the asymptotic behavior of γ_{rtn} for $n \to \infty$ when the difference between l and n can be neglected. To begin, let the graph $(rt\mathbf{0})$ be composed of only one block, that is, contain no conjunction points. Carrying out the summation in Eq. (3.121), we obtain the integral, with the denominator containing $u-1$ products of functions, each behaving at small q as q^2, whereas the numerator behaves as $\sim \exp(-lq^2)$. By performing the change of variables $\mathbf{q}'_\alpha = l^{1/2}\mathbf{q}_\alpha$, the asymptotic dependence on $l \sim n$ is taken out of the integral symbol, thus allowing us to estimate the behavior of γ_{rtn} for $n \to \infty$,

$$\gamma_{rtn} \sim (1-\tau)^n n^{c-1}, \quad (3.122)$$

where the topological structure of the cyclic graph determines the value of the exponent c,

$$c = u - 3r/2. \quad (3.123)$$

Formulas (3.122) and (3.123) can be shown to hold for the arbitrary cycle with any number of cut points but containing no loops. For example, let the cycle consist of two blocks having the exponents c_1 and c_2. Since the integral $J(l)$ in Eq. (3.121) is factorized into the product of integrals $J_1(k)J_2(l-k)$ of individual blocks, it can be estimated by using the Laplace transformation,[179] which, with the known asymptotic behavior of each of the factors $J_i(l)$ for $l \to \infty$, allows us to find the asymptotics of the sum of their products:

$$J_i(l) \sim l^{c_i-1}, \qquad J(l) = \sum_k J_1(k)J_2(l-k) \sim l^{c-1}, \qquad c = c_1 + c_2. \quad (3.124)$$

Since upon conjunction of blocks into a single graph their cyclic ranks and the number of edges are added, the calculation of the exponent c for the graph via Eq. (3.123) and via the last of Eqs. (3.124) leads to identical results.

The only exception is the case when one of the blocks is the loop ($c_i = -0.5$). For such exponents c_i the Laplace transformation can no longer be employed, and the change of order of summation and integration in Eq. (3.121) leads to divergent integrals. The estimate of the sum (3.124), obtained by other methods, showed that in calculating the complexity of the graph with loops half the number of the graph's loops should be added to the quantity (3.123). For the "eight"-graph ($r = 2, t = 2$) the complexity turns out to be -0.5.

The distribution of cycles in sizes (3.122) for $c > 1$ has a maximum, which with increasing complexity becomes more acute. Thus, among the complex cycles of the growing gel the predominant are those whose size coincides with the number-mean size $P_N^{(c)}$ of cycles of complexity c[61]:

$$P_N^{(c>0)} \simeq cP_N^{(1)} = c/[1 - (1 - \tau)]. \tag{3.125}$$

For $\varepsilon \ll 1$, in approaching the gelation point, $\tau \to \tau^*(\varepsilon) \sim \varepsilon^{2/3}$, and the sizes of cycles grow rapidly, as expected.

The approach to the description of configurational–conformational statistics of macromolecules suggested in this section is carried out the clear physicochemical model, starting from the general principle of statistical mechanics. As an example, we have considered model III of f-functional homopolycondensation (Section III.A), which can be generalized in several directions. First, the system consisting of several types of monomers can be calculated within the same model. In this case, instead of two parameters M and L, characterizing the g.f. Ψ, there appear $M_1, \ldots, M_\nu, \ldots$ for different types of units, and the set $L_{11}, \ldots, L_{ij}, \ldots$ associated with different types of pairs (i, j) of functional groups A_i and A_j reacting with each other. If these groups are "linked" to the unit by different linear chains, there arise a set of functions $\lambda_{ij}(\mathbf{r}_i - \mathbf{r}_{ij})$. Accordingly, the form of the graphs is changed, and for the labeling of nodes, pendant vertices, and edges of different types, different colors should be used. The energy of the molecule is now characterized not only by its composition $\mathbf{l} = \{l_1, l_2, \ldots, l_\nu, \ldots\}$ but also by the bond matrix $\mathbf{B}$ with elements $b_{ij} = b_{ji}$, which equal the number of intramolecular bonds formed in the course of condensation of groups A_i and A_j. At last, the characterization of cycles also changes. Besides the cyclic rank r and topology t, the composition of each edge and the types of branching nodes should be specified, since, in general, they affect conformational entropy upon cycle formation. Apart from some technical problems concerning the choice of convenient subscripts and superscripts, no problems of principle arise.

The second possible generalization of model III consists in its combination with model II, that is, the consideration of substitution effects. In the case of homopolycondensation, it is merely sufficient to take into account the kind of monomeric units (Section I.D) by replacing the parameter M with $M_0, M_1, \ldots, M_f$. In case of multicomponent systems the kind of unit becomes the matrix characteristics.[24] In addition, the peculiarities of transition to multicomponent systems, pointed out in the preceding for the combination of models I and III, should be taken into account.

The most complicated generalization is for model III supplemented with physical interactions between macromolecular fragments. One of the versions of model IV arising in this way is considered in the next section using field-theoretic methods.

IV. DIAGRAM TECHNIQUE AND FIELD THEORY

A. Stochastic Fields and Functional Integration

Diagram technique refers to a set of rules and graphical symbols enabling a spectacular representation of equalities and making it possible to dispense with cumbersome algebra. For instance, Eqs. (1.12) describe a classical branching process illustrated in Section I.C with Fig. 1.10. Each group in the root unit either escaped reaction or produced a new branch; we cut the latter off and indicate the cut with a mark (Fig. 4.1). The number of selections of k groups involved in the reaction among f groups of the unit C_f^k equals the number of graphs having just k cutting marks (cf. the Newton binomial expansion in Fig. 1.9). Such diagrams are depicted with a single representative, and the total number of equivalent (analogue) diagrams is indicated in parentheses in Fig. 4.1. Figure 4.1*b* displays the branch cut in Fig. 4.1*a*. Bonds are again indicated with cutting marks, to which similar branches can be "implanted."

Let us formulate now the diagram rules for the probability generating function $G_W(s)$. Each monomer unit corresponds to a counter s. A group evading reaction is associated with a probability $1-p$, where p is the probability that the group is involved; a bond is associated with the product of p and the generating function for a branch, $u(s)$. According to these rules, the equalities displayed in Fig. 4.1 are

$$\begin{aligned} G_W(s) &= s(1-p)^3 + 3s(1-p)^2pu + 3s(1-p)(pu)^2 + s(pu)^3, \\ U &= s(1-p)^2 + 2s(1-p)pu + s(pu)^2. \end{aligned} \tag{4.1}$$

This is a particular case (for $f=3$) of the general formulas (1.12) describing a branching process.

A special feature of polymer systems is the fact that in their statistical description each polymer molecule can be associated with a specific diagram.

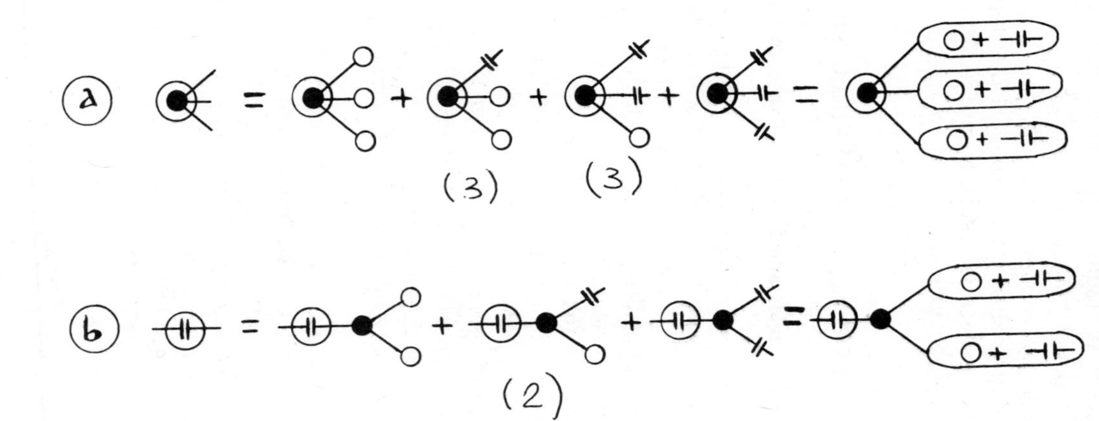

Figure 4.1. Graphical representation of Eqs. (4.1) describing branching process. Numbers of analogous graphs indicated in parentheses (cf. Fig. 1.9).

The corresponding Ω potential (3.12) for the equilibrium ensemble of macromolecules can be considered as an infinite sum of such diagrams. The sum is a perturbational series convergent for sufficiently high temperatures, which can be obtained by expansion of the logarithm of the partition function represented in terms of the functional integral. Here we have an evident analogy to the expansion in Feynman diagrams, which is employed in the field-theoretic approach to statistical mechanics of many-particle systems.[180–183] By Freed's metaphor,[2] "a box containing branching and coupled polymer molecules is filled with Feynman diagrams."

In view of the analogy, to sum up diagrams, that is, to calculate generating functionals and correlators, one can use advanced methods of field theory.[182,183] For this purpose, one has to reformulate the basic relations (3.6) and (3.7) in terms of the field theory, that is, to replace averaging over point distributions with the Gibbs measure for that over a stochastic field $\varphi(\mathbf{r})$. In the set of field configurations, a probability measure is defined, and it is used in integration of functionals depending on the field $\varphi(\mathbf{r})$.

Let us derive, for example, an expression for the partition function of the grand canonical ensemble (3.6) in terms of a functional integral. To this end, consider a set of graphs obtained by various couplings of vertices corresponding to functional groups. The chemical bond resulting from the coupling of groups with coordinates $\mathbf{r}_{ij}$ and $\mathbf{r}_{pq}$ is represented by a factor $L\exp[(h_b(\mathbf{r}_{ij}) - h_f(\mathbf{r}_{ij}) - h_f(\mathbf{r}_{pq})]\,\delta(\mathbf{r}_{ij} - \mathbf{r}_{pq})$, and the total graph $\mathscr{G}_N$ is represented by a product of such factors, in agreement with the complete set of bonds. The sum of the contribution from all the graphs $\mathscr{G}_N$ can be written as a functional integral,

$$\sum_{\mathscr{G}_N}\prod\{L\exp[h_b(\mathbf{r}_{ij}) - h_f(\mathbf{r}_{ij}) - h_f(\mathbf{r}_{pq})]\delta(\mathbf{r}_{ij} - \mathbf{r}_{pq})\}$$

$$= \frac{\int D\varphi \exp[-\mathscr{L}\{\varphi\}]\prod_{i,j}[1+\varphi(\mathbf{r}_{ij})]}{\int D\varphi \exp[-\mathscr{L}\{\varphi\}]} \equiv \left\langle \prod_{i,j}[1+\varphi(\mathbf{r}_{ij})] \right\rangle_{\varphi}. \tag{4.2}$$

Here angular brackets stand for averaging over the stochastic field φ, and the probability measure is given by a quadratic Lagrangian,

$$\mathscr{L}\{\varphi\} = \frac{1}{2L}\int \exp[2h_f(\mathbf{r}) - h_b(\mathbf{r})]\varphi^2(\mathbf{r})\,d\mathbf{r}. \tag{4.3}$$

If we sum over all the graphs $\mathscr{G}_N\{\mathbf{r}\}$ before integration over the vertex coordinates, then, according to (4.2),

$$\exp\left(-\frac{\Omega\{\mathbf{h}\}}{T}\right) = \sum_{N=0}^{\infty} \frac{1}{N!} \int \cdots \int \prod_{i=1}^{N} \left[M \exp h_u(\mathbf{r}_i)\, d\mathbf{r}_i \right.$$
$$\left. \times \prod_{j=1}^{f} \exp h_f(\mathbf{r}_{ij}) \lambda(\mathbf{r}_i - \mathbf{r}_{ij})\, d\mathbf{r}_{ij} \right] \left\langle \prod_{i,j} [1 + \varphi(\mathbf{r}_{ij})] \right\rangle_{\varphi}. \tag{4.4}$$

Interchanging the ordinary and functional integrals and using the fact that the multidimensional integral is a product of N identical integrals over coordinates of individual vertices, one can rewrite the expression in (4.4), after summation over N, as

$$\exp[-\Omega\{\mathbf{h}\}/T] = \langle \exp \mathscr{M}\{\varphi\} \rangle_{\varphi}, \tag{4.5}$$

$$\mathscr{M}\{\varphi\} = \mathscr{M} \int d\mathbf{r} \exp h_u(\mathbf{r}) \left[\int d\mathbf{r}' \exp h_f(\mathbf{r}')(1 + \varphi(\mathbf{r}'))\lambda(\mathbf{r} - \mathbf{r}') \right]^f \tag{4.6}$$

Having the thermodynamic Ω potential as a functional of the external field h, one can introduce a virtual field h^u and get a generating functional for the density correlators of the molecules, according to (3.19). Consequently, Eqs. (4.5), (4.6), and (4.3), combined with (3.19), provide, formally, an exact solution to the problem of detailed statistical description of the system.

The functional integral in (4.5) can be calculated approximately by means of the steepest descent method, since the presence of a large parameter, $\varepsilon^{-1} \gg 1$, singles out the contribution from a particular field configuration, $\bar{\varphi}(\mathbf{r})$, as compared with others. To this approximation, the generating functional of (3.19) is given simply by

$$\Psi\{\mathbf{s}\} = \mathscr{M}\{\bar{\varphi}\} - \mathscr{L}\{\bar{\varphi}\}, \tag{4.7}$$

and the condition $\delta\Psi/\delta\bar{\varphi} = 0$, corresponding to a maximum of the functional Ψ, leads to an equation for the "mean field" $\bar{\varphi}(\mathbf{r})$,

$$s_f^2(\mathbf{r})\bar{\varphi}(\mathbf{r})/L\tilde{s}_b(\mathbf{r}) = fM\tilde{s}_f(\mathbf{r}) \int d\mathbf{r}' \tilde{s}_u(\mathbf{r}')\lambda(\mathbf{r}' - \mathbf{r})$$
$$\times \left\{ \int d\mathbf{r}'' \tilde{s}_f(\mathbf{r}'')[1 + \bar{\varphi}(\mathbf{r}'')]\lambda(\mathbf{r}' - \mathbf{r}'') \right\}^{f-1}. \tag{4.8}$$

The mean field is also called the *self-consistent field* (SF), since equation (4.8) is a self-consistency condition that determines an extremal field configuration $\bar{\varphi}(\mathbf{r})$.

It is not difficult to show that using the steepest descent method for evaluation of the integral in (4.5) leads to the same results as were obtained in model I. Actually, after a change of variables in (4.7) and (4.8),

$$\bar{\varphi}(\mathbf{r}) = L\Psi^f(\mathbf{r}, \{\mathbf{s}\})\tilde{s}_b(\mathbf{r})/\tilde{s}_f^2(\mathbf{r}), \tag{4.9}$$

we reduce them to expressions that follow from the general formulas (3.42) and (3.41) if we retain only the first term (3.44) with $r = 0$ in the infinite sum, representing $\Gamma\{\mathbf{s}\}$ in (3.38). Thus, model I, where all polymer molecules have tree structures, is equivalent to the self-consistent field approximation. In the SF approximation, when calculating the generating functional Ψ, fluctuations of field $\bar{\varphi}(\mathbf{r})$ are neglected completely, and its mean value $\bar{\varphi}(\mathbf{r})$ must be obtained from the self-consistency condition (4.8). As $\Psi(\mathbf{r}, \{1\}) = \rho(\mathbf{r})$, according to (4.9), the field $\varphi(\mathbf{r})$ coincides, up to a factor of L, with the density distribution for groups not involved in the reaction.

To account for intramolecular reactions, one should have in mind that the density $\rho(\mathbf{r})$ is not constant throughout the volume but is fluctuating in separate molecules, according to their sizes, configurations, and conformations. Consequently, the incorporation of cycle-producing reactions into the theory is equivalent to an account for contributions from the field fluctuations in the integral (4.5), which can be performed by means of special methods.[182,183]

B. Diagram Technique

To represent analytical expressions in graphical terms, one needs some rules representing a correspondence between mathematical and graphical symbols. For model III those are given in Fig. 4.2. As in the preceding section, black and white vertices correspond to monomer units and functional groups, but some of the latter may be involved in the reaction. If this is the case, the group is included in a bond, which is shown as a tie. Edges in the diagram correspond to linear chains connecting units and groups. Besides, new vertices are introduced. These are marked with crosses and correspond to analytical expressions for the functional derivatives of Ψ^f and Ψ^u [Eq. (3.41)]. By Fig. 4.2, each element of a diagram corresponds to a factor, and the resulting product is to be integrated over coordinates of all the vertices, except for the root, which is encircled to distinguish it from other elements.

According to these rules, the graphical equality in Fig. 4.3 is written analytically as

$$\Psi^f(\mathbf{r}, \{\mathbf{s}\}) = fM\tilde{s}_f(\mathbf{r}) \int d\mathbf{r}'\tilde{s}_u(\mathbf{r}')\lambda(\mathbf{r} - \mathbf{r}')$$
$$\times \{ \textstyle\int d\mathbf{r}''[\tilde{s}_f(\mathbf{r}'') + \tilde{s}_b(\mathbf{r}'')L\Psi^f(\mathbf{r}'', \{\mathbf{s}\})/\tilde{s}_f(\mathbf{r}'')]\lambda(\mathbf{r}' - \mathbf{r}'')\}^{f-1}. \tag{4.10}$$

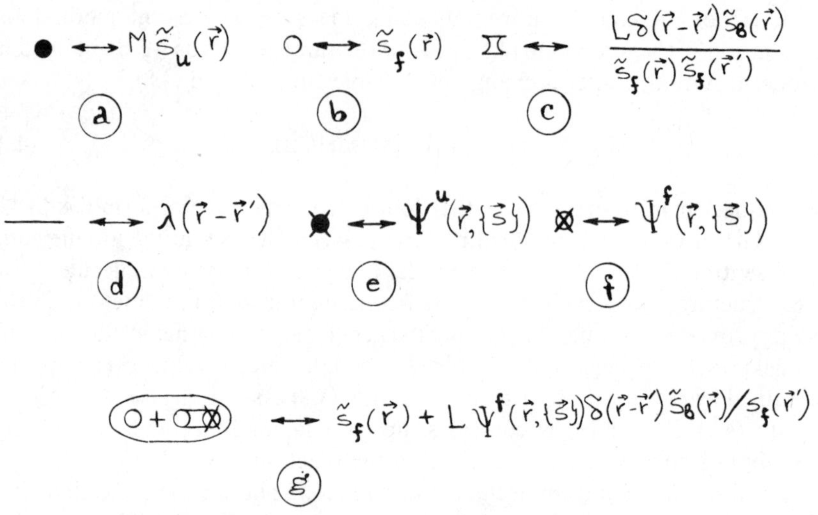

(a) ● $\leftrightarrow M\tilde{s}_u(\vec{r})$ (b) ○ $\leftrightarrow \tilde{s}_f(\vec{r})$ (c) ⌶ $\leftrightarrow \dfrac{L\delta(\vec{r}-\vec{r}\,')\tilde{s}_B(\vec{r})}{\tilde{s}_f(\vec{r})\tilde{s}_f(\vec{r}\,')}$

(d) — $\leftrightarrow \lambda(\vec{r}-\vec{r}\,')$ (e) $\leftrightarrow \Psi^u(\vec{r},\{\vec{\xi}\})$ (f) $\leftrightarrow \Psi^f(\vec{r},\{\vec{\xi}\})$

(g) (○+⌶⊗) $\leftrightarrow \tilde{s}_f(\vec{r}) + L\Psi^f(\vec{r},\{\vec{\xi}\})\delta(\vec{r}-\vec{r}\,')\tilde{s}_B(\vec{r})/s_f(\vec{r}\,')$

Figure 4.2. Elements of diagrams and corresponding analytical factors.

This equality is obtained from (4.8) after substitution of $\bar{\varphi}(\mathbf{r})$ for the derivative of Ψ, as given in (4.9).

The graphical equality shown in Fig. 4.3 can be obtained in the SF approximation by applying the general rules of the diagram technique[180–182] directly for the functionals $\mathscr{L}\{\varphi\}$ and $\mathscr{M}(\varphi)$, which are present in the integral in Eq. (4.5). These rules enable one to "discard" provisionally the analytical expressions and to deal instead with diagrams. For instance, the equality given in Fig. 4.3 could be derived first from diagrams and then translated to formulas, as in Eq. (4.10).

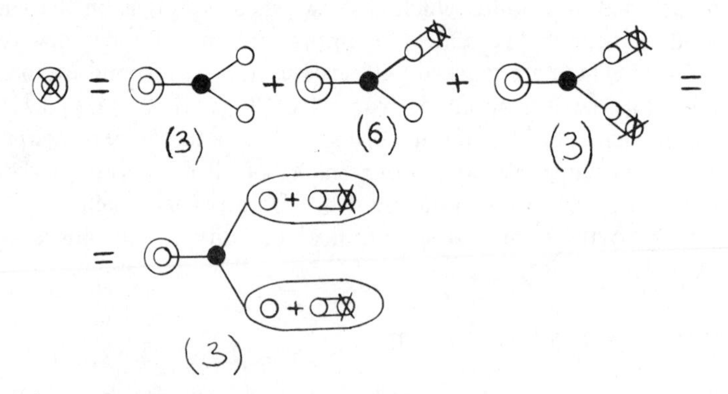

Figure 4.3. Graphical representation of Eq. (4.10) for $f = 3$.

Consider the derivation of graphical equalities. As we have already mentioned, for $\{\mathbf{s}\} = \{\mathbf{1}\}$ the derivative of Ψ^{v} is the density ρ_v, and for arbitrary s_v it bears an additional information on the relative contributions from various l-mers to the density. In particular, to get $\Psi(\mathbf{r}, \{\mathbf{s}\})$, one has to sum contributions from all molecules, groups of which get into the point $\mathbf{r}$. Each term of the sum corresponds to a graph, whereas the root vertex, which is not asso-

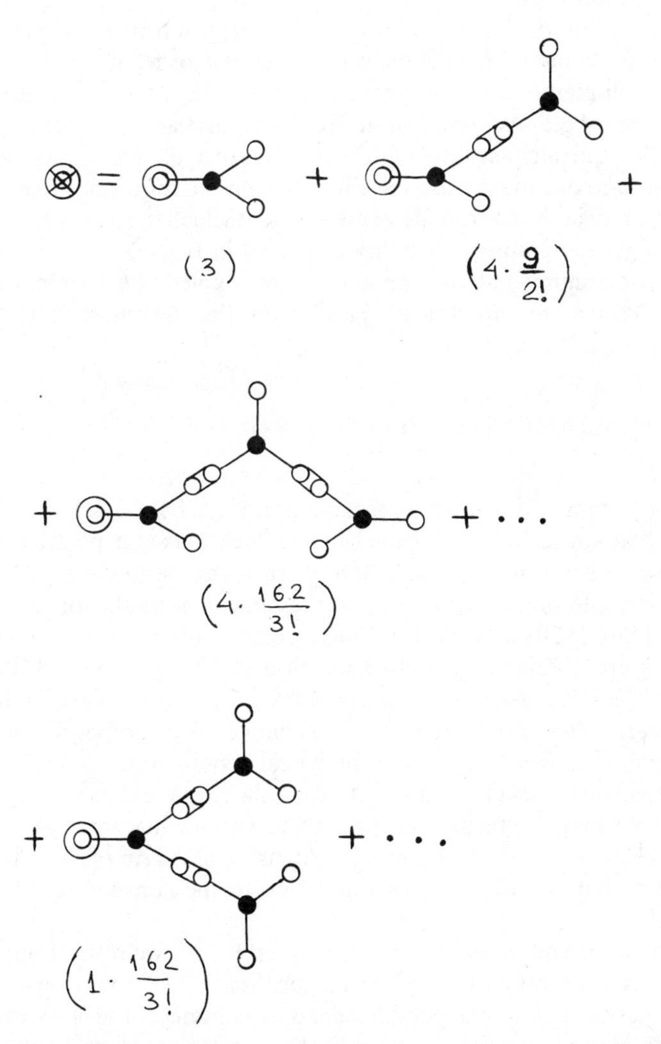

Figure 4.4. Equations for density of nonreacted groups; summing one gets equation in Fig. 4.3. Coefficient in parentheses at fraction equals number of different choices of root; numerator is number of possible bonds; denominator is factorial of number of units.

ciated with the integral, corresponds to every functional group in turn. For example, in Fig. 4.4 we give a single analogue for a chosen molecule, and the total number of the relevant graphs (taking into account the permutations of monomer units and functional groups) is given in parentheses. The exhaustion of all analogue diagrams is related to their symmetry and is equivalent to the appearance of combinatorial factors in the corresponding analytical expressions.

An inspection of Fig. 4.4 shows that to any group belonging to the unit closest to the root and having reacted, all possible ordered trees are coupled, which are implanted in a group bound to it. The set of those trees is the complete set of graphs presented in Fig. 4.4, so instead of the total series one can put the equivalent symbol of the derivative of Ψ after the bond symbol. In this procedure one must erase the circle around the root, since the integral is now necessary in the coordinate of the vertex at which the trees are implanted. Thus, we get the graphical equality displayed in Fig. 4.3.

Likewise, summing all the implanted trees coupled to a certain (root) unit, one can derive the equality of Fig. 4.5 for the derivative $\Psi^u(\mathbf{r}, \{\mathbf{s}\})$; the analytical equality is

$$\Psi^u(\mathbf{r}, \{\mathbf{s}\}) = \tilde{s}_u(\mathbf{r}) M \left\{ \int d\mathbf{r}' [\tilde{s}_f(\mathbf{r}') + \tilde{s}_b(\mathbf{r}') L \Psi^f(\mathbf{r}, \{\mathbf{s}\}) / \tilde{s}_f(\mathbf{r}')] \lambda(\mathbf{r} - \mathbf{r}') \right\}^f . \quad (4.11)$$

It can be obtained in the SF approximation from Eqs. (3.41) and (3.45).

The diagram technique exposed here, which makes it possible to avoid cumbersome problems of exhaustion of trees, can be modified slightly and used as an alternative derivation of the general formula for a branching process [Eqs. (3.50) and (3.51)]. Since, because of (3.46) and (3.47), those equalities are obtained by renormalization of Eq. (4.11) and (4.10) for the derivatives of the generating functional Ψ, they correspond to the same diagram sets. The only thing to do is to change the correspondence between the graphical elements and their analytical representatives. Instead of the correspondence rules given in Fig. 4.2, one should take those of Fig. 4.6 as a basis for the diagram technique applicable to branching processes (3.50) and (3.51). The modified rules are a natural extension of the simplest version of the diagram technique (Fig. 4.1) corresponding to the conventional branching process.[1]

In obtaining equations for the derivatives Ψ^u, Ψ^f, we restrict ourselves to tree graphs. According to the general algorithm,[180–182] to go beyond the SF approximation, one should complement the diagrams of Fig. 4.4 with various diagrams containing cycles, where a group that did not react is chosen as the root. As in the SF approximation, series of all possible diagrams can be connected to the groups nearest to the root and having reacted, and each series

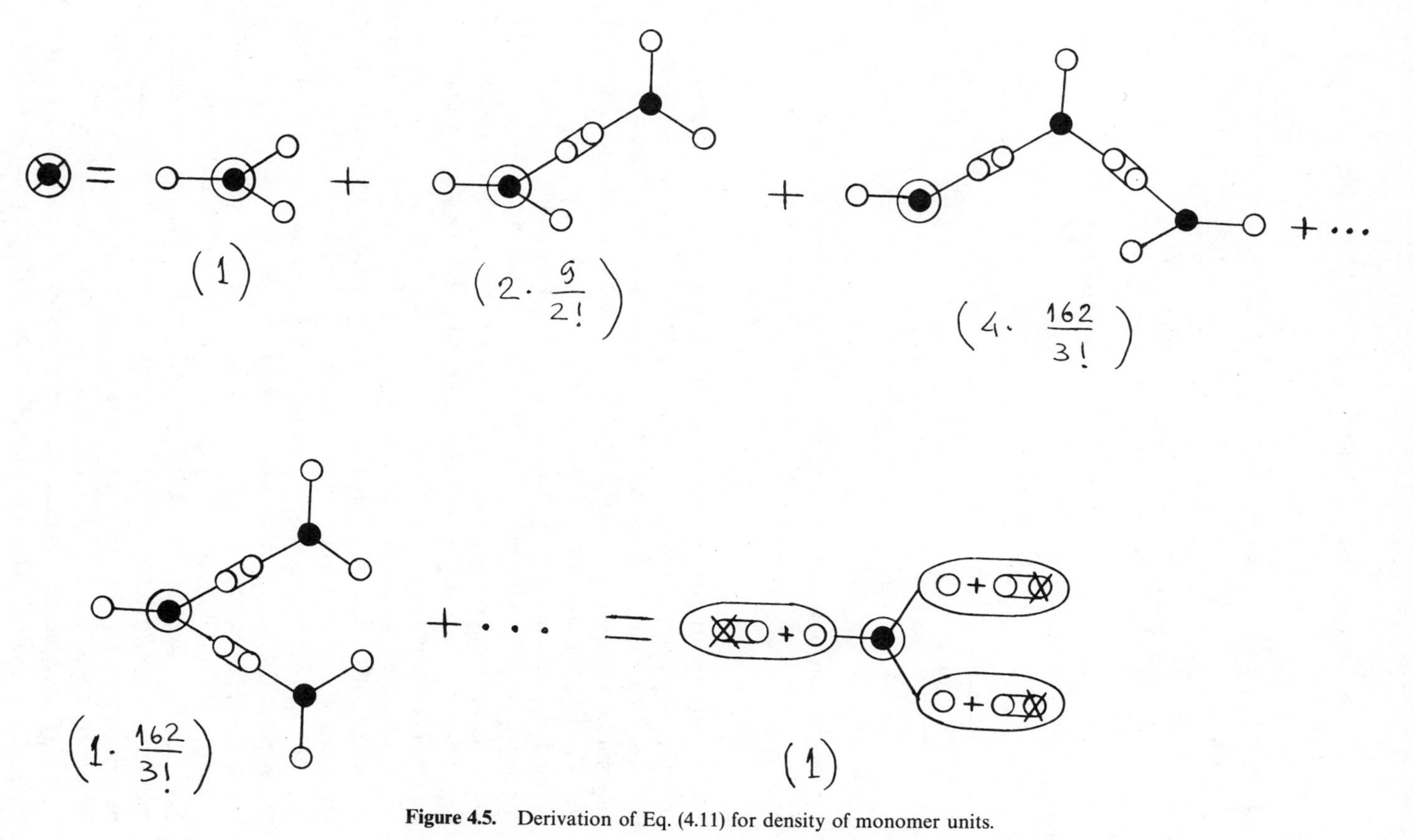

Figure 4.5. Derivation of Eq. (4.11) for density of monomer units.

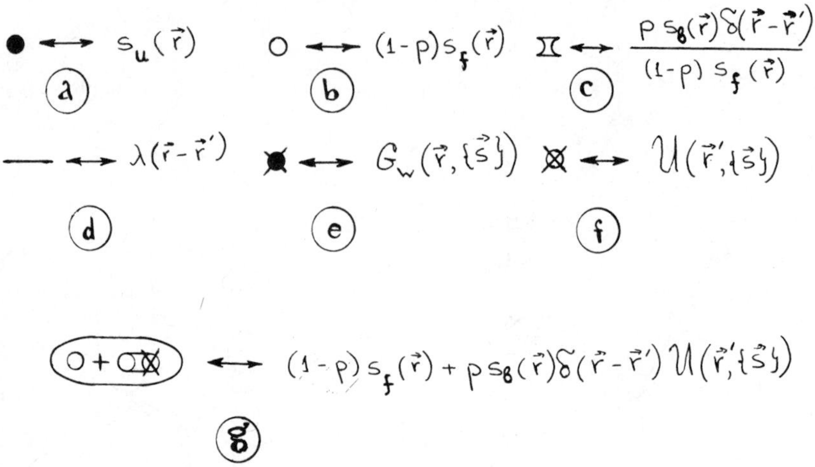

$\bullet \leftrightarrow s_u(\vec{r})$ (a) $\quad \circ \leftrightarrow (1-p)s_f(\vec{r})$ (b) $\quad \leftrightarrow \dfrac{p\, s_b(\vec{r})\delta(\vec{r}-\vec{r}')}{(1-p)\, s_f(\vec{r})}$ (c)

$\text{---} \leftrightarrow \lambda(\vec{r}-\vec{r}')$ (d) $\quad \leftrightarrow G_w(\vec{r},\{\vec{s}\})$ (e) $\quad \leftrightarrow \mathcal{U}(\vec{r}',\{\vec{s}\})$ (f)

$\leftrightarrow (1-p)s_f(\vec{r}) + p\, s_b(\vec{r})\delta(\vec{r}-\vec{r}')\,\mathcal{U}(\vec{r}',\{\vec{s}\})$ (g)

Figure 4.6. Graphical rules describing correspondence between Fig. 3.5 and analytical formulas (3.50) and (3.51) for branching process.

may be replaced by the derivative symbol Ψ^f(Fig. 4.7). It it noteworthy, however, that the replacement is possible only for bonds that are not involved in cycles. As a result, all diagrams with bridges disappear, and the remaining diagrams are sets of quasi-monomers, where effective groups are displayed with the diagram unit of Fig. 4.2*e*.

Similarly, adding cyclic diagrams to Fig. 4.5, the equality of Fig. 4.8 for the derivative Ψ^u is obtained. By means of the same renormalization, as in the SF approximation, analytical equivalents of the graphical equalities of Figs. 4.8 and 4.7 are transformed to (3.46) and (3.47), which are valid for a general branching process with cycles.

The diagram technique provides a spectacular interpretation of the important equations (3.41). For this purpose we need a representation for the series of the functional $\Gamma\{\mathbf{s}\}$, according to definition (3.38) (Fig. 4.9). In terms of the diagram technique, the functional differentiation with respect to function $s_v(\mathbf{r})$ is the selection of an arbitrary black node ($v = u$) or suspended vertex ($v = f$) as a root at point $\mathbf{r}$. The resulting series of root diagrams corresponds to the derivative Γ^v [Eq. (3.39)], where one has to substitute $\tilde{s}_f \to \tilde{s}_f + \tilde{s}_b L\Psi^f/s_f$ according to (3.41). Following the rules of the diagram technique (Fig. 4.2), this procedure is equivalent to substitution of a suspended nonroot vertex for the graph unit of Fig. 4.2d. As a result, the graphical equalities of Figs. 4.7 and 4.8 are obtained from the series of Fig. 4.9.

A diagram representation can be used instead of some algebraic transformations. For example, consider the application of the diagram technique to

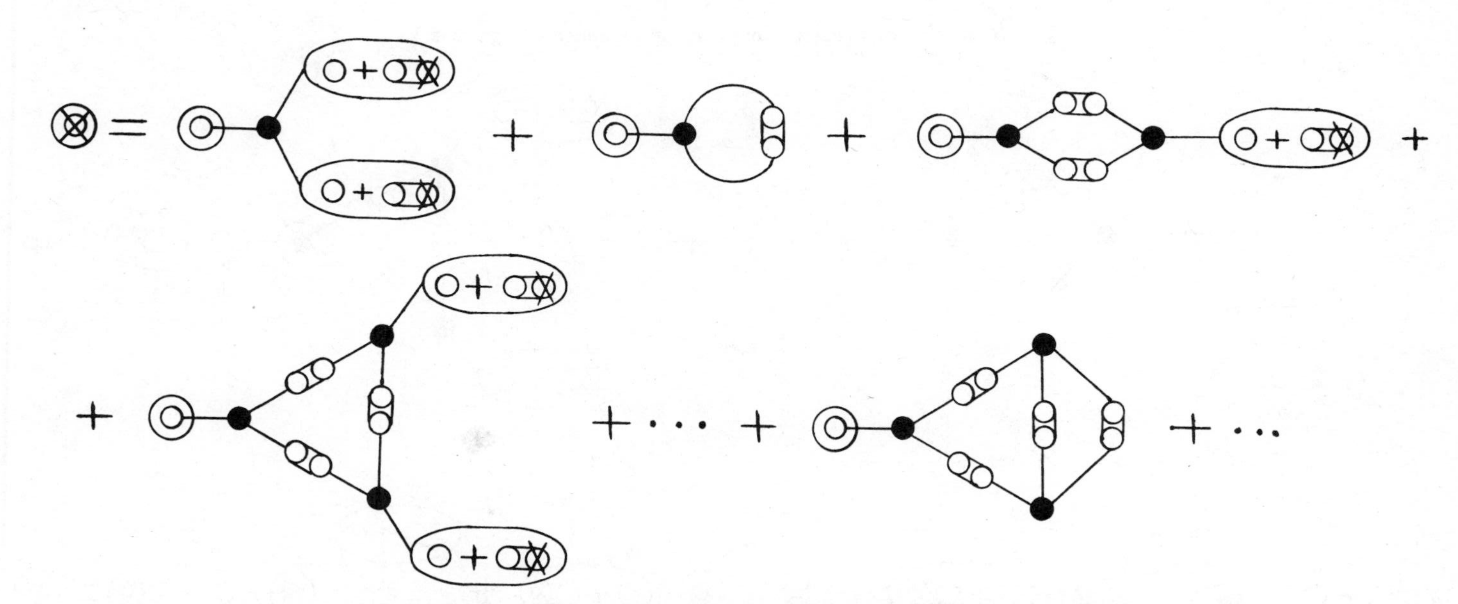

Figure 4.7. Equation for group density taking into account cycles. Numbers of analogues derived from diagram unambiguously are not indicated here and on the rest of the figures.

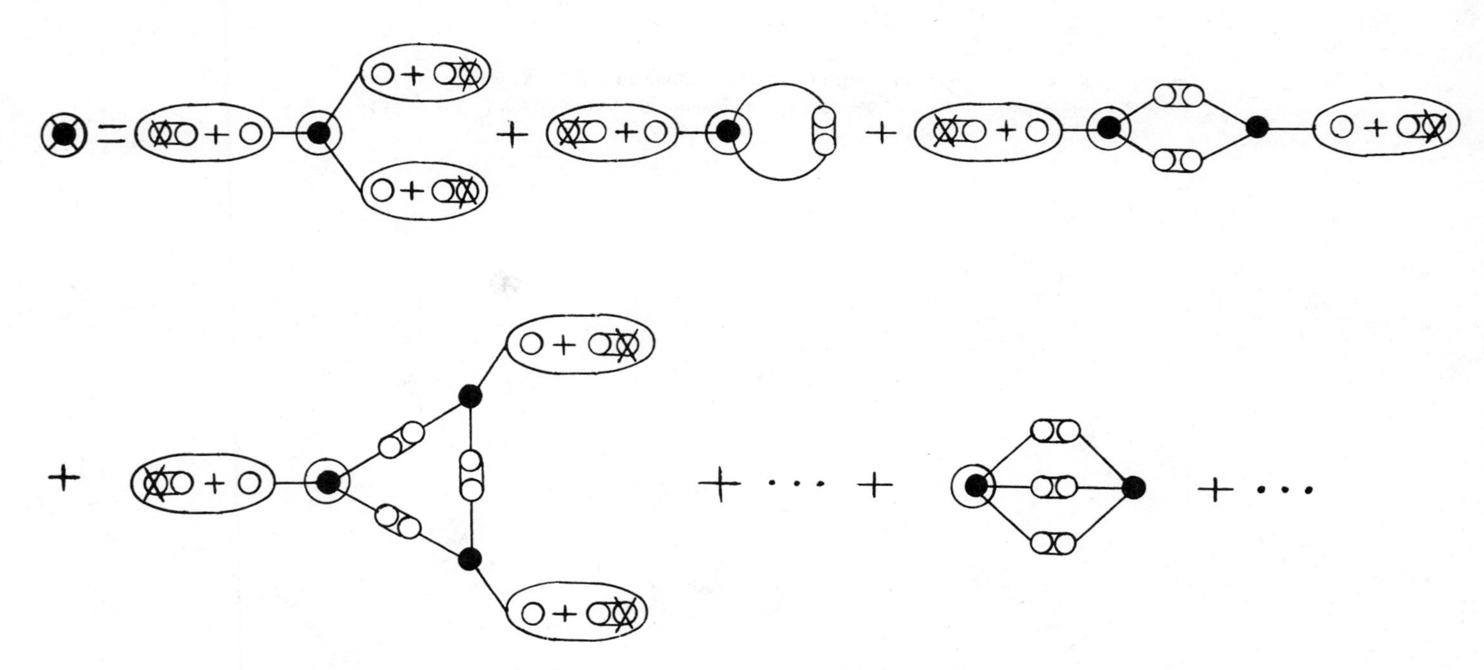

Figure 4.8. Equation for density of units taking into account cycles.

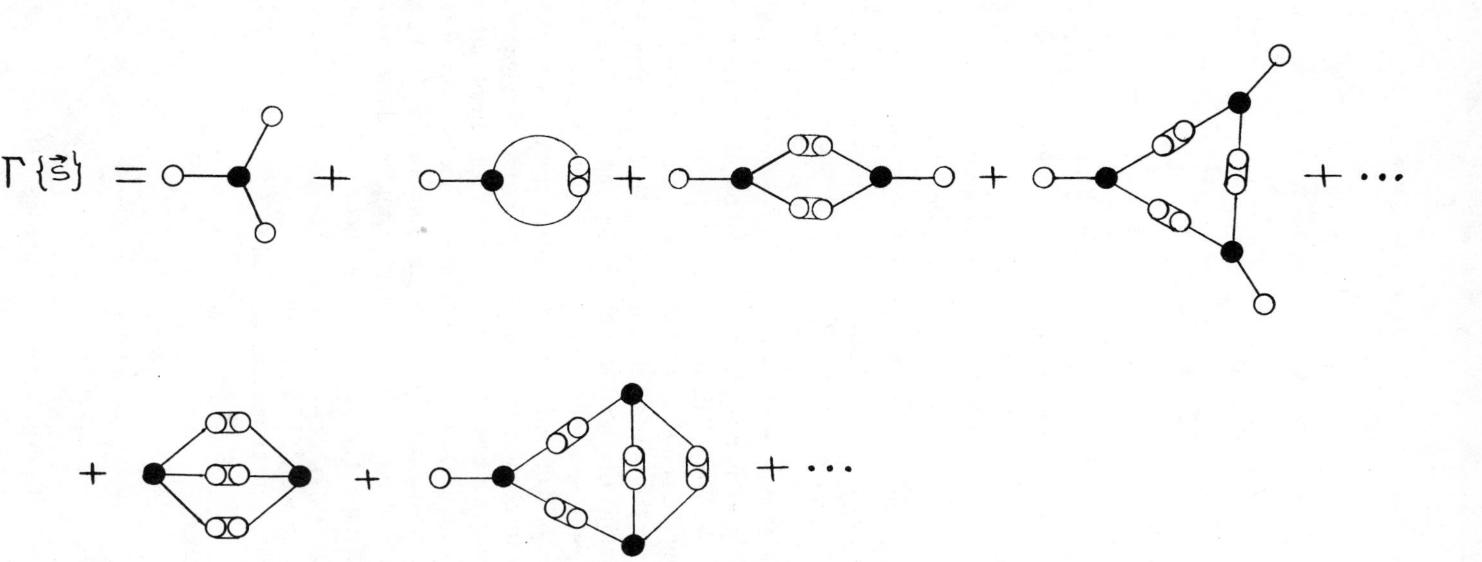

Figure 4.9. Functional Γ; equations of Figs. 4.7 and 4.8 for densities of groups and units obtained from its derivatives.

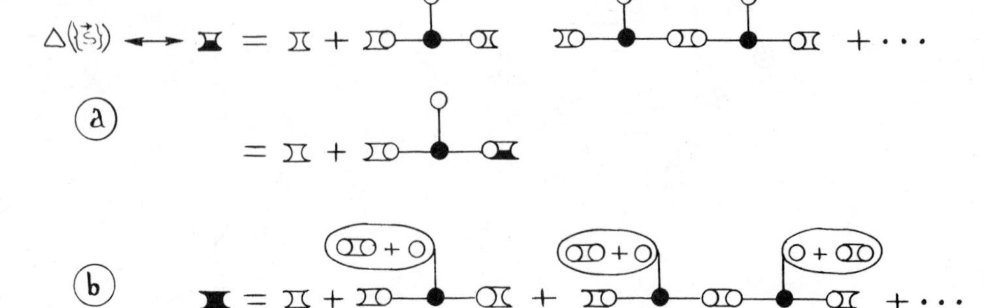

Figure 4.10. Definitions of symbols of half-shaded and shaded bonds and equations for former obtained by summation of diagrams.

summation of contributions from graphs having identical topological structures, which was treated with analytical methods in Section III.H for the case where external fields were absent. All the homeomorphic graphs are obtained from their elementary representatives by substitution of each edge for a linear chain of some length. A new notation is suitable for the sum of all the chains; we shall shade half of the bond symbol; the graphical equation is shown in Fig. 4.10*a*. The solution of its analytical counterpart is determined in Eq. (3.87). Using the half-shaded bond symbol, one can sum partially the series in Fig. 4.9 for the functional $\Gamma\{\mathbf{s}\}$ (Fig. 4.11). The graphical differentiation of the resulting diagrams (the elementary representatives; Fig. 4.12) implies, besides the above-mentioned selection of their vertices taken as roots, the differentiation of half-shaded bonds since every vertex of the corresponding chains can become the root. This procedure changes the graph topology somewhat, as a new node appears between a pair of vertices in the elementary representative, which corresponds to one of links of the chain with two groups having reacted. Either the node or one of the adjacent suspended vertices is chosen as the root (Fig. 4.12*a*). Two differentiable bonds appear instead of one. The analytical reason for this is that the factor $(\tilde{\lambda}(\mathbf{q}_\alpha)^{-2} - 1 + \tau)^{-1}$, corresponding to the αth edge, is squared in its differentiation in the derivatives $H_{rt}^{(1)}$ [Eq. (3.102)].

To get the resulting diagram equations of Fig. 4.13, one should make the

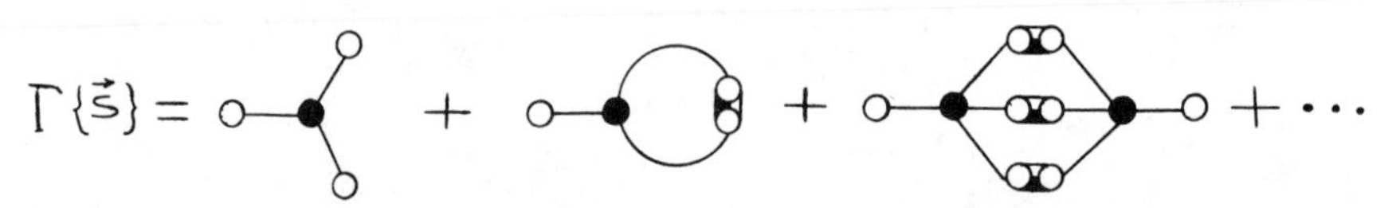

Figure 4.11. Partial summation of diagrams in Fig. 4.9.

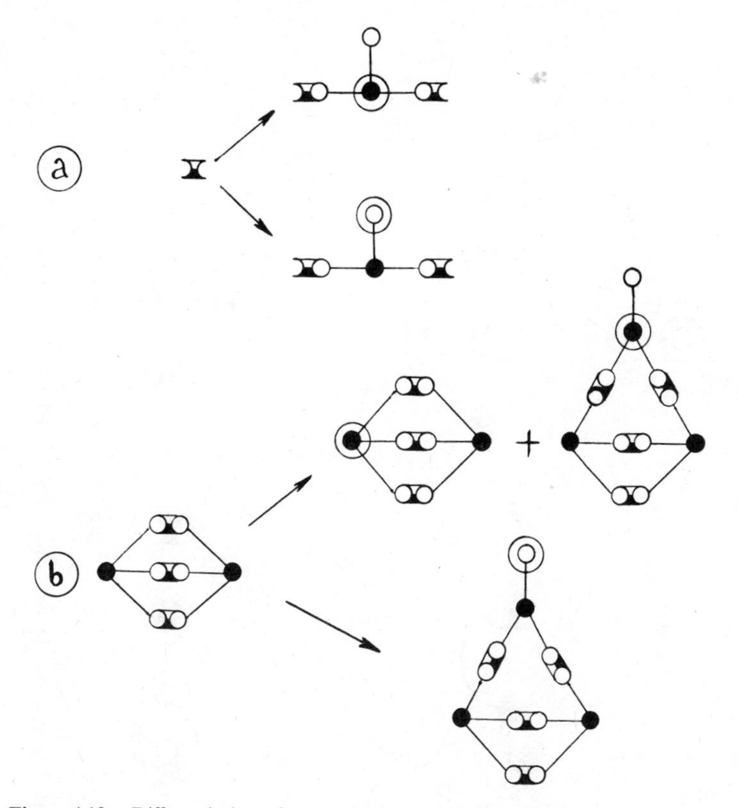

Figure 4.12. Differentiation of (*a*) semishaded bond and (*b*) one of the diagrams.

substitution $s \to \hat{s}$, according to (3.41), so all the nonreacted groups, except for the root group, are replaced by the graph of Fig. 4.2*d*, and one must use the symbol of Fig. 4.10*b*, corresponding to the functional $\Delta\{s\}$, instead of the half-shaded bond determined by Fig. 4.10*a*.

In obtaining the equations for the derivatives Ψ^u, Ψ^f by means of the diagram technique, one of the diagram vertices was chosen as the root. Similarly, the calculation of the kth derivative of the generating functional Ψ, (3.24), results in k-root diagrams. The most important among them are those for $k = 2$,

$$\Psi^{\nu\mu}(\mathbf{r}_1, \mathbf{r}_2; \{\mathbf{s}\}) \equiv \delta^2\Psi\{\mathbf{s}\}/\delta \ln s_\nu(\mathbf{r}_1)\delta \ln s_\mu(\mathbf{r}_2). \qquad (4.12)$$

For $s_\nu = 1$ they are reduced to the two-point correlators $\chi^{\nu\mu}$, (3.27). Every derivative in (4.12) is a sum of a series of all the two-root diagrams, the root

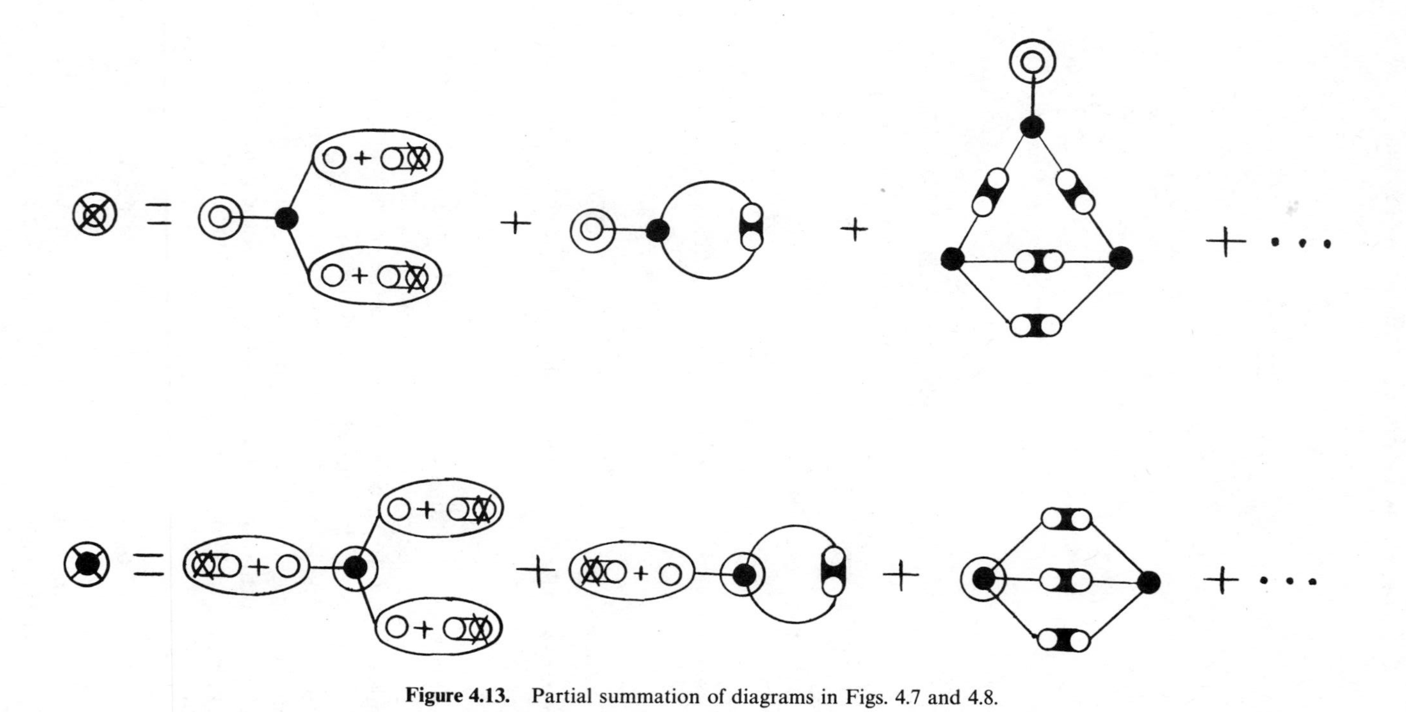

Figure 4.13. Partial summation of diagrams in Figs. 4.7 and 4.8.

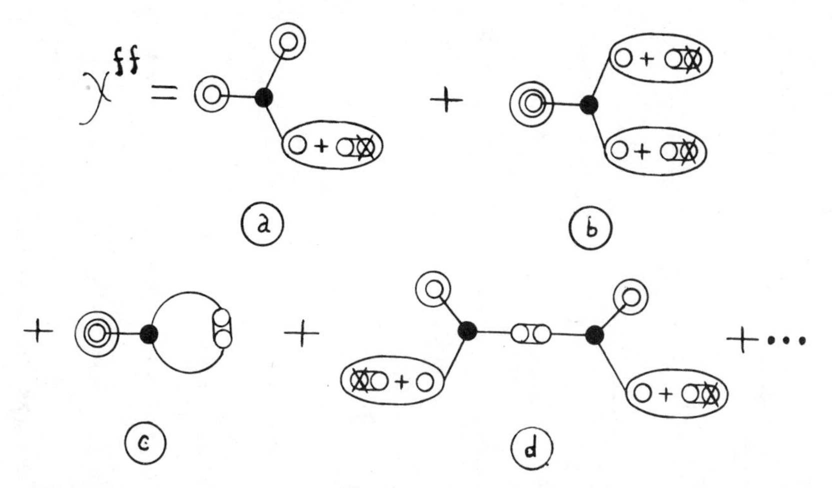

Figure 4.14. Series for correlator χ^{ff} containing, in particular, diagrams of Fig. 4.7; root is circled.

coordinates being $\mathbf{r}_1$ and $\mathbf{r}_2$. If $\mathbf{r}_1 = \mathbf{r}_2$ and $\nu = \mu$, a single vertex is taken for both the roots (Figs. 4.14*b*, *c*). The sum of the diagrams with the double roots reproduces the series for Ψ^ν, which can be avoided by substituting $\hat{\Psi}^{\nu\mu}$ for the new generating functions for the correlators,

$$\begin{aligned}\hat{\Psi}^{\nu\mu}(\mathbf{r}_1, \mathbf{r}_2; \{\mathbf{s}\}) &= \Psi^{\nu\mu}(\mathbf{r}_1, \mathbf{r}_2; \{\mathbf{s}\}) - \Psi^{\nu}(\mathbf{r}_1; \{\mathbf{s}\})\delta_{\nu\mu}\delta(\mathbf{r}_1 - \mathbf{r}_2) \\ &= s_\nu(\mathbf{r}_1)s_\mu(\mathbf{r}_2)\delta^2\Psi\{\mathbf{s}\}/\delta\mathbf{s}_\nu(\mathbf{r}_1)\delta\mathbf{s}_\mu(\mathbf{r}_2).\end{aligned} \tag{4.13}$$

The corresponding graphical notations are given in Fig. 4.15.

Some diagrams present in the series for $\Psi^{\nu\mu}$ or $\chi^{\nu\mu}$ can be obtained as combinations of other diagrams. For instance, the diagram in Fig. 4.14*d* is obtained by coupling its two parts with a central bond, integrating over the bond coordinate. To get the generating functions for the correlators $\hat{\Psi}^{\nu\mu}$, determine first the set of basic two-root diagrams having both their roots in the same cycle (Fig. 4.15). It is not difficult to see that these diagrams can

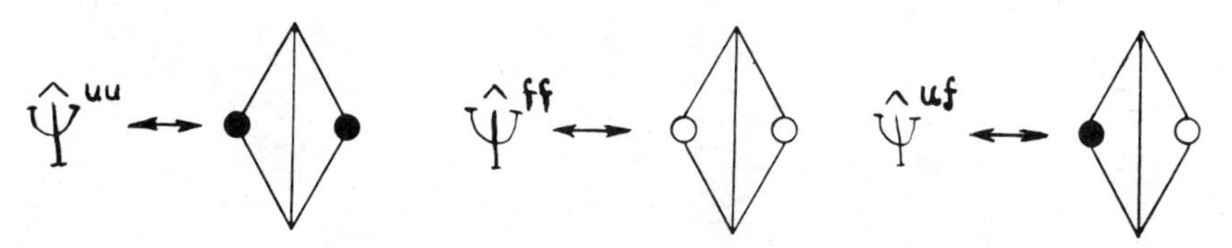

Figure 4.15. Graphical notations for correlators $\hat{\Psi}^{\nu\mu}$.

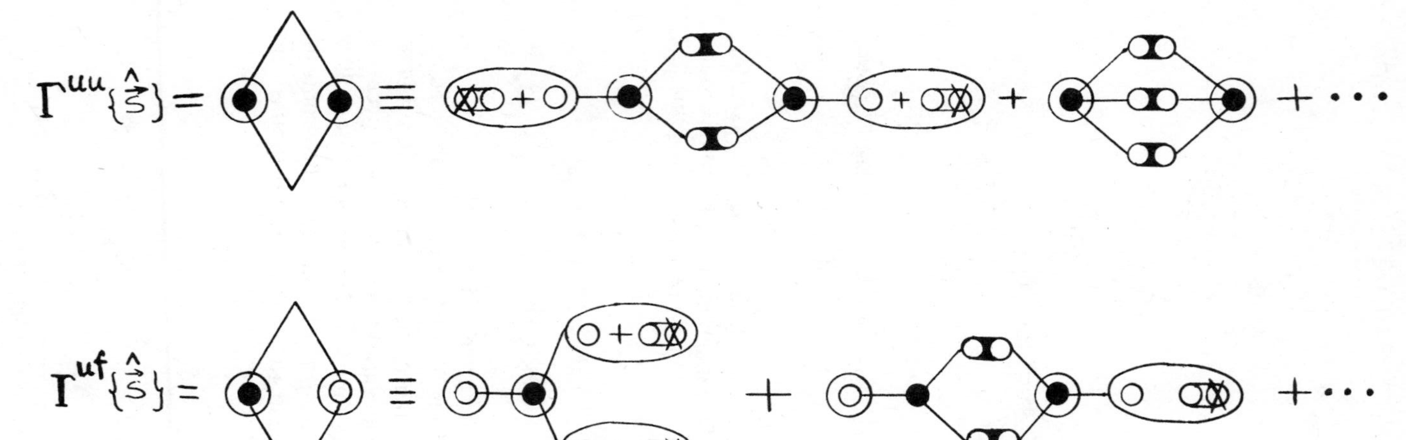

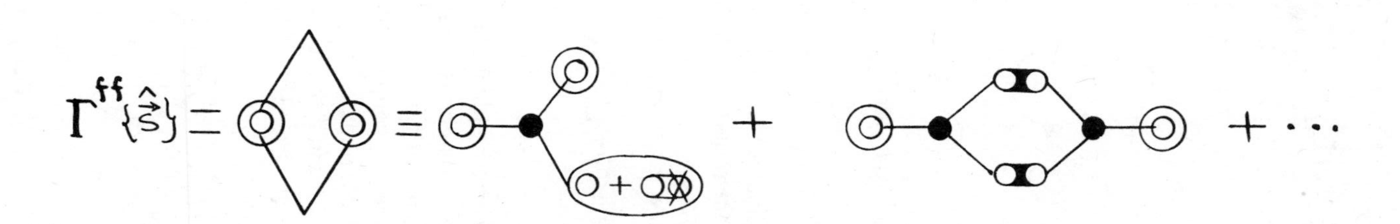

Figure 4.16. Graphical notations for sums of basic diagrams, both roots of which belong to single cycle. If vertices of these symbols are not encircled (cf. Fig. 4.17), one should integrate over their coordinates.

also be obtained by means of the procedure proposed for the graphical calculation of functional derivatives and the change of variables (Fig. 4.2*g*). Connecting the roots of basic graphs, as described for Fig. 4.14*d*, one can get any diagram with roots at different leaves. Those diagrams, combined with basic diagrams. constitute the complete set; summing these diagrams (Fig. 4.16), one gets equations for $\hat{\Psi}^{\nu\mu}$. If the system is spatially homogeneous, the solutions of the equations (Fig. 4.17) are obtained by the Fourier transformation,

$$\begin{aligned}\hat{\tilde{\Psi}}^{uu}\{\hat{s}\} &= \tilde{\Gamma}^{uu}\{\hat{s}\} + [\tilde{\Gamma}^{uf}\{\hat{s}\}]^2/[L^{-1} - \tilde{\Gamma}^{ff}\{\hat{s}\}],\\ \hat{\tilde{\Psi}}^{uf}\{\hat{s}\} &= \tilde{\Gamma}^{uf}\{\hat{s}\}/[1 - L\tilde{\Gamma}^{ff}\{\hat{s}\}],\\ \hat{\tilde{\Psi}}^{ff}\{\hat{s}\} &= \tilde{\Gamma}^{ff}\{\hat{s}\}/[1 - L\tilde{\Gamma}^{ff}\{\hat{s}\}].\end{aligned} \tag{4.14}$$

Note that the first equality has been derived in the preceding section [Eq. (3.43)].

C. Spatial Physical Interactions

To this point, we have been concerned only with the models where interactions between units were chemical reactions between their functional groups. The produced chemical bonds, equivalent to an effective attraction between the units, results in correlations of their positions in space. Besides, there are, of course, some forces of a physical origin responsible for the so-called *spatial interactions* between the units, which are similar to intermolecular interactions in gases or liquids. To take them into account in the theory, we start from model IV, describing equilibrium polycondensation of monomer RA^f, the R units of which are treated as point like particles. The pair interaction between the units is described with a potential $V(\mathbf{r}_i - \mathbf{r}_j)$, having a standard dependence on the distance between the units (Fig. 1.23).

In the model considered, the system state is determined by a spatial graph $\mathscr{G}_N\{\mathbf{r}_i\}$, specified with a set of coordinates for all the vertices corresponding to monomer units and a set of bonds connecting the vertices. The probability Gibbs measure is introduced in the set of graphs,

$$\mathscr{P}(\mathscr{G}_N\{\mathbf{r}_i\}) = \mathscr{P}^{(1)}\mathscr{P}^{(2)}, \qquad \mathscr{P}^{(2)} = \prod_{\langle i,j\rangle}^{N_c} L\lambda_{uu}(\mathbf{r}_i - \mathbf{r}_j),$$

$$\mathscr{P}^{(1)} = M^N \exp\left(\left[\Omega - \sum_{i=1}^{N} H(\mathbf{r}_i) - \sum_{\langle i,j\rangle} V(\mathbf{r}_i - \mathbf{r}_j)\right]\Big/ T\right), \tag{4.15}$$

where the first factor is due to the Gibbs distribution of vertices in space, and the second factor determines the network of bonds. The model determined

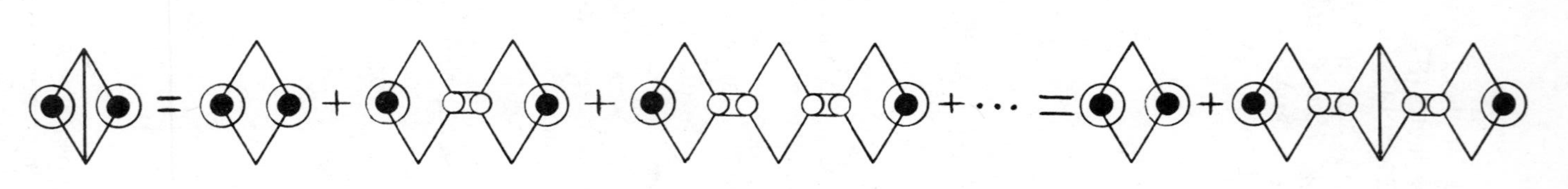

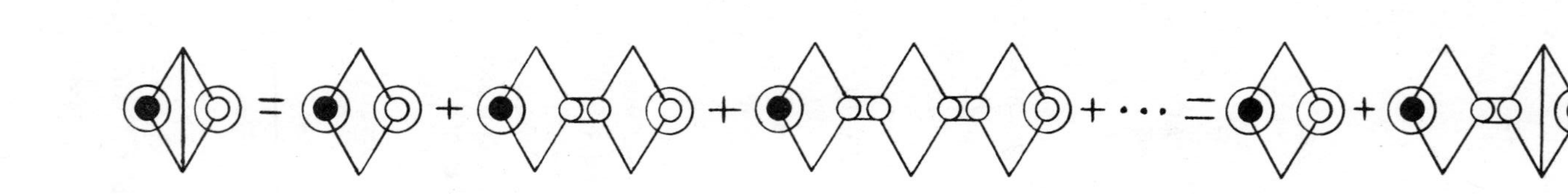

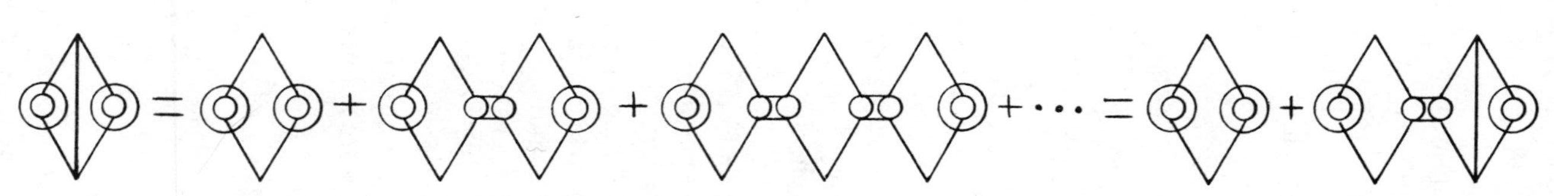

Figure 4.17. Graphical equations for two-point correlators.

by (4.15), where the probability is given by two factors, was originally proposed by Lifshitz[184] as a tool for the description of linear polymer molecules and was further extended by Erukhimovich,[133] for branching polymers. A remarkable feature of the Lifshitz–Erukhimovich model is that it produces results similar to superpositions of contributions from two simpler subsystems.

The first subsystem, which is called here a *system of chemical bonds*, was described in some detail in Section III. In that approach we have taken into account the appearance of chemical bonds resulting in the formation of polymer molecules, but spatial interactions of units were neglected completely. The probability measure for this system is

$$\mathscr{P}_{\mathrm{cb}}(\mathscr{G}_N\{\mathbf{r}_i\}) = M^N \exp\left(\left[\Omega_{\mathrm{cb}} - \sum_{i=1}^{N} H(\mathbf{r}_i)\right]\Big/ T\right)\prod_{\langle i,j\rangle} L\lambda_{uu}(\mathbf{r}_i - \mathbf{r}_j). \quad (4.16)$$

It is obtained from the general formula (4.15) for $V(\mathbf{r}_i - \mathbf{r}_j) = 0$, or from (3.4), setting $h_f = h_b = 0$ and integrating over coordinates of functional groups and bonds.

The second subsystem is called a *system of separate units*, as named by Lifshitz[184] and Erukhimovich.[133] It is an equilibrium ensemble of monomer units, interacting with a physical force potential $V(\mathbf{r}_i - \mathbf{r}_j)$ but producing no chemical bonds. The combined probability distribution for the unit coordinate in space is given by the Gibbs formula,

$$\mathscr{P}_{\mathrm{su}}\{\mathbf{r}_i\} = M^N \exp\left(\left[\Omega_{\mathrm{su}} - \sum_{i=1}^{N} H(\mathbf{r}_i) - \sum_{\langle i,j\rangle} V(\mathbf{r}_i - \mathbf{r}_j)\right]\Big/ T\right). \quad (4.17)$$

The partition function, which is given by normalization of this distribution, is given by the average of a functional,[183]

$$\exp[-\Omega_{\mathrm{su}}\{z\}/T] = \left\langle \exp\left[\left[\int d\mathbf{r} z(\mathbf{r}) \exp(-v(\mathbf{r})/T)\right]\right]\right\rangle,$$
$$z(\mathbf{r}) = M \exp[-H(\mathbf{r})/T] \quad (4.18)$$

The averaging is over configurations of a stochastic field $v(\mathbf{r})$; the corresponding probability measure is determined by a quadratic Lagrangian,

$$\mathscr{L}_{\mathrm{su}}\{v\} = \int\!\!\int d\mathbf{r}'\, d\mathbf{r}''\, J(\mathbf{r}' - \mathbf{r}'')v(\mathbf{r}')v(\mathbf{r}'')/2T, \int d\mathbf{r}\, V(\mathbf{r}' - \mathbf{r})J(\mathbf{r} - \mathbf{r}'') = \delta(\mathbf{r}' - \mathbf{r}''). \quad (4.19)$$

In obtaining Eq. (4.18), we have used the identity

$$\exp\left[-\sum_{\langle i,j\rangle} V(\mathbf{r}_i-\mathbf{r}_j)/T\right]=\left\langle \exp\left[-\sum_{i=1}^{N} v(\mathbf{r}_i)/T\right]\right\rangle_v, \tag{4.20}$$

which is verified by a change of the variable function, $v(\mathbf{r})\to v(\mathbf{r})-\sum_{i=1}^{N} V(\mathbf{r}-\mathbf{r}_i)$ in the functional integral (4.20).

The representation of the partition function in terms of a functional integral, Eq. (4.18), is suitable for constructing the diagram technique for the perturbation theory. The expression in brackets on the r.h.s. of Eq. (4.18) is reduced after expansion of the exponential in the power series to a polynomial form familiar in standard field theory. Any term of the series corresponds to a diagram that represents its analytical form by the following rules. Each diagram vertex with a coordinate $\mathbf{r}_i$ represents a factor $z(\mathbf{r}_i)$, and the edge connecting vertices with the coordinates $\mathbf{r}_i$ and $\mathbf{r}_j$ is associated with a factor $[-V(\mathbf{r}_i-\mathbf{r}_j)/T]$. Integration over all the vertex coordinates is implied.

The Mayer diagram technique,[185,186] where vertices corresponding to particles, $\mathbf{r}_i$ and $\mathbf{r}_j$, can be connected with no more than a single edge, is widely exploited in statistical mechanics of nonideal gases. In that technique, the edge corresponds to a factor $f(\mathbf{r}_i-\mathbf{r}_j)$, which is shown graphically in the figures with a thin dashed line. The factor is obtained as a partial sum of contributions from field-theoretic diagrams (4.18) and (4.19), which are different only in the edge multiplicities n,

$$f(\mathbf{r}_i-\mathbf{r}_j)=\sum_{n=1}^{\infty}\frac{1}{n!}\left[-\frac{V(\mathbf{r}_i-\mathbf{r}_j)}{T}\right]^n=\exp\left[-\frac{V(\mathbf{r}_i-\mathbf{r}_j)}{T}\right]-1. \tag{4.21}$$

Thus, the Mayer diagram technique can be reconstructed by means of a simple rearrangement of the diagram series for the field theory, (4.18) and (4.19). A subsequent partial summation of these series, resulting in, first, connected and then irreducible sets of diagrams, can be carried out on the basis of, respectively, the first and the second theorems by Mayer.[186] The latter theorem enables one to represent an expression for pressure as Taylor's series in powers of density ρ, where coefficients at ρ^n depend solely on temperature. The magnitude of the virial coefficient is determined by the nth irreducible integral $\beta_n(T)$, which corresponds to the complete set of analogue Mayer diagrams with n vertices.

A variational differential technique is appropriate for the calculation of the density correlators in the system of separate units acted on by an arbitrary external field. Let us introduce a functional $\mu^*(\mathbf{r};\{\rho\})$ as follows:

$$-\delta\Omega_{\text{su}}\{z\}/\delta z(\mathbf{r})T=\exp[-\mu^*(\mathbf{r};\{\rho\})/T], \tag{4.22}$$

where the function $z(r)$ is related to the density by

$$\rho(\mathbf{r}) = -\delta\Omega_{su}\{z\}/\delta \ln z(\mathbf{r})T = z(\mathbf{r})\exp[-\mu^*(\mathbf{r};\{\rho\})/T]. \quad (4.23)$$

Here and in the following, the argument of the Ω potential is written as $z(\mathbf{r}) = M\exp[-H(\mathbf{r})/T]$ or as $H(\mathbf{r})$ (the external field), choosing the argument more suitable for our purpose. In the theory of nonideal gases, Eq. (4.23) is well known for spatially uniform systems[186]; there is an expansion of μ^* in powers of density,

$$\mu^*(\rho) \equiv \mu - \mu_{ig} = -\sum_{n=1}^{\infty} \beta_n(T)\rho^n, \quad (4.24)$$

where μ_{ig} is the chemical potential for the ideal gas. Having the functional μ^* known, one can calculate the total correlation function for the system of separate units,[185] which is

$$h(\mathbf{r},\mathbf{r}') \equiv \frac{\hat{\Xi}(\mathbf{r},\mathbf{r}')}{\rho(\mathbf{r})\rho(\mathbf{r}')} - 1 \equiv \frac{\hat{\theta}(\mathbf{r},\mathbf{r}')}{\rho(\mathbf{r})\rho(\mathbf{r}')} = -\exp\left[\frac{\mu^*(\mathbf{r};\{\rho\})}{T}\right]\int C(\mathbf{r},\mathbf{r}'';\{\rho\})\frac{\delta\rho(\mathbf{r}'')}{\delta z(\mathbf{r}')}\,d\mathbf{r}''. \quad (4.25)$$

The latter equality is obtained by means of the variational differentiation of the generating functional according to the definition in (3.21), taking into account Eq. (4.22). We have introduced the notation

$$-C(\mathbf{r}',\mathbf{r}'';\{\rho\}) = \delta\mu^*(\mathbf{r}',\{\rho\})/\delta\rho(\mathbf{r}'')T. \quad (4.26)$$

The dimensionless functional C is reduced to the *direct correlation function* if the system is spatially uniform.[185] The known Ornstein–Zernike integral equation[185] relates that functional to the total correlation function,

$$h(\mathbf{r},\mathbf{r}') = C(\mathbf{r},\mathbf{r}') + \int C(\mathbf{r},\mathbf{r}'')\rho(\mathbf{r}'')h(\mathbf{r}'',\mathbf{r}')\,d\mathbf{r}''. \quad (4.27)$$

To derive this equation, one must first obtain an equation for the derivative $\delta\rho/\delta z$, calculating the term-by-term derivative of (4.23) with respect to $z(r)$, and then use its relation to the correlation function h, as given by (4.25).

Equation (4.27) has a simple graphical interpretaton if we introduce diagram symbols for the direct and total correlation functions (cf. Fig. 4.18*a*). In terms of the diagram technique, the equation is displayed in Fig. 4.18*b*, where the symbol of the total density is shown as in model III considered in the preceding. By definition, it is the sum of all one-root Mayer diagrams

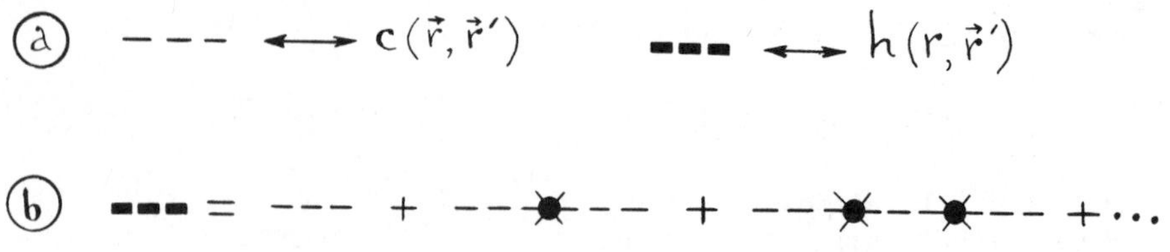

(a) $- - - \longleftrightarrow c(\vec{r}, \vec{r}')$ $\quad$ ■■■ $\longleftrightarrow h(r, \vec{r}')$

(b) ■■■ = --- + --✱-- + --✱--✱-- + ...

Figure 4.18. Symbols of (*a*) correlation functions, direct, *C*, and total, *h* and (*b*) equation representing relation between them.

(Fig. 4.19), and their partial summation leads to the graphical equation corresponding to the analytical equivalent (4.23). The functional C is defined as the sum of all the two-root connected Mayer diagrams that are not reduced to nonconnected components containing only one of two roots if any vertex is removed from them.

The exact calculation of the functional Ω_{su} in (4.18) is possible only for simplest model systems. Among them, the most interesting model is that of a lattice gas, which makes it possible to account for excluded volume effects (Section I.H). In that model, where every lattice point can be occupied at most by a single particle, the potential in (4.18) is

$$\Omega_{su}\{z\} = -T\sum_{\mathbf{r}} \ln[1 + z(\mathbf{r})]. \tag{4.28}$$

Here the sum is over all the lattice points, the coordinates of which constitute a discrete set, unlike continuum, which takes place in nonlattice models.

To account for effects due to particle interactions at large distances (cf. Fig. 1.23), the potential $V(\mathbf{r}_i - \mathbf{r}_j)$ is represented as a sum of its short-range and long-range components, $V = V_s + V_l$. Correspondingly, one can draw diagrams with edges of two types.[186] The sum of contributions from diagrams with various dispositions (and numbers) of the long-range-type edges can be interpreted exactly as an average over configurations of the stochastic field $v(\mathbf{r})$. The corresponding probability measure is given by the Lagrangian in (4.19), where the long-range component V_l stands for V. Summing the contributions from the diagrams with the V_s edges before averaging, one gets

$$\exp[-\Omega_{su}\{H\}/T] = \langle \exp[-\Omega_s\{H + v\}/T] \rangle_v. \tag{4.29}$$

The functional $\Omega_s\{H\}$ equals the thermodynamic potential for a system of particles under the action of an external field and interacting with potential V_s. A relevant example is the expression in (4.28), where the excluded volume effects are taken into account exactly in the framework of the lattice gas model. The evaluation of the functional integral in (4.29), by means of the

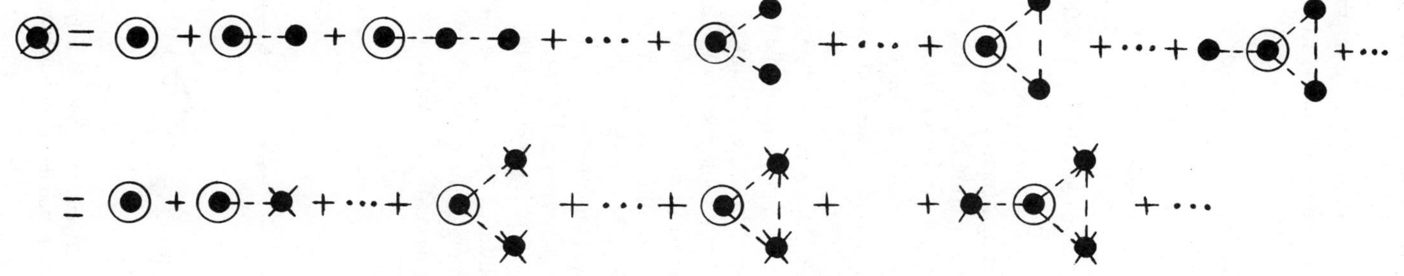

Figure 4.19. Series of Mayer diagrams that determines symbol of total density.

steepest descent method, produces an estimate for the contribution from the long-range interaction to the system thermodynamic potential,

$$\frac{1}{2}\iint V_l(\mathbf{r}' - \mathbf{r}'')\rho(\mathbf{r}')\rho(\mathbf{r}'')\,d\mathbf{r}'\,d\mathbf{r}''. \tag{4.30}$$

In the considered approximation, this contribution is independent of the excluded volume effects.

D. Diagram Technique in Lifshitz–Erukhimovich Model

Because in the Lifshitz–Erukhinovich model, we have the Gibbs distribution (4.1), the diagrams contain a single type of vertex corresponding to a monomer unit and two types of edges connecting the vertices. Edges of the first type (solid lines) correspond to chemical bonds; edges of the second type (dashed lines) are those of the Mayer diagram technique for the system of separate units (Fig. 4.20*a*). The edges correspond to the factors $L\lambda_{uu}(\mathbf{r}_i - \mathbf{r}_j)$ and $f(\mathbf{r}_i - \mathbf{r}_j)$ [Eq. (4.21)]. A sample diagram is shown in Fig. 4.20*b*.

Following Erukhimovich,[133] we call a "branch" a subset of diagram vertices connected by solid lines only, and marked branches (i.e., those which are different from others) will be called "trunks." Evidently, every branch corresponds to an (l, q)-isomer molecule. A combination of all branches specifies a graph $\mathcal{G}_N\{\mathbf{r}_i\}$, that is, a state of the considered polymer system, and the Gibbs probability measure (4.15) is determined in the set of graphs. A generic diagram is constructed by connecting the vertices of the graph $\mathcal{G}_N\{\mathbf{r}_i\}$ with dashed lines. It is called "connected" if any pair of its vertices can be connected with a line including edges of arbitrary types. The concentration of the (l, q)-isomer molecules, $c(l, q)$, is the sum of the contributions from all the one-trunk connected diagrams, where the trunk is the (l, q) branch[133] (Fig. 4.21). Similarly, summing all the one-root connected diagrams, one gets the total unit density ρ, which is indicated with a symbol, the one introduced in Fig. 4.2.

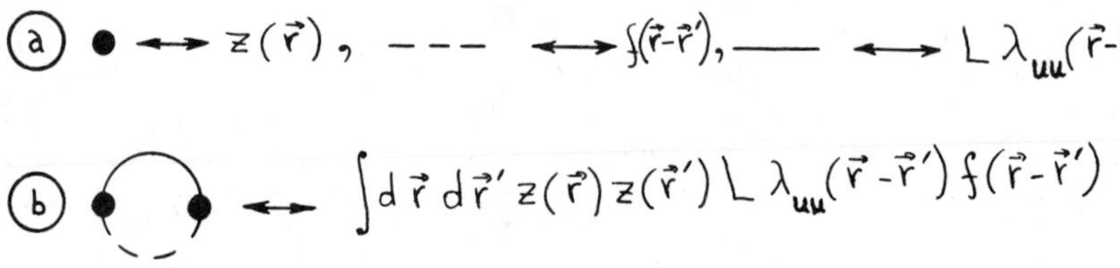

(a) $\bullet \longleftrightarrow z(\vec{r})$, $---\longleftrightarrow f(\vec{r}-\vec{r}')$, $\text{——} \longleftrightarrow L\lambda_{uu}(\vec{r}-\vec{r}')$

(b) $\longleftrightarrow \int d\vec{r}\,d\vec{r}'\,z(\vec{r})z(\vec{r}')L\lambda_{uu}(\vec{r}-\vec{r}')f(\vec{r}-\vec{r}')$

Figure 4.20. (*a*) Elements of diagram technique for polymer system with physical interactions and (*b*) simplest diagram.

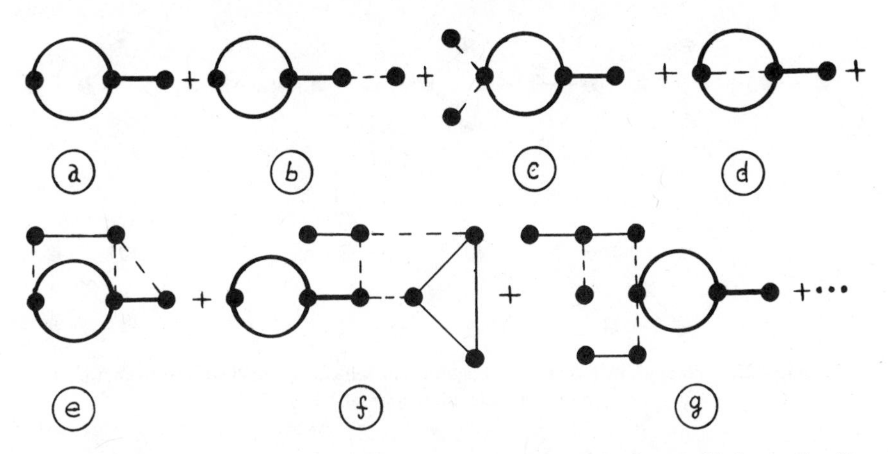

Figure 4.21. Sum of one-trunk connected diagrams (trunk is shown with heavy line for clarity) that determine trimer concentration.

In cases where the specification of system states would require the data on the space positions of functional groups and chemical bonds, besides coordinates of the monomer units, the graphical presentation of branches must be somewhat more complicated, as discussed in Section IV.B. To reduce the system considered there to the system of chemical bonds, (4.16), one has to integrate over the coordinates of groups and bonds in the corresponding formulas of Section IV.B, that is, to "erase" white vertices and couplings in the diagrams of Figs. 4.4–4.8. This procedure enables us to get a relation between the diagrams of model III and model IV.

The usefulness of the diagram technique is due to the fact that it enables one to obtain the desired approximate quantities by means of a partial summation of some particular classes of the diagrams. Each approximation corresponds to certain rules of diagram selection. In polymer systems having high substance concentrations, units of various molecules are present near any given unit, so it is reasonable to neglect in the leading approximation the physical interactions between units belonging to the same polymer molecule. In terms of the diagram technique, one has to drop diagrams, similar to those in Figs. 4.21*d–f*, where a pair of vertices (units) of the same branch (molecule) is connected by a route passing by a dashed line. The sum of the remaining graphs is expressed in terms of the thermodynamic functions for the system of separate units.

To verify this statement, let us consider first the one-root graphs, the sum of which equals the total density of the units. The partial summation of these graphs is carried out as in the system of separate units (Fig. 4.22). It was found that the resulting effect owing to the physical interaction is the sub-

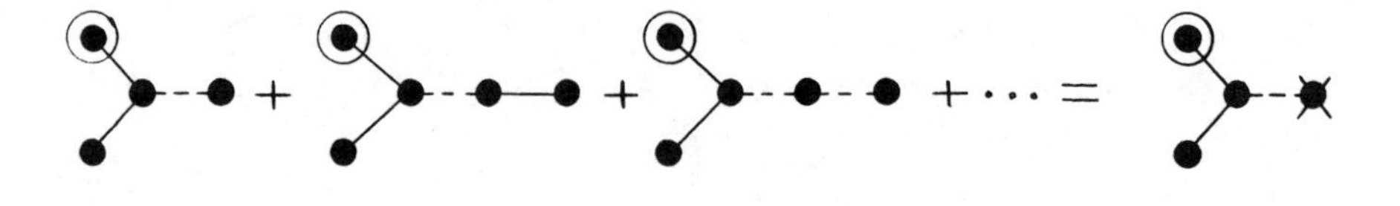

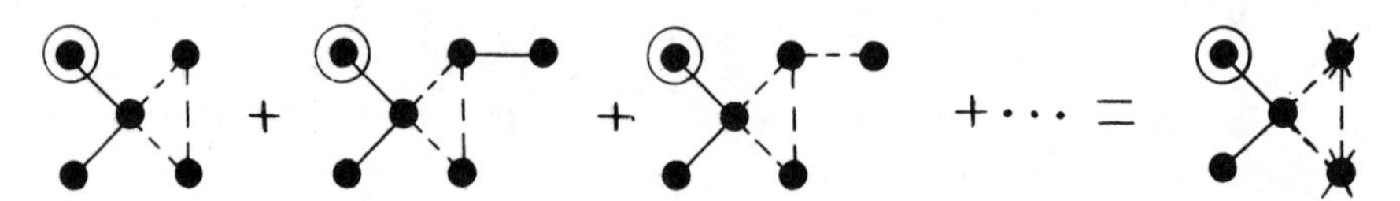

Figure 4.22. Summation of one-root diagrams in self-consistent field approximation for physical interactions.

stitution of the factor $z(\mathbf{r})$ associated with a vertex assigned to a monomer unit in the framework of the diagram technique of chemical bonds for a modified function $z^*(\mathbf{r})$,

$$z^*(\mathbf{r}) = z(\mathbf{r}) \exp[-\mu^*(\mathbf{r}; \{\rho\})/T]. \tag{4.31}$$

The quantity μ^* has the meaning of the self-consistent field of physical interactions; it equals the chemical potential for the system of separate units minus the chemical potential for the ideal gas having the same density distribution.[184] Thus, in the approximation concerned, the physical interactions are taken into account by adding μ^* to the external field $H(\mathbf{r})$. Within the approximation of the self-consistent field of physical interactions (SFPI), Eq. (3.24), which relates the Ω potential for the system of chemical bonds to the generating functional of the correlators, is still valid. Consequently, in this approximation we have the relation

$$\Psi\{s\} = -\Omega_{\text{cb}}\{H + \mu^* - T \ln s\}/T, \tag{4.32}$$

which suggests that taking into account physical interactions does not change the MSD, as compared with the system of chemical bonds; it is reduced just to a renormalization of some parameters. Therefore, in that case all the results obtained model III in describing statistical properties of separate molecules remain valid up to renormalization.

This conclusion is not extended, of course, to the thermodynamic relations (e.g., the equation of state), which are different in the SFPI approximation of model IV and in model III. So to obtain an exhaustive statistical description of the system, one has to get its thermodynamic potential $\Omega\{z\}$, which in the approximation in view depends only on the density distribution for the units,

$\rho(\mathbf{r})$. This distributions is the solution of the self-consistency equation,

$$\rho(\mathbf{r}) = \left. \frac{\delta \Psi\{s\}}{\delta \ln s(\mathbf{r})} \right|_{s(\mathbf{r})=1}. \tag{4.33}$$

The dependence of the r.h.s. on $\rho(\mathbf{r})$ is determined by the functional (4.32), which has been obtained from the analysis of the system chemical bonds. The resulting function $\rho(\mathbf{r})$ makes it possible to get the self-consistent field $H + \mu^*$, present in (4.32), thus determining completely the renormalization of the parameters in the MSD. The evident thermodynamic equality $\rho(\mathbf{r}) = \partial\Omega\{H\}/\delta H(\mathbf{r})$, combined with Eq. (4.33), leads to a reconstruction of the thermodynamic potential for the polymer system,

$$\Omega\{H\} = -T\Psi_{cb}\{H + \mu^*\} - \int d\mathbf{r}\, P^*(\mathbf{r}; \{\rho\}). \tag{4.34}$$

Here $P^*(\mathbf{r}; \{\rho\}) = P_{su}(\mathbf{r}; \{\rho\}) - \rho(\mathbf{r})T$ is the difference of the pressures at a point r for the system of separate units and for the ideal gas of the same density $\rho(\mathbf{r})$. Note a relation of Eq. (4.34) to the so-called conditional thermodynamic potential $\Omega(\{\rho\}; \{H\})$ for a given, in general nonequilibrium, density distribution $\rho(\mathbf{r})$. The conditional potential was introduced[186] to obtain the system fluctuation characteristics, and in the SF approximation it coincides with the expression in (4.34), if one treats $\rho(\mathbf{r})$ and $H(r)$ as formally independent functional variables. Thus the thermodynamics of the Lifshitz–Erukhimovich model is determined by the characteristics of the separate unit system, P^* and μ^*, and their variations are related simply by $\delta P^*(\mathbf{r}; \{\rho\}) = \rho(\mathbf{r})\delta\mu^*(\mathbf{r}; \{\rho\})$. It is not difficult to show, using this relation and Eq. (4.33), that the equilibrium density distribution $\rho(\mathbf{r})$ in a given external field $H(\mathbf{r})$ can be obtained as the minimum of the conditional thermodynamic potential $\delta\Omega(\{\rho\}; \{H\})/\delta\rho(\mathbf{r}) = 0$. Its value for the equilibrium density $\rho(\mathbf{r})$ determines the magnitude of the thermodynamic Ω potential, and the SFPI approximation corresponds to neglecting density fluctuations.

It is suitable to rewrite (4.34) in another form,

$$\Omega\{H\} = \Omega_{cb}\{H + \mu^*\} + \Omega_{su}\{\rho\} - \Omega_{ig}\{\rho\}, \tag{4.35}$$

where $\Omega_{ig}\{\rho\}$ is the thermodynamic potential for the ideal gas of density $\rho(\mathbf{r})$. Relation (4.35) is obviously equivalent to the factorization of the partition function, which is the product of the partition functions for the chemical bond system and the separate unit system. Factorization is the basis of a modern approach to the theory of polymer solutions as proposed by Flory.[187] The most consistent treatment of this matter was done by Lifshitz[184] and

Erukhimovich;[133] the latter derived a particular case of Eq. (4.34) for a spatially uniform ensemble of tree molecules. Equality (4.34) has been obtained without such restrictions, since in the considered SFPI approximation one takes into account formation of polymer molecules having arbitrary topological structures and acted upon by arbitrary external fields.

Besides the calculation of the thermodynamic potential, the diagram technique was applied in ref. 133 to the correlation functions of the total density. The summation of the two-root diagrams, necessary for that purpose, is, however, more cumbersome, as compared with that for the one-root diagrams. The same results can be obtained in a simpler way by means of the functional differentiation (3.18) with respect to the external field applied to the thermodynamic potential Ω in Eq. (4.34),

$$\theta(\mathbf{r},\mathbf{r}') = -T\frac{\delta^2\Omega}{\delta H(\mathbf{r})\,\delta H(\mathbf{r}')} = \frac{\delta\rho(\mathbf{r}')}{\delta H(\mathbf{r})}. \tag{4.36}$$

The step-by-step differentiation of (4.33) with respect to $H(r)$ leads to an equation that determines the derivative $\delta\rho/\delta H$. Using (4.36), this equation can be written as a linear integral equation for the irreducible two-point correlator for the total density of units, $\theta = \theta^{uu}$:

$$\theta(\mathbf{r},\mathbf{r}') = \theta_{\mathrm{su}}(\mathbf{r},\mathbf{r}') - \iint d\mathbf{r}_1\, d\mathbf{r}_2\, \theta_{\mathrm{cb}}(\mathbf{r},\mathbf{r}_1)C(\mathbf{r}_1,\mathbf{r}_2)\theta(\mathbf{r}_2,\mathbf{r}'). \tag{4.37}$$

Here C is the direct correlation function for the system of separate units, as defined in Eq. (4.26). For a spatially uniform system, the Fourier transform can be applied to (4.37), and we get a simple relation, which holds for quantities inverse to the Fourier transforms of the correlator of the total density of units, θ, and the structure function for the chemical bond system, g,

$$\tilde{\theta}^{-1}(\mathbf{q}) = \tilde{g}^{-1}(\mathbf{q}) - \tilde{C}(\mathbf{q}), \qquad \tilde{g}(\mathbf{q}) = \tilde{\theta}_{\mathrm{cb}}(\mathbf{q}). \tag{4.38}$$

This important relation was obtained originally in ref. 188. Equality (4.27) makes it possible to express C in Eq. (4.38) in terms of the total correlation function for the total correlation function of the system of separate units; the resulting Fourier transform of the total density correlator is

$$\tilde{\theta}(\mathbf{q}) = \tilde{g}(\mathbf{q}) + \tilde{g}(\mathbf{q})\tilde{\beta}(\mathbf{q})\tilde{g}(\mathbf{q}), \tag{4.39}$$

where the function β is related to h as follows:

$$[\tilde{\beta}(\mathbf{q})]^{-1} = [\tilde{h}(\mathbf{q})]^{-1} - [\tilde{g}(\mathbf{q}) - \rho]. \tag{4.40}$$

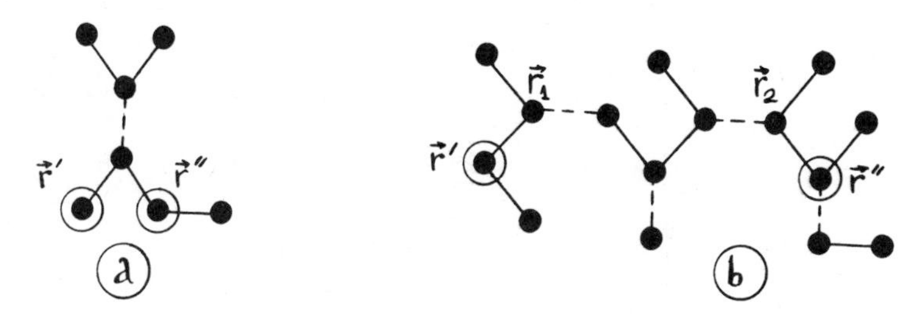

Figure 4.23. Two types of diagrams corresponding to two terms in Eq. (4.39).

It is not difficult to interpret Eqs. (4.39) and (4.40) using the diagrams. The correlation function in (4.39) is the sum of the contributions from all two-root diagrams taken into account in the approximation concerned. The roots at points $\mathbf{r}'$ and $\mathbf{r}''$ may belong to the same branch (Fig. 4.23*a*) or to different branches (Fig. 4.23*b*). The sum of all the diagrams of each type equals the corresponding term in (4.39). Actually, as in the SFPI approximation, the spatial interactions do not change the MSD at a fixed density, and the summed contributions from the diagrams of Fig. 4.23*b* are equal to the structure function, which is determined completely by the MSD for the ensemble of polymer molecules. For the same reason, the factors $g(\mathbf{r}' - \mathbf{r}_1)$ and $g(\mathbf{r}_2 - \mathbf{r}'')$ appear from summation of the contributions from the diagrams of Fig. 4.23*b*. Each factor corresponds to the part of the route connecting the diagram roots at the points $\mathbf{r}'$ and $\mathbf{r}''$, which passes the solid lines (chemical bonds) of the trunk (the polymer molecule) that grows from the corresponding root. The remaining third part of the route corresponds to the factor $\beta(\mathbf{r}_1 - \mathbf{r}_2)$, which is due to the spatial interaction of units belonging to different molecules and having coordinates $\mathbf{r}_1$ and $\mathbf{r}_2$. Those units can interact both directly and via units of other molecules (Fig. 4.10), which produces their mutual repulsion because of the excluded volume factor. A specific scale of these forces is the mean size of the polymer molecule in the system. As the mean size is increasing with conversion, the effective repulsion is suppressed, and the resulting total density correlator in (4.39) has no singularity at the gel formation point.

The diagram series that determines β is presented in Fig. 4.24. Its partial summation leads to a diagram equation for β (Fig. 4.24). As the structure is determined by the diagram series of Fig. 4.25, Eq. (4.40) appears in the spatially uniform system after the Fourier transformation.

Going beyond the SFPI approximation considered in the preceding, one has to account for diagrams where vertices (units) of the same branch (molecule) may be connected with a route passing a dashed line. As the order of the perturbation theory is increasing, the summation of the diagrams becomes rapidly more and more complicated. Complications arise also when the

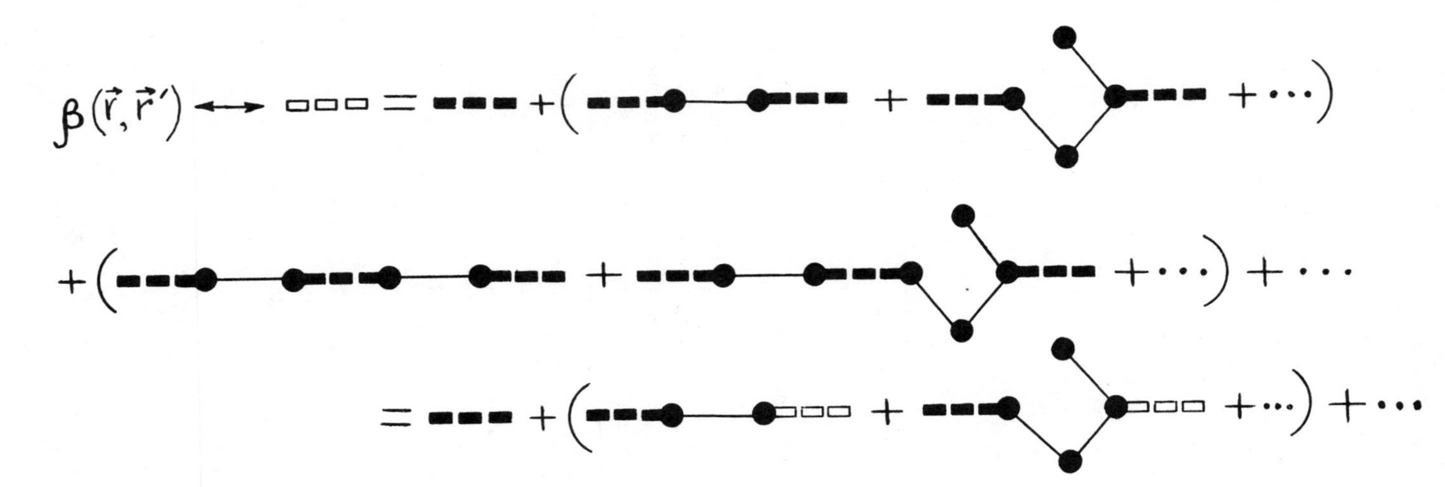

Figure 4.24. Graphical equation that determines function $\beta(\mathbf{r}, \mathbf{r}')$.

Figure 4.25. Diagram expansion of structure function.

infinite gel (condensate) molecule formed above the gel point is to be concerned; the latter is neglected completely in this section. Analytical methods of field theory are fairly effective in overcoming the difficulties. These methods also permit an analysis of the behavior of polymer systems in the presence of strong critical fluctuations.

E. Analytical Methods of Field Theory

It was shown in Section IV.A that in the framework of model III one gets an expression for the generating functional of the density correlatiors of separate molecules in terms of the functional integral (4.5) over the stochastic field $\varphi(\mathbf{r})$. To get from that integral, the generating functional for the analogous correlators in the system of chemical bonds, Eq. (4.16), one has to set $h_f = h_b = 0$ in functionals (4.3) and (4.6) and introduce a new field φ', which is the convolution of the old field φ and the function λ. As a result, (4.3) and (4.6) are transformed into

$$\mathscr{L}\{\varphi\} = \frac{1}{2L}\iint \Lambda_{uu}(\mathbf{r}' - \mathbf{r}'')\varphi(\mathbf{r}')\varphi(\mathbf{r}'')\, d\mathbf{r}'\, d\mathbf{r}'',$$

$$\int \Lambda_{uu}(\mathbf{r}' - \mathbf{r})\lambda_{uu}(\mathbf{r} - \mathbf{r}'')\, d\mathbf{r} = \delta(\mathbf{r}' - \mathbf{r}''), \tag{4.41}$$

$$\mathscr{M}\{\varphi\} = \int d\mathbf{r}\, z(\mathbf{r})[1 + \varphi(\mathbf{r})]^f, \qquad z(\mathbf{r}) = M \exp[-H(\mathbf{r})/T], \tag{4.42}$$

where for brevity we omit the prime on the new field with respect to which averaging is implied in (4.5). Thus, using (3.24), we get the generating functional for the system of chemical bonds,

$$\Psi_{\text{cb}}\{s\} = -\Omega_{\text{cb}}\{H - T\ln s\}/T = \ln\langle \exp[\textstyle\int d\mathbf{r}\, z(\mathbf{r})s(\mathbf{r})[1 + \varphi(\mathbf{r})]^f]\rangle_\varphi. \tag{4.43}$$

The derivation of an analogous result in the theory taking into account the physical interactions between the units is a more difficult problem, since the first equality in (4.43) does not hold in this case. Therefore, in model IV, unlike the previous case, Eq. (4.43), the evaluation of the generating functional $\Psi\{s\}$ is not reduced to calculation of the Ω potential for the system in the effective external field $H(\mathbf{r}) = -T\ln\tilde{s}(\mathbf{r}) = H(\mathbf{r}) - T\ln s(\mathbf{r})$. In the presence of spatial interactions, as well as without them, $-\Omega\{H\}/T$ is, by definition, the generating functional for the irreducible correlators of the total density. The expression for the generating functional $\Psi\{s\}$ in terms of the functional integral for model IV was obtained originally by Panyukov,[134] who, for that purpose, used the replics method[189] well known in the theory of disordered systems. The method we propose here to derive an expression for $\Psi\{s\}$ is in essence similar to that in ref. 134, but it permits one to reveal an important analogy between the descriptions of polymer and percolation systems.[135]

An important formula stems from combining (4.15) with (4.20);

$$\mathscr{P}(\mathscr{G}_N\{\mathbf{r}_i\}|\{H\}) = \exp\left[\frac{\Omega(H)}{T}\right]\left\langle \exp\left[-\frac{\Omega_{\text{cb}}\{H+v\}}{T}\right]\mathscr{P}_{\text{cb}}(\mathscr{G}_N\{\mathbf{r}_i\}|\{H+v\})\right\rangle_v \tag{4.44}$$

Thus, the problem for the system with spatial interactions is reduced to another problem where such interactions are absent but there is a stochastic Gaussian field $v(\mathbf{r})$. In view of Eq. (4.44), any stochastic functional $W(\mathscr{G}_N\{\mathbf{r}_i\})$ determined in the set of the system states $\mathscr{G}_N\{\mathbf{r}_i\}$ can be averaged with the probability measure of model IV in two stages. First, W can be averaged with the Gibbs distribution for the system of chemical bonds [Eq. (4.16)]; afterward, it can be averaged over configurations of the stochastic field v, with the probability measure specified by the Lagrangian of Eq. (4.19). Setting $W = 1$, from Eq. (4.44) and taking into account Eqs. (4.43) and (4.18), one gets an expression for the partition function,

$$\begin{aligned}\exp\left[-\frac{\Omega\{H\}}{T}\right] &= \left\langle \exp\left[-\frac{\Omega_{\text{cb}}\{H+v\}}{T}\right]\right\rangle_v \\ &= \left\langle \left\langle \exp\left[\int z(\mathbf{r})\exp\left(-\frac{v(\mathbf{r})}{T}\right)(1+\varphi)^f\,d\mathbf{r}\right]\right\rangle_\varphi\right\rangle_v \\ &= \left\langle \exp\left[-\frac{\Omega_{\text{su}}\{z(\mathbf{r})(1+\varphi)^f\}}{T}\right]\right\rangle_\varphi .\end{aligned} \tag{4.45}$$

The latter equality, presented originally by Erukhimovich[133] with no derivation, results from the possibility of permuting the averagings over the fields φ and v.

By definition, the generating functional $\Psi\{s\}$ equals the average over the Gibbs ensemble, Eq. (4.15), of the connectedness functional,

$$W(\mathscr{G}_N\{\mathbf{r}_i\}|\{s\}) = \sum \exp\left[\int \ln s(\mathbf{r})\rho^M_{(l,q)}(\mathbf{r})\,d\mathbf{r}\right], \tag{4.46}$$

where the sum is over all the (l, q)-isomers, their microscopic densities being $\rho^M_{(l,q)}(\mathbf{r})$. According to Eq. (4.43), averaging this functional over the Gibbs distribution $\mathscr{P}_{\text{cb}}$ leads to the quantity $[-\Omega_{\text{cb}}\{H + v - T\ln s\}/T]$, so that in view of (4.44) one can get an expression for the desired generating functional,

$$\begin{aligned}\Psi\{s\} &= \left\langle -\frac{\Omega_{\text{cb}}\{H+v-T\ln s\}}{T}\exp\left[-\frac{\Omega_{\text{cb}}\{H+v\}}{T}\right]\right\rangle_v \Bigg/ \\ &\quad \cdot\left\langle \exp\left[-\frac{\Omega_{\text{cb}}\{H+v\}}{T}\right]\right\rangle_v \\ &= -\frac{1}{T}\frac{d\Omega_n}{dn}\bigg|_{n=0}, \qquad \Omega_n = -T\ln \Xi_n,\end{aligned} \tag{4.47}$$

where we have used the following notation;

$$\Xi_n = \left\langle \exp\left[-\frac{(\Omega_{\text{cb}}\{H+v\} + n\Omega_{\text{cb}}\{H+v-T\ln s\})}{T}\right]\right\rangle_v. \tag{4.48}$$

Following the replics method,[189] we assume in the following that n is a positive integer, setting it equal to zero only in the final formulas. For an integer n, the expression in angular brackets in Eq. (4.48) is the product of $1+n$ partition functions for the system of chemical bonds in the presence of the corresponding external field. Each ith partition function can be represented, according to (4.43), as an exponential functional averaged over the field φ_i, so that it is possible to write the product of the partition functions in the same form;

$$\left\langle \exp\left\{\int d\mathbf{r}\, z(\mathbf{r})\exp\left(-\frac{v}{T}\right)\left[(1+\varphi_0)^f + s(\mathbf{r})\sum_{i=1}^{n}(1+\varphi_i)^f\right]\right\}\right\rangle_{\varphi}. \tag{4.49}$$

Unlike (4.43), here we average over the $(1+n)$-component stochastic replic field $\boldsymbol{\varphi} = (\varphi_0, \varphi_1, \ldots, \varphi_i, \ldots, \varphi_n)$, and the corresponding probability measure is given by the Lagrangian,

$$\mathscr{L}_r\{\boldsymbol{\varphi}\} = \sum_{i=0}^{n}\mathscr{L}\{\varphi_i\}, \tag{4.50}$$

where the functional $\mathscr{L}$ was defined in Eq. (4.41). Setting the expression of (4.49) into the angular bracket in (4.48) and changing the order of averaging over stochastic fields, one gets, taking into account (4.18), the basic formula

$$\Xi_n = \left\langle \exp\left[-\frac{\Omega_{su}\{\hat{z}(\mathbf{r})\}}{T} \right] \right\rangle_{\varphi}, \quad \hat{z}(\mathbf{r}) = z(\mathbf{r}) \left[(1+\varphi_0)^f + s(\mathbf{r}) \sum_{i=1}^{n} (1+\varphi_i)^f \right] \tag{4.51}$$

Combined with (4.47), this is formally an exact solution for the generating functional of the molecular density correlators in the framework of the Lifshitz–Erukhimovich model.

In cases where one has to label the polymer molecules not only with the number of the constituent monomer units but also with the numbers of the functional groups and chemical bonds, one can introduce the corresponding counters s_f and s_b into the functional (4.51). For this purpose, it is sufficient to replace $(1+\varphi_i)$ for $(s_f+\varphi_i)$ in the expressions for z and to multiply the parameter L [Eq. (4.41)] by s_b in all the terms in (4.50), except for the first one, $\mathscr{L}\{\varphi_0\}$.[134]

F. Self-Consistent Field Approximation

As we have already mentioned in Section IV.A, this approximation corresponds to the evaluation of the field-theoretic functional integral by means of the steepest descent method. In model IV, it is sufficient to deal with integral (4.45), which determines the system Ω potential, to obtain the equation of state and correlators of the total density of units. In the framework of the considered SF approximation, it is

$$\Omega\{z\} = \Omega_{su}\{z_{su}\} + T\mathscr{L}\{\bar{\varphi}\}, \qquad z_{su}(\mathbf{r}) = z(\mathbf{r})(1+\bar{\varphi})^f, \tag{4.52}$$

where the extremal field configuration $\bar{\varphi}(\mathbf{r})$ provides a minimum for the thermodynamic potential,

$$\frac{\delta\Omega\{z\}}{\delta\bar{\varphi}(\mathbf{r})} = \frac{\delta\Omega_{su}\{z_{su}\}}{\delta z_{su}(\mathbf{r})} z(\mathbf{r}) f(1+\bar{\varphi})^{f-1} + \frac{T}{L}\int \Lambda_{uu}(\mathbf{r}-\mathbf{r}')\bar{\varphi}(\mathbf{r}')\,d\mathbf{r}' = 0. \tag{4.53}$$

This equation, taking into account (4.22), coincides exactly with the analogous equation for the system of chemical bonds, Eq. (4.41), where the self-consistent field for the physical interactions, μ^*, is added to the external field H. The self-consistent density distribution in the system is determined by the thermodynamic equality,

$$-T\rho(\mathbf{r}) = z(\mathbf{r})\delta\Omega\{z\}/\delta z(\mathbf{r}) = z_{su}(\mathbf{r})\delta\Omega_{su}\{z_{su}\}/\delta z_{su}(\mathbf{r}), \tag{4.54}$$

which is used to get z_{su} and, consequently, the dependence $\Omega_{su}\{\rho\}$. It is not difficult to see that Eq. (4.52) coincides with Eq. (4.35) for the Ω potential, which was obtained within the SFPI approximation, since the second term in (4.52) is just the difference $\Omega_{cb} - \Omega_{ig}$. The ρ dependence of this term, TN_b (which is, up to a factor T, the total number of chemical bonds in the system) is given by Eq. (4.53), which can be represented in another form because of Eq. (4.54):

$$f\rho(\mathbf{r}) = [1 + \bar{\varphi}(\mathbf{r})] \int \Lambda_{uu}(\mathbf{r} - \mathbf{r}')\bar{\varphi}(\mathbf{r}')\, d\mathbf{r}'/L. \tag{4.55}$$

This is the local analogue of the stoichiometric formula (3.2), to which Eq. (4.55) is reduced after both its parts are integrated over the system volume.

Thus, the equalities (4.53) and (4.54), together with conditions (3.9) for determination of parameters M and L, are a closed set of equations for the thermodynamic characteristics and correlators of the total density of units. The problem is especially easy for a spatially uniform system concerned in what follows.

For the uniform system, Eqs. (4.55) and (3.9) lead to simple expressions for the densities of units, bonds, and conversion (3.10),

$$\rho = \bar{\varphi}(1 + \bar{\varphi})/fL, \qquad \rho_b = \bar{\varphi}^2/2L, \qquad p = \bar{\varphi}/(1 + \bar{\varphi}), \tag{4.56}$$

which make it possible to write Eq. (4.52) as an equation of state for the considered equilibrium polymer system,

$$P(\rho) = P_{su}(\rho) - T\rho f p/2, \qquad L = p/f\rho(1 - p)^2. \tag{4.57}$$

Here the second equality is obtained from (4.55) and (4.56); it is the mass action law for the reaction between the functional groups, and parameter L coincides with the equilibrium constant. It is clear from the equation of state, Eq. (4.57), that the chemical binding of groups is equivalent to the effective attraction of units and can induce a phase transition even under conditions where it is absent in the system of separate units. Condition $\delta P/\delta\rho = 0$, which determines the boundary of the system thermodynamic stability, when applied to (4.57), results in a simple expression for conversion at the spinodal,

$$p_{sp} = \frac{1}{\nu f - 1}, \qquad \frac{1}{\nu} = \frac{\partial P_{su}}{T\partial\rho} = \frac{\rho}{T}\frac{\partial\mu_{su}}{\partial\rho}. \tag{4.58}$$

In the absence of spatial interactions, $P_{su} = \rho T$, $\nu \equiv 1$, so that $p_{sp} = 1/(f - 1) = p^*$, and the thermodynamic stability in the SF approximation is lost at the gel point. With an account of the physical interactions between the units depending on their character [i.e., on the form of the function $P_{su}(\rho)$], gel formation may happen both below and above the phase separation point.

The variational differentiation of the thermodynamic potential in (4.52) with respect to the field $H(\mathbf{r})$ (by analogy to the procedure described in Section IV.D) enables one to get an integral equation for the total density correlator $\theta(\mathbf{r}-\mathbf{r}')$. Its Fourier transform, obtained from this equation, is

$$\tilde{\theta}^{-1}(\mathbf{q}) = \tilde{g}^{-1}(\mathbf{q}) + \tilde{\theta}_{su}^{-1}(\mathbf{q}) - \rho^{-1}, \tag{4.59}$$

where $g(\mathbf{q})$, which coincides with the structure function below the gel formation point, is given by

$$\tilde{g}(\mathbf{q}) = \rho + \rho p f / [\tilde{\lambda}_{uu}^{-1}(\mathbf{q}) - p(f-1)]. \tag{4.60}$$

The correlation function in (4.59), obtained originally in ref. 188, coincides with that in Eq. (4.38); the fact is quite natural since Eq. (4.52), which determines the Ω potential, coincides with Eq. (4.35). In the presence of the spatial interactions, in general, the Fourier transform at its zero argument, $\tilde{\theta}(0)$, has no singularity at the gel formation point but increases infinitely as the system approaches the spinodal, where the r.h.s. of Eq. (4.59) is vanishing.

To describe gel formation and MWD for polymers, one should consider the generating functional for the molecular density correlators, Eq. (4.51). Even working within the SF approximation, one is able to obtain a number of important results,[134] in particular, to verify the validity of the Flory theory (both below and above the gel point) in the presence of spatial interactions. Among the results of this type, one should mention the extraction of a contribution from the gel fraction out of sol molecules, performed in ref. 134, which enabled that author to obtain for the first time the density correlation function for units of a polymer mesh. It was found that this function is described with "asymmetric" solutions of equations of the mean-field theory, which are different from the solution that is symmetric with respect to replics and corresponds to sol molecules. The equations are

$$\frac{\delta \phi_n}{\delta \varphi_i(\mathbf{r})} = \frac{\delta \Omega_{su}\{\hat{z}\}}{T \delta \hat{z}(\mathbf{r})} z(\mathbf{r}) f (1+\varphi_i)^{f-1} \sigma_i + \frac{1}{L} \int \Lambda_{uu}(\mathbf{r}-\mathbf{r}') \varphi_i(\mathbf{r}') \, d\mathbf{r}' = 0; \tag{4.61}$$

$$\sigma_i = \delta_{i0} + (1-\delta_{i0}) s \quad (i = 0, 1, \ldots, n);$$

$$-T\phi_n = \Omega_{su}\{\hat{z}\} + T \sum_{i=0}^{n} \mathscr{L}\{\varphi_i\}, \tag{4.62}$$

They are maximum conditions for the functional ϕ_n in the set of various configurations of the $(n+1)$-component replic field and have always a symmetric spatially uniform solution,

$$\varphi_0^{(0)} = \bar{\varphi}^s, \qquad \varphi_i^{(0)} = \bar{\varphi}_-^s \qquad (i = 1, 2, \ldots, n). \tag{4.63}$$

The solution $\varphi^{(0)}$, where all the components of the replic field are identical, except for the zeroth component, corresponds for $s \neq 1$ to the symmetry group of the functional ϕ_n with respect to component permutations. If the conversions are low enough (i.e., under the condition $\rho L \ll 1$), the symmetric solution (4.63) is unique, since in that case all $\varphi_i \ll 1$ are determined from the solution of linear equations that stem from (4.61) as $L \to 0$,

As the conversion increases and attains a "critical level," $p = p^*$, a "bifurcation" takes place, giving rise to n asymmetric solutions separating from the solution (4.63), symmetric in replics,

$$\varphi_0^{(j)} = \bar{\varphi}^a, \qquad \varphi_i^{(j)} = \bar{\varphi}_+^a \delta_{ij} + \bar{\varphi}_-^a (1 - \delta_{ij}) \qquad (i, j = 1, 2, \ldots, n). \tag{4.64}$$

In each solution there is a component $\bar{\varphi}_+^a$ that is larger than all the other components, φ_-^a. The appearance of the solution (4.64), having a symmetry group smaller than that of the functional ϕ_n, at the gel point is called "spontaneous symmetry breaking," a phenomenon well known in statistical physics.[85] It is a phase transition in a system with an order parameter proportional to the difference $\bar{\varphi}_+ - \bar{\varphi}_-$, which is zero identically for the symmetric solution. For equations (4.61) at $n = 0$ the latter is

$$\bar{\varphi}^s(s) = \frac{p}{1-p}, \qquad \bar{\varphi}_+^s(s) = \bar{\varphi}_-^s(s) = \frac{pu(s)}{1-p}. \tag{4.65}$$

Here $u(s)$ is a root of Eq. (1.12) satisfying $u(0) = 0$; its magnitude for $s = 1$ will be denoted simply by u.

The order parameter is nonzero for the asymmetric solution of Eqs. (4.61),

$$\bar{\varphi}_{(s)}^a = \frac{\beta}{1-\beta}, \qquad \bar{\varphi}_+^a(s) = \frac{\beta V_+(s)}{1-\beta}, \qquad \bar{\varphi}_-^a(s) = \frac{\beta V_-(s)}{1-\beta}, \tag{4.66}$$

where the dependencies of three functions β, V_+, V_- $(V_+ > V_-)$ on two variables s and p are given by

$$V_\pm = s(1 - \beta + \beta V_\pm)^{f-1}, \quad 1 + s[(1 - \beta + \beta V_+)^f + (n-1)(1 - \beta + \beta V_-)^f]$$
$$= p(1-\beta)^2/\beta(1-p)^2. \tag{4.67}$$

For $s = 1$ and $n = 0$, the solution of Eqs. (4.67).

$$\beta = pu/(1 - p + pu), \qquad V_+ = 1/u, V_- = 1, \tag{4.68}$$

being set into (4.66), provides the important relations

$$\bar{\varphi}^a(1) = \frac{pu}{1-p}, \qquad \bar{\varphi}^a_+(1) = \frac{p}{1-p}, \qquad \bar{\varphi}^a_-(1) = \frac{pu}{1-p}, \tag{4.69}$$

which are necessary for the calculation of the correlation functions for the gel.

We now turn to the evaluation of the generating functional (4.47) in the SF approximation. Calculating the functional integral in (4.51) by means of the steepest descent method, one gets

$$-\frac{\Omega_n}{T} \equiv \ln \Xi_n = \ln[\exp \phi^s_n + n \exp \phi^a_n], \tag{4.70}$$

where the functional defined in (4.62) is given by

$$\begin{aligned} \phi_n &= -\Omega_{\text{su}}\{\hat{z}\}/T - \mathscr{L}\{\bar{\varphi}\} - \mathscr{L}\{\bar{\varphi}_+\} - (n-1)\mathscr{L}\{\bar{\varphi}_-\}, \\ \hat{z} &= z[(1+\bar{\varphi})^f + s(1+\bar{\varphi}_+)^f + s(n-1)(1+\bar{\varphi}_-)^f, \end{aligned} \tag{4.71}$$

and the superscript indicates the solution of (4.61) for which the functional is being evaluated. Setting (4.70) into (4.47), we obtain expressions for the generating functionals of the density coorelators for sol molecules, $\Psi^{(s)}$, and gel molecules, $\Psi^{(g)}$,

$$\Psi\{s\} = \Psi^{(s)} + \Psi^{(g)}, \qquad \Psi^{(s)}\{s\} = \left.\frac{d\phi^s_n}{dn}\right|_{n=0}, \qquad \Psi^{(g)}\{s\} = \exp(\phi^a_0 - \phi^s_0). \tag{4.72}$$

Naturally, $\Psi^{(g)}$ is zero below the gel formation point, since there are no asymmetric solutions at that stage of polycondensation. It is possible, however, to use formally the expressions in (4.70) and (4.72) also below the gel point, adopting there a definition, $\bar{\varphi}^a_+ = \bar{\varphi}^a_-$, for the asymmetric solutions so that they coincide with the symmetric solutions.

When differentiating with respect to n in calculating $\Psi^{(s)}\{s\}$, Eq. (4.72), one should take into account only the explicit n dependence of the functional ϕ^s_n, since to solve Eq. (4.61), the partial derivatives with respect to φ_i vanish. The unit counters s in the functional $\Psi^{(s)}\{s\}$ obtained in this way must be treated as coordinate independent, and the bond counters s_b are introduced into the functional and into Eq. (4.61) for $\bar{\varphi}_+$ and $\bar{\varphi}_-$ by substituting L for $s_b L$. The result,

$$\Psi^{(s)}(s, s_b) = N[s(1 - p + pu(s, s_b))^f - pfu^2(s, s_b)/2s_b],$$
$$u(s, s_b) = ss_b[1 - p + pu(s, s_b)]^{f-1}, \quad (4.73)$$

coincides, up to a normalization factor, with the generating functional for numerical MWDs, (1.14). In view of (4.73), in the considered SF approximation, there are only tree molecules, and in each of them the number of bonds equals the number of units minus 1. Equality (4.73) is valid both below and above the gelformation point, according to the method we employed for its derivation. Calculating the derivatives of (4.73) with respect to s and s_b at the point $s = s_b = 1$, one gets the known relations[2] between the density $\rho^{(s)}$ and conversion $p^{(s)}$ for the sol and for the whole polycondensation system,

$$\rho^{(s)} = \rho(1 - p + pu)^f, \qquad p^{(s)} = pu^{(f-2)/(f-1)}. \quad (4.74)$$

Thus, in the Lifshitz–Erukhimovich model the SF approximation produces exactly the same results as those of the Flory theory throughout the range of conversion values. In the framework of this approximation, not only the molecular mass characteristics but also correlations between separate polymer molecules coincide with those evaluated in model I. This important result is due to the fact that the probability measure in the set of configurations and conformations of separate molecules in the ensemble we deal with in the SF approximation for model IV coincides with the measure for the ensemble corresponding to ideal polycondensation (model I). This fact is verified if we calculate, say, the second variational derivative of the generating functional $\Psi^{(s)}\{s\}$ with respect to the virtual field $h^u(\mathbf{r}) = \ln s(\mathbf{r})$. The resulting Fourier transform of the generating functional for the two-point correlators, (1.25), for the density of sol molecule units is given by

$$\tilde{\chi}^{(s)}(\mathbf{q}, s) = \rho s \xi^f + \frac{\rho \lambda_{uu}(\mathbf{q}) s^2 pf \xi^{2f-2}}{1 - \tilde{\lambda}_{uu}(\mathbf{q}) sp(f-1)\xi^{f-2}}, \qquad \xi = 1 - p + pu(s). \quad (4.75)$$

This formula can be rewritten in a form equivalent to (3.59). Setting $s = 1$, we get from (4.75) the sol structure function $g^{(s)}(\mathbf{q})$, which coincides with (4.60) up to the gel formation point. It has a singularity at the gel point and is different from the correlator for the total density of units in the system, Eq. (4.59), in the presence of spatial interactions.

Expression for the generating functional $\Psi^{(g)}\{s\}$ in Eq. (4.72) provides the correlator for the gel, which consists of a single giant molecule occupying the total volume of the reaction system. Actually, as for $s = 1$, the condition holds that $\phi_0^a = \phi_0^s$, meaning that the pressures in the sol and gel are equal, and we have $\Psi^{(g)}\{1\} = 1$. This condition is valid because of Eq. (4.71) and Eqs. (4.65) and (4.69). The thermodynamic limit (where the number of particles in the system and its volume are going to infinity for a finite density) is inadequate in Eq. (4.72), since for $s < 1$ the gel contribution to the generating functional $\Psi\{s\}$ is lost. For all derivatives of the functional, however, the limit can be calculated in the standard manner, a fact that is important if we are concerned with gel correlators. The one-point correlators provide the known expressions, Eq. (1.29),[134,190] for the numbers of units and independent cycles in the gel. One should bear in mind that the cycles in the gel are taken into account, even though we work within the mean-field theory, since the term $\Psi^{(g)}$ in Eq. (4.72) cannot be obtained, in principle, by means of the high-temperature perturbative expansion. Calculating the second derivative of its logarithm with respect to $\ln s(\mathbf{r})$, we get the Fourier transform of the irreducible correlator of the density of units,

$$\tilde{\theta}^{(g)}(\mathbf{q}) = \tilde{g}^{(g)}(\mathbf{q}) + \tilde{g}^{(g)}(\mathbf{q})\tilde{\beta}(\mathbf{q})\tilde{g}^{(g)}(\mathbf{q}), \qquad \tilde{g}^{(g)}(\mathbf{q}) = \tilde{g}(\mathbf{q}) - \tilde{g}^{(s)}(\mathbf{q}), \tag{4.76}$$

where the function β is given in Eq. (4.40). The sol structure function is, by definition, $g^{(s)}(\mathbf{q}) = \chi^{(s)}(\mathbf{q}, 1)$; it is obtained from Eq. (4.75), and $g(\mathbf{q})$ is given by Eq. (4.60). In obtaining $\theta^{(g)}$ by differentiation of the functional $\Psi^{(g)}$, we must take into account both the explicit dependence on $s(r)$ and the implicit dependence via the fields $\bar{\varphi}$, $\bar{\varphi}_+$, and $\bar{\varphi}_-$. The arising variational derivatives are determined from a set of three linear equations, which are obtained by the term-by-term differentiation of Eqs. (4.61).

It is not difficult to see that Eq. (4.76) is identical to Eq. (4.39), solely with g substituted for $g^{(g)}$. It may be supposed that the similarlity is inexplicable, since Eq. (4.39) has been derived by summing the contributions from the tree graphs only, whereas Eq. (4.76) describes gel-containing cyclic elements. Nevertheless, the similarity is not accidental; it can be explained by means of field-theoretic methods for systems containing a condensate,[180] and the gel plays the role of condensate in this case. It is shown in the preceding approach that calculating correlation functions for such systems in the framework of the SF approximation, one can consider, instead of a cyclic diagram having a complicated topology corresponding to the condensate in the thermodynamic limit, an equivalent set of infinite chord tree graphs. The latter are obtained from the graph by cutting all cyclic edges in all possible ways. Thus, the problem of calculating $\theta^{(g)}$ is reduced to calculating the correlational function of density of units for infinite trees in an ensemble containing only finite and infinite tree sol molecules. For that ensemble, the role of the structure function is played by g in Eq. (4.60), which is the sum of the sol structure function $g^{(s)}$ and the function $g^{(g)}$, relevant to the chord trees characterizing gel. The same arguments suggest that Eq. (4.39) is valid even above the gel point. Because g describes the total density correlations, it is involved instead of $g^{(g)}$. By the meaning of the function β in (4.40), which determines the screening of interaction between any pair of units from different molecules, whether these molecules are finite or infinite. Therefore, the function β is present in the same manner in both the equalities (4.39) and (4.76).

Note that the expression for the inverse correlator of the density of the gel units can be written in a form similar to (4.59),

$$[\tilde{\theta}^{(g)}(\mathbf{q})]^{-1} = [\tilde{g}^{(g)}(\mathbf{q})]^{-1} + [\tilde{\theta}^{(g)}_{\rm su}(\mathbf{q})]^{-1} - (\rho^{(g)})^{-1}, \tag{4.77}$$

where $\theta^{(g)}_{\rm su}$ is given by

$$[\tilde{\theta}^{(g)}_{\rm su}(\mathbf{q})]^{-1} = (\rho^{(g)})^{-1} + [\tilde{h}^{-1}(\mathbf{q}) + \rho - \tilde{g}^{(s)}(\mathbf{q})]^{-1}. \tag{4.78}$$

It was noted[191] that this function has the meaning of the correlator for the density of monomers produced from breaking all chemical bonds in the gel, the molecules of which have physical interactions with each other and with sol molecules. Below the gelformation point $\theta^{(g)}$ is zero, and for $p \to 1$ it tends toward the total density correlator for units, Eq. (4.59). The correlation function (4.76) has singularities at the spinodal, $p = p_{\rm sp}$, and at the gel point, $p = p^*$. In the vicinity of the latter, $\tau \equiv |1 - p/p^*| \ll 1$, but sufficiently far from the former, $|\tau| \gg 1 - p/p_{\rm sp}$, the correlator $\theta^{(g)}$ is independent of a particular form of the physical interaction potential; it is given by an asymptotic formula in the three-dimensional space,

$$\theta^{(g)}(\mathbf{r}) = \frac{f}{4\pi(f-1)} \frac{\rho|\tau|^{1/2}}{a^3} \exp\left(-\frac{r}{\xi_c}\right). \tag{4.79}$$

At a scale substantially smaller than the correlation length $\xi_c = a/|\tau|^{1/2}$ (i.e., $r \ll \xi_c$), the function in Eq. (4.79) has a rather unconventional behavior. Unlike the standard $r \to 0$ asymptotics, $\theta(\mathbf{r}) \sim r^{-1}$, given by the Ornstein–Zernike formula,[185,186] the expression in Eq. (4.79) has a constant limit. It is responsible for the smallness of the relative gel density fluctuations for $r \lesssim \xi_c$,

$$\frac{\langle(\delta\rho^{(g)})^2\rangle}{(\rho^{(g)})^2} = \left(\frac{Gi}{|\tau|}\right) \ll 1, \qquad Gi = \left[\frac{(f-2)^2}{16\pi f(f-1)\rho a^3}\right]^{2/3}, \tag{4.80}$$

which characterizes the domain of applicability of the SF approximation. As the Ginzburg number Gi in Eq. (4.80) has a small numerical factor, the fluctuational region of conversions near the gel point, where the SF approximation is invalid, is sufficiently narrow even for $\rho a^3 \sim 1$.

Equations (4.61) for the field components are necessary conditions for the maximum of the functional ϕ_n in (4.62). The relevant sufficient condition is the positive definiteness of an integral operator with the kernel,

$$\begin{aligned}\mathscr{K}_{ij}(\mathbf{r}'-\mathbf{r}) &\equiv L\delta^2\phi_n/\delta\varphi_i(\mathbf{r}')\delta\varphi_j(\mathbf{r}) = \gamma_i(\mathbf{r}'-\mathbf{r})\delta_{ij} - \varphi_i\varphi_j h_{\mathrm{su}}(\mathbf{r}'-\mathbf{r})/L, \\ \gamma_i(\mathbf{r}'-\mathbf{r}) &= \Lambda_{uu}(\mathbf{r}'-\mathbf{r}) - \delta(\mathbf{r}'-\mathbf{r})(f-1)\varphi_i/(1+\varphi_i),\end{aligned} \tag{4.81}$$

on the solutions of Eq. (4.61). In a spatially uniform system, the eigenfunctions of this operator are plane waves, $\tilde{\varphi}_j(\mathbf{q})\exp(i\mathbf{qr})$, and the corresponding continuous spectrum of the eigenvalues, $\lambda(\mathbf{q})$, is determined by the roots of a characteristic equation,

$$D_\lambda(\mathbf{q}) \equiv |\tilde{\mathscr{K}}_{ij} - \lambda\delta_{ij}| = 0. \tag{4.82}$$

As the operator with the kernel of Eq. (4.81) is Hermitian, all of its eigenvalues are real. They are all positive at small conversions; that is, the operator is positive definite. As the conversion p increases, however, the minimum eigenvalue vanishes at the bifurcation point. The latter belongs to one of two types, depending on whether the positive definiteness is violated when p passes the bifurcation value. If this is the case, we have a spinodal transition; otherwise, the transition is the gelformation. To verify this fact, one has to calculate the determinant in (4.82),

$$D_\lambda(\mathbf{q}) = \prod_{i=0}^{n}[\tilde{\gamma}_i(\mathbf{q}) - \lambda]\left[1 - \frac{\tilde{h}_{\mathrm{su}}(\mathbf{q})}{L}\sum_{i=0}^{n}\frac{\varphi_i^2}{\tilde{\gamma}_i(\mathbf{q}) - \lambda}\right]. \tag{4.83}$$

As soon as we consider only spatially uniform solutions, it is sufficient to set $\mathbf{q} = 0$.

Consider first the stability of the symmetric solution (4.63), for which we get the determinant in (4.83) at integer n,

$$D_\lambda = (\tilde{\gamma}_1 - \lambda)^{n-1}\{(\tilde{\gamma}_0 - \lambda)(\tilde{\gamma}_1 - \lambda) - [\bar{\varphi}^2(\tilde{\gamma}_1 - \lambda) + n\bar{\varphi}_-^2(\tilde{\gamma}_0 - \lambda)]\tilde{h}_{\mathrm{su}}/L\}. \tag{4.84}$$

According to (4.84), Eq. (4.82) has two n-dependent simple roots λ_{I} and λ_{II}, which correspond to vanishing the quadratic trinomial in braces, and (for $n > 1$) an extra $(n-1)$-tuple degenerate root $\lambda_{\mathrm{III}} = \tilde{\gamma}_1$. The latter must be discarded since we are concerned with the limit $n \to 0$, as the method of replics implies. In the limit, the determinant (4.82) for $s = 1$ does vanish only for $\lambda = \lambda_{\mathrm{I}} = 1 - p^{(s)}/p^*$ and $\lambda = \lambda_{\mathrm{II}} = 1 - p/p_{sp}$. The first root, λ_{I}, is positive for all conversion values, except for the gel point, where it vanishes. Therefore, the positive definiteness of operator (4.81) is violated only at the spinodal, $p = p_{sp}$. In the case where $p^* < p_{sp}$, the bifurcation point at $\lambda_{\mathrm{I}} = 0$ corresponds to the formation of infinite gel molecules; in mathematical terms, it is the appearance of asymmetric solutions (4.69). Equality (4.84), according to which the gel cannot undergo a transition until the spinodal point is attained, is still valid for those solutions. In the inverse situation, $p_{sp} < p^* = 1/(f-1)$, the spinodal transition to a new phase having its specific values of ρ and p takes place even before gel formation. The investigation of gel formation and thermodynamic transitions in that phase can be performed along exactly the same line as that exposed in the preceding. For $0 < s < 1$, $\lambda_{\mathrm{I}}(s) > 0$ on the symmetric solution, so that this solution does indeed provide the maximum of the functional ϕ_n, Eq. (4.62), until $\lambda_{\mathrm{II}} = 1 - p/p_{sp}$ vanishes.

G. Multicomponent Systems and Extension of Potts Model

In model IV, the system state is specified by a spatial graph $\mathscr{G}_N\{\mathbf{r}_i\}$, namely, a set of numbers $N = \{N_\nu\}$ of different type-ν vertices having coordinates $\{\mathbf{r}_i\}$ and a set of bonds connecting the vertices. The probability Gibbs measure is determined in the set of the graphs,

$$\mathscr{P}(\mathscr{G}_N\{\mathbf{r}_i\}) = \mathscr{P}^{(1)}\mathscr{P}^{(2)}, \qquad \mathscr{P}^{(2)} = \prod_{(\alpha,\beta)}\prod_{(i,j)} L_{\alpha\alpha}\lambda_{\alpha\beta}(\mathbf{r}_i - \mathbf{r}_j),$$

$$\mathscr{P}^{(1)} = \prod_{\nu} M_\nu^{N_\nu} \exp\left(\left[\Omega - \sum_{\nu}\sum_{i=1}^{N_\nu} H_\nu(\mathbf{r}_i) - \sum_{(\nu,\mu)}\sum_{(i,j)} V_{\nu\mu}(\mathbf{r}_i - \mathbf{r}_j)\right]\Big/ T\right). \tag{4.85}$$

Here the first factor represents positions of vertices in space, and the second factor is determined by the chemical bonds coupling the vertices. Equality (4.85) is the Lifshitz–Erukhimovich description of a chemical equilibrium for an ensemble of macromolecules produced in the process of polycondensation of an arbitrary mixture of monomers $R_\nu A_1^{f_{\nu_1}} \cdots\ A_\alpha^{f_{\nu_\alpha}} \cdots$ ($\nu = 1, \ldots$), each containing $f_{\nu_1}, \ldots, f_{\nu_\alpha}, \ldots$ functional groups $A_1, \ldots, A_\alpha, \ldots$, respectively, besides the unit R_ν, and amounts to a fraction of α_ν in the whole mixture. A chemical bond coupling two groups A_α and A_β in Eq. (4.85) that appears in the course of the reaction corresponds to the product of its equilibrium constant $k_{\alpha\beta} = L_{\alpha\beta}$ and a function $\lambda_{\alpha\beta}$, which accounts for a spatial correlation between coordinates of two units coupled by the bond. The units R_ν having activities M_ν are under the action of an external field H_ν and interact with a pair potential $V_{\nu\mu}$.

An expression for the thermodynamic potential $\Omega\{z(\mathbf{r})\}$ (where $z_\nu(\mathbf{r}) = M_\nu \exp[-H_\nu(\mathbf{r})/T]$) was found for the equilibrium system in the functional integral representation.[133] Its variational derivatives with respect to the external field provide the total density correlators of units,

$$\rho_\nu(\mathbf{r}) = \frac{\delta\Omega}{\delta H_\nu(\mathbf{r})}, \qquad \theta_{\nu\mu}(\mathbf{r}', \mathbf{r}'') = \langle \delta\rho_\nu(\mathbf{r}')\delta\rho_\mu(\mathbf{r}'')\rangle = -T\frac{\delta^2\Omega}{\delta H_\nu(\mathbf{r}')\delta H_\mu(\mathbf{r}'')} \tag{4.86}$$

The first unit is a vector; the second is a matrix. The matrix can be obtained from solution of the integral equation (1.37), which has the same form for single- and multicomponent systems, whereas for the latter matrix, functions are involved. In particular, a simple equality that is an extension of Eq. (1.26) is valid for matrix elements of the direct correlation function,

$$-C_{\nu\mu}(\mathbf{r}', \mathbf{r}'') = \delta\mu_\nu^*(\mathbf{r}', \{\rho_1, \ldots, \rho_\alpha, \ldots\})/\delta\rho_\mu(\mathbf{r}'')T. \tag{4.87}$$

The functional μ_ν^* present here has the meaning of a self-consistent field acting on a monomer unit R_ν. In a spatially homogeneous system, matrices $\mathbf{C}$ and $\mathbf{g}$ depend on coordinate differences, $\mathbf{r}' - \mathbf{r}''$, only, so it is possible to get an extension of Eq. (1.38) starting from Eq. (1.37),

$$\tilde{\boldsymbol{\theta}}^{-1}(\mathbf{q}) = \tilde{\mathbf{g}}^{-1}(\mathbf{q}) - \tilde{\mathbf{C}}(\mathbf{q}), \tag{4.88}$$

which is a relation between the inverse correlation matrices for the systems with physical interactions and without them.

The generating functional $\Psi\{\mathbf{s}\}$ for the density correlations of units in separate molecules is, by definition, the connection functional averaged with the Gibbs measure (4.85),

$$W(\mathcal{G}_N\{\mathbf{r}_i\}|\{\mathbf{s}\}) = \sum \exp\left[\sum_\nu \int \ln s_\nu(\mathbf{r})\rho^m_{\nu(l,q)}(\mathbf{r})\, d\mathbf{r}\right], \tag{4.89}$$

where the summation over all molecules is implied, the microscopic density of units R_ν being $\rho^m_{\nu(l,q)}(\mathbf{r})$. The functional was calculated for a single-component system[134,191] by means of the replic method. The extension to an arbitrary multicomponent system[135] gives

$$\Psi\{\mathbf{s}\} \equiv \langle W\rangle = -\frac{1}{T}\frac{d\Omega_n}{dn}\bigg|_{n=0}. \tag{4.90}$$

The functional Ω_n is evaluated by averaging over the replic field $\boldsymbol{\varphi} = (\varphi_0, \ldots, \varphi_n)$

with the Lagrangian

$$\mathcal{L}_r\{\boldsymbol{\varphi}\} = \frac{1}{2}\sum_{i=0}^{n}\sum_{\alpha,\beta} L_{\alpha\beta}^{(-1)} \iint \Lambda_{\alpha\beta}(\mathbf{r}' - \mathbf{r}'')\varphi_{i\alpha}(\mathbf{r}')\varphi_{i\beta}(\mathbf{r}'')\, d\mathbf{r}'\, d\mathbf{r}'',$$
$$\sum_{\gamma} L_{\alpha\gamma} \int \lambda_{\alpha\gamma}(\mathbf{r}' - \mathbf{r})\Lambda_{\gamma\beta}(\mathbf{r} - \mathbf{r}'')\, d\mathbf{r}\, L_{\gamma\beta}^{(-1)} = \delta_{\alpha\beta}\delta(\mathbf{r}' - \mathbf{r}'') \tag{4.91}$$

for the partition function of the system of separate units,

$$-\frac{\Omega_n}{T} = \ln\left\langle \exp\left[-\frac{\Omega_{su}\{\hat{\mathbf{z}}\}}{T}\right]\right\rangle_{\varphi}, \tag{4.92}$$

with modified activities and external fields,

$$\hat{z}_v(\mathbf{r}) = M_v \sum_{i=0}^{n} \exp[-\hat{H}_{iv}(\mathbf{r})/T] \prod_{\alpha} [1 + \varphi_{i\alpha}(\mathbf{r})]^{f_{v\alpha}},$$
$$\hat{H}_{iv}(\mathbf{r}) = H_v(\mathbf{r}) - (1 - \delta_{i0})T \ln s_v(\mathbf{r}). \tag{4.93}$$

The elements of the matrix $\mathbf{L}^{-1}$, which is the inverse of the matrix $\mathbf{L}$ with elements $L_{\alpha\beta}$, are denoted by $L_{\alpha\beta}^{(-1)}$ in Eqs. (4.91).

The functional Ω_n is the thermodynamic potential of a model[135] described in the following, which is an extension of the Potts model, Eq. (1.60). Each unit R_v of type v is in one of $1 + n$ states specified by its number ("color"), $i = 0, 1, \ldots, n$. The unit number (color) i is assigned to all adjacent groups. In this model, only groups of an identical color can be coupled by a chemical bond, whereas the bond probability is independent of i, as well as the unit interaction energy $V_{v\mu}$. All the colored units, except for the "white" units (those with $i = 0$), are acted upon by a virtual field $-T \ln s_v(\mathbf{r})$, except for $H_v(\mathbf{r})$. Calculating the partition function, Eq. (4.92), for the model of colored units and "erasing" subsequently their colors with Eq. (4.90), one gets the desired functional $\Psi\{\mathbf{s}\}$.

A similar procedure, Eq. (1.61), revealed a relation between the probabilistic percolation model and the statistical Potts lattice model with $1 + n$ states.[118,119] The derivative of the free energy with respect to n per lattice site, taken at $n = 0$, is the generating function for the cluster distribution in the number of their sites, and the percolation transition corresponds to the second-kind phase transition in the Potts model. This correspondence enables one to employ the advanced methods of the theory of phase transitions and apply them to the solution of combinatorial geometry problems in percolation theory; a number of important results have been obtained along this line.[192,193]

An analogous correspondence was found in ref. 135 for equilibrium polymer systems consisting of any number of components that have arbitrary physical and chemical interactions in the framework of the Lifshitz–Erukhimovich model. Besides these extensions, particles are not necessarily put at lattice sites, unlike the Potts model.

To calculate the density correlators for units of separate molecules, one must evaluate the corresponding derivatives of the generating functional in Eq. (4.46) with respect to $\ln s_\nu(\mathbf{r})$. Permuting the differentiations of Ω_n with respect to these arguments [virtual fields, Eq. (4.93)] and n, one draws an interesting conclusion. It is found that to calculate the k-point correlator of the molecular density, one can get the corresponding correlator in the system of colored units and then erase their colors taking the derivative with respect to n for $n \to 0$. The same method is used in percolation theory to express the pair connection in terms of the derivative of the two-point correlation function of the Potts model with respect to the number of its components q for $q \to 1$.[120]

In the framework of the SF approximation, Eqs. (4.72) are still applicable to multicomponent systems; it is sufficient to increase the number of equations in the set (4.61) in proportion to the number of different types of functional groups in the reactive mixture. The calculation of one-point correlators yields known formulas for the generating function of its weight SCD,

$$G_W(\mathbf{s}) \equiv \sum_{l_\nu=1}^{\infty} f_W(\mathbf{l}) \prod_\nu s_\nu^{l_\nu} = \sum_\nu \frac{s_\nu}{\rho} \frac{\delta \Psi^{(u)}}{\delta s_\nu} = \sum_\nu \alpha_\nu s_\nu \prod_i \xi_i^{f_{i\nu}}, \tag{4.94}$$

where the s dependence of ξ_i is determined from

$$\xi_i = 1 - p_i + \sum_j \sum_\nu p_{ij} \sigma_{j\nu} s_\nu \prod_k \xi_k^{f_{k\nu} - \delta_{kj}}. \tag{4.95}$$

Here p_{ij} is the ratio of concentration of all chemical bonds produced in the reaction between groups A_i and A_j to the initial concentration of groups A_i. Their conversion p_i equals the sum of p_{ij} over all j, and $\sigma_{j\nu}$ is the fraction of j-type groups that were initially at the ν-type monomer; δ_{kj} is the Kronecker delta. Equations (4.94) and (4.95) are the extension of (1.12) to arbitrary multicomponent systems; they are valid also above the gelformation point, describing the sol fraction there. To calculate its SCD, one must find those solutions ξ_i of the set of equations (4.95) at $s = 1$ that are less than 1, and then introduce the modified generating functions. These functions have the same form as before the modification, yet their parameters p_i, p_{ij}, $\sigma_{j\nu}$ must be substituted for the corresponding sol values $p_i^{(s)}$, $p_{ij}^{(s)}$, $\sigma_{j\nu}$ according to the formulas of ref. 2,

$$p_{ij}^{(s)}=\frac{p_{ij}}{\xi_i\xi_j}\sum_{v}\sigma_{jv}\prod_{k}\xi_k^{f_{kv}-\delta_{kj}},\qquad p_i^{(s)}=\sum_j p_{ij}^{(s)},$$

$$\sigma_{jv}^{(s)}=\frac{\sigma_{jv}\prod_k \xi_k^{f_{kv}-\delta_{kj}}}{\sum_\mu \sigma_{j\mu}\prod_k \xi_k^{f_{k\mu}-\delta_{kj}}}. \tag{4.96}$$

The contents of the sol fraction, $\alpha^{(s)}$, and the ratio of all monomer units in it. $\rho^{(s)}/\rho$, can be obtained from

$$\alpha_v^{(s)}=\alpha_v\prod_i \xi_i^{f_{iv}}\rho/\rho^{(s)},\qquad \rho^{(s)}/\rho=\sum_v\alpha_v\prod_i\xi_i^{f_{iv}}. \tag{4.97}$$

For multicomponent systems we can get matrix elements $\chi_{v\mu}^{(s)}(\mathbf{r}'-\mathbf{r}'';\mathbf{s})$ of the generating function $\chi^{(s)}$ for two-point density correlators of units in separate sol molecules. By definition, they are the second variational derivatives, $\delta^2\Psi^{(s)}/\delta\ln s_v(\mathbf{r}')\,\delta\ln s_\mu(\mathbf{r}'')$, for coordinate independent counters $s(\mathbf{r})\equiv s$. Calculating the derivatives,[135] one gets

$$\begin{aligned}\hat{\chi}_{v\mu}^{(s)}(\mathbf{q},\mathbf{s})&=\rho s_v\alpha_v\prod_k \xi_k^{f_{kv}}\delta_{v\mu}\\&+\rho s_v s_\mu\sum_{i,j,m}\alpha_v f_{vi}\prod_k\xi_k^{f_{vk}-\delta_{ik}}\Delta_{ij}(\mathbf{q},\mathbf{s})\tilde{\lambda}_{jm}(\mathbf{q})p_{jm}\sigma_{m\mu}\prod_k\xi_k^{f_{\mu k}-\delta_{mk}},\end{aligned} \tag{4.98}$$

where $\Delta_{ij}(\mathbf{q},\mathbf{s})$ are matrix elements of $\mathbf{\Delta}(\mathbf{q},\mathbf{s})=[\mathbf{E}-\mathbf{M}(\mathbf{q},\mathbf{s})]^{-1}$, and the matrix elements of $\mathbf{E}$ and $\mathbf{M}(\mathbf{q},\mathbf{s})$ are, respectively, δ_{ij} and

$$M_{ij}(\mathbf{q},\mathbf{s})=\sum_{v,m}\tilde{\lambda}_{im}(\mathbf{q})p_{im}\sigma_{mv}s_v(f_{vj}-\delta_{mj})\prod_k\xi_k^{f_{vk}-\delta_{mk}-\delta_{jk}}. \tag{4.99}$$

Using the matrices $\mathbf{M}=\mathbf{M}(\mathbf{0},\mathbf{1})$ and $\mathbf{\Delta}=\mathbf{\Delta}(\mathbf{0},\mathbf{1})$, one can get a simple expression for the weighted average degree of polymerization,

$$P_W=1+\sum_{i,j,v}\alpha_v f_{vi}\Delta_{ij}p_j. \tag{4.100}$$

The condition under which it goes to infinity, that is, by definition, at the gel point, is

$$|\mathbf{E}-\mathbf{M}|=0. \tag{4.101}$$

Clearly, it is the same as the appearance of a divergence in the Fourier transform of molecular density at zero, $\hat{\chi}^{(s)}(\mathbf{0}, \mathbf{1})$, Eq. (4.98). Equalities (4.98) and (4.99) are an extension of (4.75); they are reduced to the latter at the polycondensation of the monomer RA^f. Meanwhile, below the gel point, the quantities $\tilde{\chi}^{(s)}_{\nu\mu}(\mathbf{q}, \mathbf{1}) \equiv \tilde{g}^{(s)}_{\nu\mu}(\mathbf{q})$ coincide with the Fourier-transformed elements of the structure matrix $\mathbf{g}(\mathbf{r}' - \mathbf{r}'')$, so that after substitution of Eq. (4.88) into (4.98) and (4.99), one is able to get the Fourier-transformed matrix $\tilde{\boldsymbol{\theta}}(\mathbf{q})$ of the total density correlation functions for units of various types. The condition of the spinodal transition in the multicomponent system according to (1.88) is

$$|\tilde{\mathbf{g}}^{-1}(\mathbf{0}) - \tilde{\mathbf{C}}(\mathbf{0})| = 0, \tag{4.102}$$

which determines the boundary of its thermodynamic equilibrium. Above the gel formation point, matrix $\tilde{\mathbf{g}}(\mathbf{q})$ is different from $\tilde{\mathbf{g}}^{(s)}$. Its elements are obtained from Eqs. (4.98) and (4.99), where one should set $\boldsymbol{\xi} = \mathbf{1}$ for $\mathbf{s} = \mathbf{1}$ on the r.h.s., and the solution of Eq. (4.95) with $\xi_i < 1$ is taken for sol structure matrix elements, $\tilde{g}^{(s)}_{\nu\mu}(\mathbf{q}) \equiv \tilde{\chi}^{(s)}_{\nu\mu}(\mathbf{q}, \mathbf{1})$.

Simple analytical expressions describing arbitrary multicomponent system can also be obtained for the statistical characteristics of the gel. An extension of Eqs. (1.28) and (1.29) are the formulas

$$1 - p_i^{(g)} = (1 - p_i)\left[1 - \sum_\nu \sigma_{i\nu} \prod_k \xi_k^{f_{k\nu} - \delta_{ki}}\right] \bigg/ \left[1 - \sum_\nu \sigma_{i\nu} \prod_k \xi_k^{f_{k\nu}}\right], \tag{4.103}$$

$$r^{(g)} = N \sum_\nu \alpha_\nu \left\{ \sum_i \frac{f_{i\nu}}{2}\left[\prod_k \xi_k^{f_{k\nu} - \delta_{ki}}(1 - \xi_i) + p_i\left(1 - \prod_k \xi_k^{f_{k\nu} - \delta_{ki}}\right)\right] - \left(1 - \prod_k \xi_k^{f_{k\nu}}\right)\right\}, \tag{4.104}$$

and the Fourier-transformed elements $\theta^{(g)}_{\nu\mu}(\mathbf{r})$ of the gel correlation matrix $\boldsymbol{\theta}^{(g)}(\mathbf{r})$ are obtained from

$$\tilde{\boldsymbol{\theta}}^{(g)}(\mathbf{q}) = \tilde{\mathbf{g}}^{(g)}(\mathbf{q}) + \tilde{\mathbf{g}}^{(g)}(\mathbf{q})\tilde{\boldsymbol{\beta}}(\mathbf{q})\tilde{\mathbf{g}}^{(g)}(\mathbf{q}), \qquad \tilde{\mathbf{g}}^{(g)}(\mathbf{q}) \equiv \tilde{\mathbf{g}}(\mathbf{q}) - \tilde{\mathbf{g}}^{(s)}(\mathbf{q}), \tag{4.105}$$

$$[\tilde{\boldsymbol{\beta}}(\mathbf{q})]^{-1} = [\tilde{\mathbf{h}}(\mathbf{q})]^{-1} - [\tilde{\mathbf{g}}(\mathbf{q}) - \boldsymbol{\rho}]. \tag{4.106}$$

These equalities are, naturally, reduced to (4.39) and (4.40) for the polycondensation of a single monomer RA^f.

The formulas presented in the preceding make it possible to calculate statistical characteristics in the framework of the SF approximation and to analyze various transitions in any concrete systems having arbitrary compositions and functionalities of primary monomers.

H. Theory of Swelling for Stochastic Polymer Networks

Until now we considered only polymer systems in complete chemical equilibrium with respect to all polycondensation reactions. In that case, besides the density of monomer units, ρ, the system was characterized with an equilibrium constant $k = L$ for a chemical reaction between functional groups, and their conversion p was determined from the mass action law for the reaction, Eq. (3.49). Certain polycondensation regimes are also possible, however, in which exchange reactions between the polymer molecules are rapid enough, so that an equilibrium MSD is established for them, where the distribution parameter (conversion p) may have a nonequilibrium magnitude. This is the case if the equilibrium between the condensation of the functional groups and the destruction of the resulting chemical bond is established to slowly as compared with exchange reactions. For such systems one should specify, besides ρ, not the equilibrium constant but the conversion of the functional groups, p, or the density of the chemical bonds, $\rho_b = f\rho p/2$.

As already mentioned in Section I.B, conditions under which polymers are exploited are substantially different from those under which they are synthesized. For instance, a polymer network produced by equilibrium polycondensation in melt (after the fixation of its topological structure by sharp cooling and the separation from sol molecules) may be put into low-molecular-weight solvent, where gel swelling would take place. Its equilibrium conformational set can be changed by variation of temperature and solvent type, but in the absence of chemical reactions the topological structure of the gel would be intact, remaining the same as that formed in the process of synthesis. If the synthesis was carried out in the equilibrium regime, the structure of all polymer gels with a fixed value of $p^{(g)}$, which were produced under the conditions where the Flory theory is valid, (4.80), would be practically identical because of ergodicity. It follows from the latter condition that for such stochastic gels with $N^{(g)}$ units, fluctuations of the structure parameters per single unit are proportional to $(N^{(g)})^{-1/2}$, so they are negligible. Consequently, the topology of meshes of this type is given unambiguously by a single parameter, $p^{(g)} = 2N_b^{(g)}/fN^{(g)}$, where $N_b^{(g)}$ is the number of bonds in the network; when combined with $N^{(g)}$, that number specifies completely a stochastic network.

To calculate the equilibrium swelling degree, the elastic properties of the mesh, and the density correlation function of its units, $\theta^{(g)}(\mathbf{r})$, it is sufficient to get an expression for the thermodynamic $\Omega^{(g)}$ potential of the gel and solvent system in the presence of an arbitrary external field $h(\mathbf{r})$ acting upon the gel units. The expression, which has been obtained by means of the replic method,[191] is

$$\Omega^{(g)}\{h\} = \Omega_{\rm su}\{\hat{z}(\mathbf{r}), z_s^{(g)}(\mathbf{r})\} + T\int d\mathbf{r}\,\frac{[\bar{\varphi}^a_+(\mathbf{r})]^2 - [\bar{\varphi}^a_-(\mathbf{r})]^2}{2L},$$
$$\hat{z}(\mathbf{r}) = z(\mathbf{r})[(1+\bar{\varphi}^a_+(\mathbf{r}))^f - (1+\bar{\varphi}^a_-(\mathbf{r}))^f], \qquad z(\mathbf{r}) = Me^{h(\mathbf{r})}. \tag{4.107}$$

Here $z_s^{(g)}$ and M are the activities of the solvent and monomer units in the gel, respectively, and the functions $\bar{\varphi}^a_{\pm}(\mathbf{r})$ are obtained from the equations

$$\int d\mathbf{r}'\,\Lambda_{uu}(\mathbf{r}-\mathbf{r}')\bar{\varphi}^a_{\pm}(\mathbf{r}') = -z(\mathbf{r})\frac{Lf}{T}\frac{\delta\Omega_{\rm su}\{\hat{z}, z_s^{(g)}\}}{\delta\hat{z}(\mathbf{r})}(1+\varphi_{\pm}(\mathbf{r}))^{f-1}. \tag{4.108}$$

The densities of units and bonds are obtained from Eqs. (4.107);

$$\rho^{(g)}(\mathbf{r}) = -\frac{1}{T}\frac{\delta\Omega^{(g)}_{\rm su}\{\hat{z}(\mathbf{r}), z_s^{(g)}\}}{\delta\ln\hat{z}(\mathbf{r})}, \qquad \rho_b(\mathbf{r}) = \frac{[\bar{\varphi}^a_+(\mathbf{r})]^2 - [\bar{\varphi}^a_-(\mathbf{r})]^2}{2L} \tag{4.109}$$

whereas M and L are, in general, different from their values for the system in a chemical equilibrium with the same distributions $\rho(\mathbf{r})$ and $\rho_b(\mathbf{r})$. They are determined by the normalization conditions

$$N^{(g)} = \int d\mathbf{r}\,\rho^{(g)}(\mathbf{r}), \qquad N_b = \int d\mathbf{r}\,\rho_b^{(g)}(\mathbf{r}). \tag{4.110}$$

In the following we restrict ourselves to an analysis of a gel in a finite-volume approximation, where $h = 0$, and $\bar{\varphi}^a_{\pm} \equiv \varphi_{\pm}$ are constants in a volume $V^{(g)}$ occupied by the gel and are equal to each other outside the volume. It is not difficult to show that equations (4.108) for $\varphi_{\pm}$ can be written as

$$\varphi_+ = \frac{p}{1-p}, \qquad \varphi_- = \frac{pu}{1-p}, \qquad u = \xi^f, \qquad \xi = 1 - p + pu. \tag{4.111}$$

The first equality is a definition of p. In the case where the network was produced from equilibrium polycondensation of f-functional monomer units, p equals the conversion of such a system and is related to the gel conversion $p^{(g)}$,[128]

$$p^{(g)} \equiv \frac{2N_b^{(g)}}{fN^{(g)}} = \frac{p(1+u)}{1+pu}. \tag{4.112}$$

Equations (4.111) and (4.112) determine $\varphi_{\pm}$ implicitly in terms of $p^{(g)}$; naturally, they are meaningful only for $p > p^* = (f-1)^{-1}$.

To get a closed set of equations for other quantities, M, L, $V^{(g)}$, and $z_s^{(g)}$, one should add two other equations to (4.110) that are determined by specific features of the physical problem concerned. Let us start with the problem of the isotropic swelling of the polymer gel in an excess solvent. In this case the additional conditions are the equalities of the pressure and chemical potential of the solvent in both the phases, corresponding to swelled gel and pure solvent,

$$P^{(g)} \equiv -\Omega^{(g)}/V^{(g)} = P, \qquad z_s^{(g)} = z_s, \tag{4.113}$$

where P is a given external pressure in the solvent phase, and z_s is its activity. The solution of this problem yields a magnitude of the maximum volume $V_{\max}^{(g)}$ that can be occupied by the given network in that particular solvent. When the gel swelling takes place in a bounded volume, $V < V_{\max}^{(g)}$, which does not confine solvent molecules, the system is an osmotic cell. In this case $\rho^{(g)}$ is fixed, and the osmotic pressure, $\pi = P^{(g)} - P$, is determined by the equality of the solvent chemical potentials,

$$\rho^{(g)} = N^{(g)}/V^{(g)}, \qquad z_s^{(g)} = z_s. \tag{4.114}$$

The third problem arises if the quantity of the solvent is not sufficient for complete swelling. The desired quantity is the volume of swelled gel $V^{(g)}$ that contains a fixed number of the solvent molecules N_s at a given magnitude of the external pressure,

$$N_s = \rho_s^{(g)} V^{(g)}, \qquad P^{(g)} = P. \tag{4.115}$$

Next we use the local approximation

$$\Omega_{\text{su}}\{z(\mathbf{r}), z_s(\mathbf{r})\} = -T \int \omega[z(\mathbf{r}), z_s(\mathbf{r})]\, d\mathbf{r}, \tag{4.116}$$

introducing the density of the Ω potential. This is correct if the specific scales at which the functions $z(\mathbf{r})$ and $z_s(\mathbf{r})$ are varying are large as compared with the correlation length of density in the system of separate units.

For the problem given in Eqs. (4.113) we get an equation for $\hat{z}$ based on Eqs. (4.107), (4.109), and (4.112),

$$\omega(\hat{z}, z_s) - \omega(0, z_s) = \frac{(f p^{(g)}/2)\partial\omega(\hat{z}, z_s)}{\partial \ln \hat{z}}, \tag{4.117}$$

where $p^{(g)}$ is the gel conversion, Eq. (4.112), and the function $\omega(z, z_s)$ is introduced in (4.116). Setting the obtained value of z into the first equality in (4.109), we evaluate $\rho^{(g)}$. In a system where $p^{(g)}$ is close to its minimum, $2/f$, the density of gel units, $\rho^{(g)}$, is small, and the pressure in the system of separate units is given by the virial expansion,

$$P_{su}(\rho^{(g)})/T = \omega(z, z_s) = \omega(0, z_s) + \rho^{(g)} + B(\rho^{(g)})^2 + C(\rho^{(g)})^3 + \cdots. \quad (4.118)$$

Hence in the first order in the small parameter, $p^{(g)} - 2/f$, we obtain an expression for $\rho^{(g)}$, and the domain of its applicability,

$$\rho^{(g)} = \frac{(1/2)fp^{(g)} - 1}{B(T)}, \qquad p^{(g)} - \frac{2}{f} \ll \min(1, B^2/C). \quad (4.119)$$

As the system approaches the θ conditions, the density of units in swelled gel increases with diminishing B, according to (4.119), in the domain of its applicability.

For the problem given in Eqs. (4.114) an equation for $\hat{z}$ is obtained from the from the first equality in (4.109), and π is determined by setting the resulting $\hat{z}$ into (4.107). Using the virial expansion (4.118) at small $\rho^{(g)} \ll \min(1/B, B/C)$, we get an expression for osmotic pressure,

$$\pi/T = B(T)(\rho^{(g)})^2 - \rho^{(g)}(fp^{(g)}/2 - 1). \quad (4.120)$$

When density of the units in the gel is higher than that corresponding to the equilibrium swelling, (4.119), the osmotic pressure in (4.120) is positive. By isotropic expansion of the network in a solvent, one can attain a situation where $\rho^{(g)}$ is less than the value $\rho^{(g)}_{\min} = N^{(g)}/V^{(g)}_{\max}$, and elastic stresses appear that correspond to negative values of π.

Finally, for the problem of (4.115), the set of equations that determines $\hat{z}$ and $\hat{z}_s^{(g)}$ is

$$\frac{P}{T} = \omega(\hat{z}, z_s^{(g)}) - \frac{(fp^{(g)}/2)\partial\omega(\hat{z}, z_s^{(g)})}{\partial \ln \hat{z}},$$
$$\frac{N_s}{N^{(g)}} = \frac{\partial\omega(\hat{z}, z_s^{(g)})/\partial \ln z_s^{(g)}}{\partial\omega(\hat{z}, z_s^{(g)})/\partial \ln \hat{z}} = -\left.\frac{d \ln \hat{z}}{d \ln z_s^{(g)}}\right|_{\omega}. \quad (4.121)$$

Solution of these equations is fixed by a concrete choice of the function $\omega(\hat{z}, \hat{z}_s)$, which describes the physical interaction of particles in the system of separate units. As in Eq. (4.120), a further simplification is possible for low-density gel units, when the density of solvent molecules is obtained as the solution of the equation

$$P = P_{su}(\rho_s^{(g)}) - \frac{N^{(g)}}{N_s}\left(\frac{fp^{(g)}}{2} - 1\right)\rho_s^{(g)}T, \tag{4.122}$$

where $P_{su}(\rho_s)$ is the equation of state for the pure solvent. The resulting solution $\rho_s^{(g)}$ determines the volume of swelled network, $V^{(g)} = N_s/\rho_s^{(g)}$.

Let us now find out the domain of applicability of the SF approximation within which the functional $\Omega^{(g)}\{h\}$ has been obtained, (4.107). To this end, we calculate the derivative of $\Omega^{(g)}$ with respect to the field h and evaluate the correlation function of gel units.[191] Its Fourier transform is given in Eq. (4.76), where the function β is, however, different from that of Eq. (4.40) and is given by

$$[\tilde{\beta}(\mathbf{q})]^{-1} = [\tilde{h}^{(g)}(\mathbf{q})]^{-1} - [\tilde{g}^{(g)}(\mathbf{q}) - \rho^{(g)}]. \tag{4.123}$$

Here $h^{(g)}(\mathbf{q})$ is the Fourier transform of the total correlation function for the system of separate monomer gel units of density $\rho^{(g)}$ that are put into a given solvent. By the definition of $g^{(g)}$ [Eq. (4.76)] and taking into account (4.75) and (4.60), it is proportional to factor ρ, which has no direct physical meaning in the present case,

$$\rho \equiv M(1-p)^{-f}\partial\omega(z, z_s)/\partial\hat{z}. \tag{4.124}$$

This quantity is related to the density of units of swelled gel, as follows from (4.108), $\rho^{(g)} = \rho(1 - \xi^f)$.

The diagram interpretation of Eq. (4.76), taking into account (4.123), is similar to that in Section IV.D (see Figs. 4.23 and 4.24). The sole difference is that one should discard the contributions from finite sol molecules in the diagrams of Fig. 4.24. In the present case, Eq. (4.77) is valid for the reciprocal Fourier transform of the density correlator of gel units, which is, however, simpler here,

$$[\tilde{\theta}^{(g)}(\mathbf{q})]^{-1} = [\tilde{g}^{(g)}(\mathbf{q})]^{-1} - \tilde{C}^{(g)}(\mathbf{q}), \quad [\tilde{C}^{(g)}(\mathbf{q})]^{-1} = [\tilde{h}^{(g)}(\mathbf{q})]^{-1} + \rho^{(g)}, \tag{4.125}$$

because $g^{(s)} = 0$ since the system does not contain sol.

An analogy exists between Eqs. (4.125) and (4.38) that must be noted since here we are concerned with a two-component system, unlike that in (4.38). Since the second component is a solvent, molecules of which are not coupled with chemical bonds, one can exclude it from consideration and treat the system as a one-component object. The effect of the solvent is a renormalization of the direct correlation function of monomer units, $C^{(g)}(\mathbf{q})$, which is determined in the second equality in (4.125). Similarly, the solvent can be discarded in the same manner for any multicomponent system for which Eqs. (4.125) and (4.38) are matrix equalities, and $h^{(g)}$ in (4.125) is the direct correlator matrix for monomer units of different types in the system of separate units,

$$h_{\nu\mu}(\mathbf{r}', \mathbf{r}'') = \theta_{\mathrm{su}\mu\nu}(\mathbf{r}', \mathbf{r}'')/\rho_\nu(\mathbf{r}')\rho_\mu(\mathbf{r}'') - [\rho_\nu(\mathbf{r}')]^{-1}\delta_{\mu\nu}\delta(\mathbf{r}' - \mathbf{r}''), \quad (4.126)$$

where the matrix $\theta_{\mathrm{su}\mu\nu}$ is defined in (4.86). Equation (4.126) is an extension of (4.25) to multicomponent systems.

As has been mentioned, the spinodal transition corresponds to a divergence in the Fourier transform of the correlator, (4.125), at a zero argument. It is noteworthy that such a transition is possible even in a strong solvent if repulsive forces dominate in the system of separate units.

Consider a stochastic gel produced at a conversion p, close to $p^* = (f-1)^{-1}$, in the case where the density of its units in the solvent is low, so that one can use the virial expansion (4.118). In the following we are concerned with distances that are large compared with $\mathbf{a}$, and an approximate expression is valid for $\tilde{\Lambda}_{uu}(\mathbf{q})$,

$$\tilde{\Lambda}_{uu}(\mathbf{q}) = 1 + 2a^2q^2, \qquad \rho^{(g)} = 2f\rho|\tau|/(f-2), \qquad \tau \equiv 1 - p/p^*. \quad (4.127)$$

The second expression systems from $\rho^{(g)} = \rho(1 - \xi^f)$ for small τ, $|\tau| \ll 1$. Setting (4.127) in (4.125), we get, for $a^2\mathbf{q}^2 \ll 1$, the following asymptotics for the density correlator of gel units[191]:

$$\tilde{\theta}^{(g)}(\mathbf{q}) = \frac{2f}{f-1}\frac{\rho|\tau|}{\tau_{\mathrm{sp}}|\tau| + 4a^4\mathbf{q}^4}, \qquad \tau_{\mathrm{sp}} = \tau + \frac{4B(T)\rho f}{f-1}. \quad (4.128)$$

At the spinodal transition point, τ_{sp} is zero, so the condition $\tau_{\mathrm{sp}} < 0$ determines the region of absolute thermodynamic instability of the swelled gel with respect of its collaps. Under the θ conditions, $B(T) = 0$, triple interactions of monomer gel units should be taken into account, and the boundary of the spinodal instability is determined by the magnitude of the third virial coefficient,

$$C(T) > (f-1)(f-2)/12f^2\rho^2. \quad (4.129)$$

An expression for the density correlator of the gel units is obtained from (4.128) by means of inverse Fourier transformation,

$$\theta^{(g)}(\mathbf{r}) = \frac{\rho f}{f-1}\left(\frac{|\tau|}{\tau_{\mathrm{sp}}}\right)^{1/2} \frac{\mathrm{e}^{-r/\xi_{\mathrm{c}}}}{4\pi a^2 r} \sin\frac{r}{\xi_{\mathrm{c}}}. \tag{4.130}$$

The specific correlation length ξ_{c} is given by

$$\xi_{\mathrm{c}} = a(\tau_{\mathrm{sp}}|\tau|)^{-1/4}. \tag{4.131}$$

It goes to infinity at the spinodal.

An analogue of Eq. (4.80) in the present case is

$$\frac{\langle(\delta\rho^{(g)})^2\rangle}{(\rho^{(g)})^2} = \left(\frac{\mathrm{Gi}}{\tau_{\mathrm{sp}}}\right)^{1/4}, \qquad \mathrm{Gi} = \frac{1}{|\tau|^5}\left[\frac{f-2}{32\pi f(f-1)\rho a^3}\right]^4. \tag{4.132}$$

Hence, one can estimate the boundary of the applicability of the employed SF approximation, $1 \gg \tau_{\mathrm{sp}} \gg \mathrm{Gi}$.

I. Other Field-Theoretic Approaches to Description of Branching Polymers

In the preceding sections we have shown how certain methods of field theory suggest a compact representation of the fundamental characteristics of polymer systems. Various probability measures in the set of stochastic field configurations can be constructed depending on the choice of the system model. In this concluding section we present a brief review of some cases from the literature where ideas and calculational methods of field theory and the theory of phase transitions were applied to the analysis of lattice, as well as continual, models of branching polymers.

The basic idea in using lattice models is to choose such a Hamiltonian that the corresponding high-temperature perturbative series and the representing diagrams reproduce exactly the diagram expansion for the concerned polymer system. In other words, the probability measure in the set of diagrams of the lattice model [edges in the diagrams are dashed lines representing physical interaction, and it is implied that the summation is performed over the internal degrees of freedom, say, the Potts variables σ_{i} in the Hamiltonian (1.60)] coincides with the probability measure in the set of polymer molecules having various configurations and conformations. The correspondence is attained when we use certain variants of the Potts model with arbitrary numbers of components, q, with a subsequent limit in the parameter q, its limiting value being determined by the specific features of the model concerned. The field-theoretic formulation is employed usually to get concrete results in the lattice models.

The most important works on the description of the statistics of branching polymers by means of various lattice models are compiled in Table III, the composition of which is as follows. Each column of the table corresponds to a polymer system: first corresponds to a dilute solution of a polymer in a strong solvent, and the second to a solution in a solvent having an arbitrary affinity for the polymer. As shown in refs. 130 and 115, the description of the latter system in the framework of the lattice models is equivalent to a consideration of the polymer melt [Eq. (1.59)], so it is possible to put both systems into the second column. The third column refers to semidiluted and strong polymer solutions. The distribution of references corresponds to the model used. The first row refers to works where a version of the Potts model was used [the Hamiltonian for one of them is presented in Eq. (1.60)]. A common feature of these models is that the interaction energy for any particles of different types is zero. The second row refers to works using lattice gas models, where each particle of n types is in one of q Potts states. Like the conventional model of lattice gases, which is equivalent to the Ising spin model, the generalized version also has an equivalent formulation in terms of spin variables. It is reasonable therefore to put ref. 113 in the second row. Note that despite the formal difference between the Hamiltonians of the Potts models and the lattice gas models, they can describe the same polymer system.

The results obtained in refs. 106–108, 110, 113, 115–117, and 124 have been mentioned in Section I.H. Therefore, here we will consider only the results of the other works cited in Table III.

In ref. 194 the statistical properties of the ensemble of branching polymers without cycles was described in the framework of a variant of the Hilhorst model. The latter is an extension of the lattice gas model to the case where there are n particle types, and each particle can be in one of q Potts states.

TABLE III References for Statistics of Branched Polymers Models

	System		
Model	Random Lattice Animals	Correlated Lattice Animals or Random Percolation	Correlated Percolation
Conventional Potts models	116	115, 117	115, 196
Extended models of lattice gas	194	113, 124, 131	106–108, 110

According to the definition of the model, only particles of the same type are interacting, so every cycle is present in the partition function with a factor of n. Correspondence to the polymer system arises in the limit $q-1 \sim n \to 0$; a field-theoretic formulation of the model[194] has been proposed for this case.

Field-theoretic methods were found fruitful also for the case where the number of particle types is $n = 2$, as for the conventional lattice gas. It was shown[130] that the free energy for the Potts model of lattice gas ($n = 2$, $q = 1$) determines the generating functional for the ensemble of lattice animals. The field theory constructed for the model enabled those authors to use the renormalization group method of ε expansion to obtain the critical exponents corresponding to each of three first regimes of the critical behavior for diluted solutions of branching polymers [Eq. (1.65)].

The critical behavior of an ensemble of lattice animals has also been investigated in the framework of the three-parameter Potts model, (1.60). It is shown[131] that this model also describes the statistical properties of the ensemble of percolation clusters. Its field-theoretic formulation using the renormalization group method provides a possibility to trace the transition in the statistical properties of clusters up to the gel formation point as the number of cluster units l is increasing from the percolation regime ($l \ll l_T$) to the regime of lattice animals ($l \gg l_T$). Above the gel point, the free energy of the extended Potts model has singularities owing to the presence of instanton solutions in the corresponding field theory. As shown in ref. 131, the instanton configurations correspond to so-called droplets (cf. Section I.H), and their contribution to free energy enables one to calculate the distribution of the droplets in the number of units contained in them.

In three works[195,107,196] polymer systems were described by means of the correlated percolation model. In the first work (for $n = 2$, $q = 1$) the author obtained the exact solution of the problem of nodes in the Bethe lattice. The solution evidences that the transition point is shifted from its value for random percolation. A more general problem of correlated percolation both in nodes and couplings ($n = 2$, $q = 1$) has been considered,[107] and the corresponding field theory has been constructed. In the framework of the ε-expansion method, a regime corresponding to the point Q in Fig. 1.29*b* has been found, besides the percolation regime. Critical exponents are calculated[107] for both these points, which are stationary with respect to the renormalization group transformations. It was shown[196] that the model can be reformulated in terms of the four-parameter Potts model with $q = 2$ (cf. also ref. 115).

The field-theoretic formulation of the polymer lattice models proposed first for the Potts Hamiltonian[197] and employed in subsequent works makes it possible to analyze the critical behavior of the system of branching macromolecules in the region of developed fluctuations. There is, however, an essential defect in its early versions, namely, the lack of correspondence between the Feynman graphs of the field theory and configurations of real polymers. As the exact solutions are known only for a number of model Hamiltonians, this correspondence is quite important both for the justification of the SF approximation and for the calculation of the fluctuational correction.

This defect is absent in continual models of polymers, which exploit an analogy between polymer molecules and Feynman graphs of field theory. The Lagrangian is chosen in such a way as to reproduce in the diagram expansion of the perturbative high-temperature series the correct values of the statistical weights (probabilities) of various molecules in the polymer ensemble.

The corresponding field theory for the Lifshitz–Erukhimovich model has been exposed in Section IV.D. Based on field-theoretic formulation, it was possible[134] to develop a perturbation theory in a small parameter, $\varepsilon = (\rho a^3)^{-1}$, for systems having high concentrations and to calculate fluctuational corrections to the equation of state (cf. also ref. 133) as well as to the generating functional of the molecule density correlators. It was also shown that in the presence of sufficiently strong physical interactions in concentrated systems the percolation regime describes only a narrow fluctuational domain, $|\tau| \ll \mathrm{Gi} \ll \tau_{\mathrm{sp}}$ $(\tau = 1 - p/p^*,\ \tau_{\mathrm{sp}} = 1 - p/p_{\mathrm{sp}})$, where the Ginzburg number, Eq. (4.80), is sufficiently small. Thus, the applicability of the lattice models for the description of the gelformation phenomenon is restricted only to the region of developed fluctuations, where the Fourier transform of the correlator of the gel unit density is[134,198]

$$\tilde{\theta}^{(g)}(\mathbf{q}) \approx \rho|\tau|^{-\gamma}[1 + \mathbf{q}^2\xi_c^2]^{-2}, \qquad \xi_c \approx a|\tau|^{-\nu}. \tag{4.133}$$

Here ξ_c is the correlation length of the systems, γ and ν are the critical exponents of the percolation theory, as introduced in Table I. Thus, a fairly complete description of polymer systems is possible in the framework of the Lifshitz–Erukhimovich model.

Another original variant of the continual model describing statistical properties of branching polymers has been proposed by Lubensky and Isaakson.[113,114] Their method was based on the consideration of linear chains coupled by semifunctional elements, and it is assumed that all groups of the chains have reacted completely, whereas the distribution of the unit coordinates is the same as that for an equilibrium ensemble with a fixed number of molecules Π and numbers N_f of the f-functional units. It has been shown[113,114] that the statistical weights of configurations and conformations of branching polymers in the model are determined by contributions from the Feynman diagrams for the field theory with a Lagrangian,

$$\mathscr{L}_k\{\psi_{ij}; h_f\} = \int d\mathbf{x}\left\{\frac{r}{2}\psi^2 + \tfrac{1}{2}(\nabla\psi)^2 + \frac{\beta}{2}(\psi^2)^2 - \sum_{f\neq 2}\frac{w_f}{f!}\left[\sum_{j=1}^{k}\psi_{1j}^f + e^{-h_f}\sum_{j=k+1}^{q}\psi_{1j}^f\right], \qquad \psi^2 \equiv \sum_{i,j}\psi_{ij}^2, \tag{4.134}$$

where the field ψ_{ij} has nq components, $i = 1,\ldots,n$, $j = 1,\ldots,q$. In the diagram expansion of the partition function Z_{nq} for the field theory with the Lagrangian of (4.134) in the perturbative series, the Feynman graph, containing Π connected parts corresponding to polymer molecules, is present with a factor of q. Therefore, q determines the activity of the molecules. The physical meaning of other parameters present in the Lagrangian (4.134) is revealed similarly. In particular, the magnitude of β is proportional to the second virial coefficient for the corresponding system of separate units. The partition function for the polymer system is Z_{nq} for $k = q$ and $n = 0$, where one should express the parameters $\{w_f\}$, r in terms of the numbers $\{N_f\}$ and Π, and the generating function for the numerical MWD is, up to a normalization,

$$\left.\frac{\partial \ln Z_{nq}}{\partial \ln q}\right|_{k=q,n=0}. \tag{4.135}$$

One can show that a similar formula is obtained for the generating functionals of correlators for individual molecules. Let us now discuss the most important results obtained in the framework of the model.

As shown in ref. 114, in the absence of physical interactions ($\beta = 0$), the SF approximation leads to results obtained in the Flory theory. For $\beta > 0$, however, it was found[114] that there is a shift in the gel formation point and a change in the shape of the MWD, as compared with Flory's results. This conclusion is wrong, as can be proved by calculations similar to those in Section IV.E. The analysis of the system behavior above the gel point given in ref. 114 is also incorrect. For $h_f = 0$, the Lagrangian (4.134) has a symmetry under permutations of fields ψ_{ij}. It is true, as noted in ref. 114, that the symmetry is broken spontaneously, but this fact was not taken into account correctly there. In fact, above the transition point the ground state of the field theory with the Lagrangian (4.134) has a q-fold degeneration,

$$\langle\psi_{ij}\rangle^{(l)} = \bar{\varphi}\delta_{i1}\delta_{jl} + \varphi\delta_{i1}(1-\delta_{jl}), \qquad j, l = 1,\ldots,q, \bar{\varphi} \geqslant \varphi. \tag{4.136}$$

In the Lubensky–Isaakson model, $\bar{\varphi}$ coincides with the total density of the monofunctional units up to a factor, and φ is the density is the sol. We can show that correct calculations for the model concerned in the SF approximation lead to results in qualitative agreement with those obtained in the framework of the Lifshitz–Erukhimovich model. For the critical region, practically the same fluctuational regimes were obtained[113,114] for the statistical behavior of molecules as those obtained in the framework of the polymer lattice model.[130]

The statistical properties of branching molecules in dilute solutions have also been considered in ref. 111 both without and with cycles. The conclusion that the critical exponents are the same for the systems of both types was corrected later.[112] A special case of polymer systems where molecules cannot have more than a single cycle is the systems obtained from the polycondensation of monomer RBA^f, when chemical bonds couple only different groups A and B. A field theory constructed for this system[199] exactly reproduces Flory's results for $\beta = 0$ in the whole region of conversions p. For $\beta > 0$, fluctuations are substantial in dilute solutions near the point $p = 1$, and the critical exponents for this stationary point are related by simple algebraic equalities with critical exponents for the problem of lattice animals.[199] The density correlation functions for such an infinite molecule have been found in ref. 191.

A common defect of most of the works discussed here, where various fluctuational regimes have been found, is uncertainty as to the domain of their applicability. This fact diminishes considerably their practical usefulness in the application to concrete polymer systems.

V. CONFORMATIONS OF GAUSSIAN MOLECULES AND SPECTRAL PROPERTIES OF THEIR GRAPHS

A. Space Metric and Graph Metric

A method widely used in the theory of polymers is the Gaussian model of macromolecules, which takes into account only their entropy elasticity and neglects spatial interactions between units. In the model, the probability $\mathscr{P}(\mathscr{G}_{lq}\{\mathbf{r}_i\})$ for an arbitrary equilibrium conformation of a particular (l, q)-mer molecule is given by the Gibbs distribution with the elastic potential, Eq. (1.32). Taking into account (1.34), the distribution is written as

$$\mathscr{P}(\mathscr{G}_{lq}\{\mathbf{r}_i\}) = \exp[-\gamma \operatorname{Tr}(\mathbf{RKR}^{\mathrm{T}})]/Q_{lq}. \tag{5.1}$$

The configurational integral Q_{lq} is determined by the normalization of the distribution $\mathscr{P}$. In the case with which we are concerned, where external fields are absent, the integral over the coordinates of the pth unit, which was omitted in Eq. (1.35), yields a trivial factor, namely, the system volume. This factor

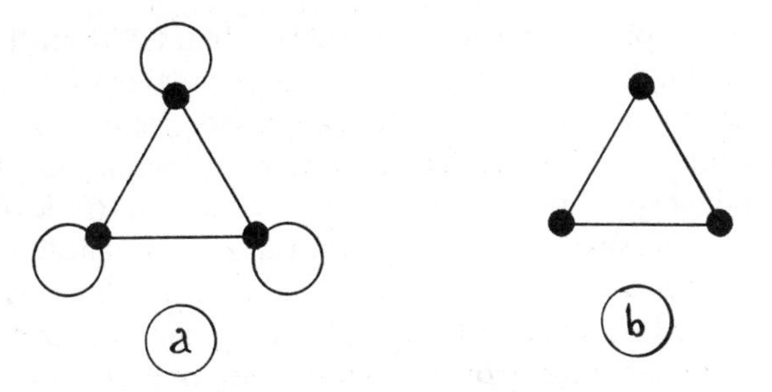

Figure 5.1. Molecules (*a*) with and (*b*) without loops having identical positions of units.

is due to the translational motion in space of the molecule as a whole, so it must be discarded in the analysis of molecule conformations. If a molecular graph for an (l, q)-mer contains loops (see Fig. 5.1), they must be "erased" when constructing the corresponding Kirchhoff matrix, since polymer chains associated with the loops are not subject to extensions, independently of any conformational rearrangements of molecular units, so they do not contribute to the elastic energy in Eq. (1.32).

To calculate the statistical characteristics of a polymer molecule, say, its radius of gyration, one has to average some functions with the probability measure (5.1). Values of the corresponding multidimensional integrals depend on the molecule configuration solely via its Kirchhoff matrix. Thus, we substitute the space (Riemannian) metric for a graph metric. According to Gordon,[200] an expert in the application of graph theory methods to the chemical physics of polymers,[200] the reduction of probability problems in continuum to problems of discrete mathematics is fundamental to a number of fields in the natural sciences.

The Gibbs distribution in Eq. (5.1) for an (l, q)-mer Gaussian molecule having an arbitrary topological structure q makes it possible to relate its various equilibrium and dynamical properties to the spectrum of Kirchhoff's matrix $\mathbf{K}$ for the corresponding molecular graph. The spectrum contains l eigenvalues (taking into account their multiplicity), $\lambda_0, \lambda_1, \ldots, \lambda_i, \ldots, \lambda_{l-1}$, which are confined in an interval $0 \leqslant \lambda_i \leqslant 2f$, where f is the functionality of the branching units and $\lambda_0 = 0$, whereas $\lambda_1 \neq 0$.

The configurational integral responsible for the normalization of the probability distribution in (5.1) is

$$Q_{lq} = \left(\frac{\pi}{\gamma}\right)^{3(l-1)/2} |\mathbf{K}_p|^{-3/2} = \left(\frac{\pi}{\gamma}\right)^{3(l-1)/2} l^{3/2} \prod_{i=1}^{l-1} \lambda_i^{-3/2}. \tag{5.2}$$

It concides with the partition function for the (l, q)-mer molecule up to an inessential factor. For the matrix $\mathbf{K}_p$, which is obtained from the Kirchhoff matrix by removing its pth row and pth column, a value of its determinant, $|\mathbf{K}_p|$, equals the number of spanning rooted trees for the molecular graph $\mathscr{G}_{lq}$ having their root at the pth vertex. This number is independent of p and equals the product of all nonzero eigenvalues of the matrix $\mathbf{K}$ divided by l.

A quantity that is usually considered is the ratio of configurational integrals for the qth topological isomer of an l-mer and a linear chain with the same number of units, l. For a fixed value of l, the ratio characterizing a decrease in the conformational entropy owing to formation of cyclic fragments in the molecule diminishes with a rise of its cyclic rank r. In particular, for any tree ($r = 0$), including a linear molecule, one has $\mathbf{K}_p = 1$, and for a single cycle ($r = 1$) one has $|\mathbf{K}_p| = 1$, since all its rooted-spanning trees are obtained by removing one of l edges. Hence, we get a familiar result by Jacobson and Stockmayer,[46] according to which the probability of a cyclization falls as $l^{-3/2}$ with the number of its units l. Equality (5.2) is a generalization of that result to molecules having an arbitrary topological structure.

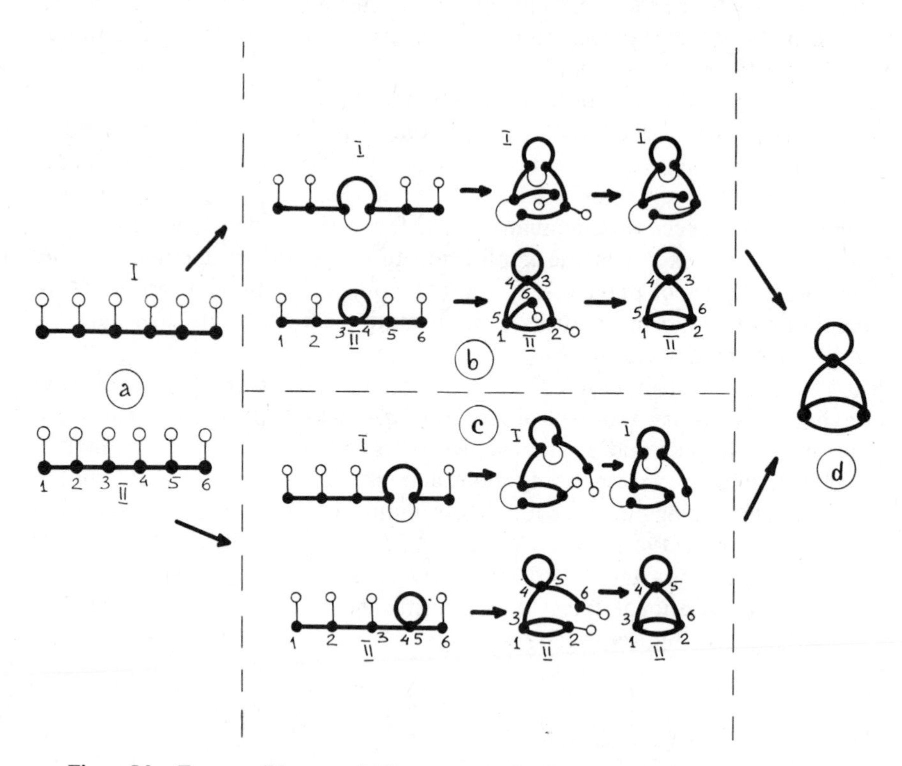

Figure 5.2. Two possible ways (I, II) to account for linear memory during two different pathways [(*b*) and (*c*)] of cyclic isomer formation.

The problem of evaluating configurational integrals arises also in calculating the topology distribution $\mathscr{G}_{l,q}$ for various (l, q)-isomers, which are formed in the course of the equilibrium cyclization of a linear macromolecule.[201] A specific feature of the system concerned is conservation of the "linear memory"[184] in all cyclization products, fixing the initial order of the connection of monomer units with stable chemical bonds. The latter are denoted by heavy lines, whereas labile bonds, which appear in the course of cyclization, are indicated by b thin lines (see Fig. 5.2). Another equivalent method to account for "linear memory" in the cyclization process is the use of enumerated graphs, where one can deal with edges of a single type. In the latter approach every pair of monomer units connected with a labile bond is represented by one common vertex in the molecular graph (Fig. 5.2). This method is a reasonable approximation for the calculation of the cyclization of molecules, for which the numbers of atoms coupled by labile and stable bonds are substantially different, so that the former can be neglected and contracted to a point.

A one-to-one correspondence between the graphs with edges of two types and those with indices holds only if the unmarked graph is symmetric (Fig. 5.2*b*). Otherwise (Fig. 5.2*c*), two equivalent indexings are possible, corresponding to the choice of an extreme unit of the asymmetric graph, which is the starting point of vertex enumeration (graphs 9 and 10 in Fig. 5.3).

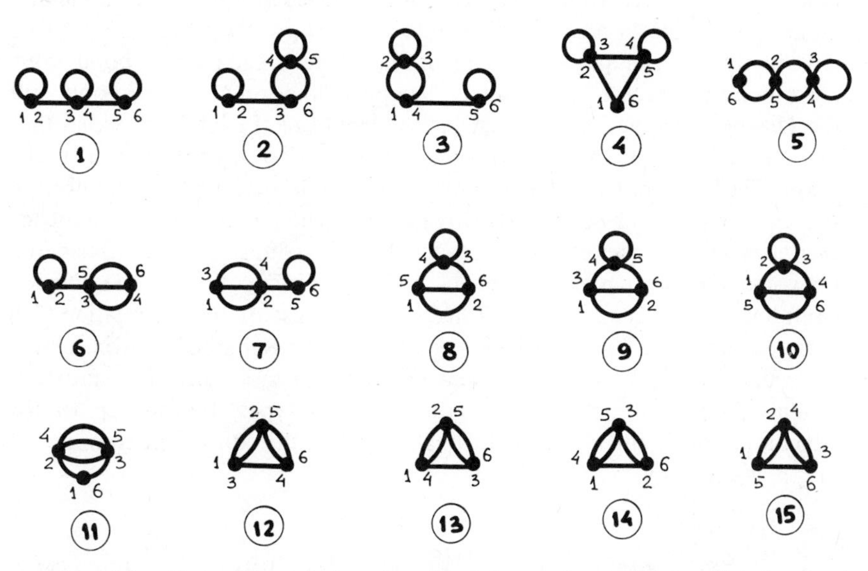

Figure 5.3. All possible methods to form bonds in complete cyclization of six-functional molecule.

Other enumeration ways, which conserve the vertex adjacency determined by linear memory, are also possible, in addition to trivial inversion. Unlike the latter, the different indexing ways correspond to different (topologically) isomers (Figs. 5.2*b*, *c* and graphs 8 and 9 in Fig. 5.3). A number of characteristics of such isomers (e.g., their geometric dimensions) are practically indistinguishable experimentally, so their separation in the reaction mixture (say, by chromatographic methods) is hardly possible. Experimental separation can be performed for polymer molecules represented by the same unlabeled graph (Fig. 5.2), their equilibrium concentration being determined, however, not by the order of the automorphic group $\mathscr{S}(l, q)$ of the graph G_{lq} but by the number and symmetry of those labeled graphs (Figs. 5.2*b*, *c*), which yield the graph after all labels are erased. Actually, the number of ways that give rise to an (l, q)-mer produced from monomers, which determines the combinatorial entropy, is related to $\mathscr{S}(l, q)$ by a simple formula, (1.5), under the condition that all these ways are permissible in that particular chemical reaction. Evidently, this condition holds for polycondensation, whereas in the cyclization reaction with which we are now concerned some (l, q)-mer formation cannot take place because of restrictions owing to the linear memory.

The probability to find each macromolecule in the considered equilibrium ensemble is evaluated similar to that in the discussion of products generated by equilibrium polycondensation, as given in Section III.B. In the present case, however, there are two types of molecular chemical bonds, stable and labile. The stable bonds are taken into account by the exponential factor (5.1) in the molecular partition function, whereas each labile bond must correspond to a factor $L\lambda_{\text{lb}} \equiv L\delta(r)$. The approximation of a labile bond with the Dirac delta function $\delta(r)$ neglects the length of the labile bond, so that any pair of coupled monomer units correspond to a single vertex in the graph. The factor of L has the same meaning as in the discussion of equilibrium polycondensation in Section III.B, where it is the equilibrium constant for the elementary reaction between functional groups. In the considered Gaussian model of macromolecules applied to evaluation of the entropy component of its free energy, $-T \ln Q_{lq}$, one can use the simple relation (5.2) between the value of Q_{lq} and the spectrum of the Kirchhoff matrix for the molecular graph. In the considered ensemble of molecules, the statistical weight Z_{lq} for an isomer with bonds of two types, that is, up to the stoichiometric factor κ_{lq}, the statistical weight of the equivalent enumerated graph, is given by

$$Z_{lq} = \kappa_{lq} \int \cdots \int \exp[-\gamma \operatorname{Tr}(\mathbf{R}\mathbf{K}_{\text{in}}\mathbf{R}^{\text{T}})] \Pi[L\lambda_{\text{lb}}(\mathbf{r}_i - \mathbf{r}_j)] d\{\mathbf{r}_i\}, \qquad (5.3)$$

where $\mathbf{K}_{\text{in}}$ is the Kirchhoff matrix for the initial linear polymer molecule, and the product is over all labile bonds of the (l, q)-mer. The factor is $\kappa_{lq} = 1$ for the symmetric (l, q)-isomer, which corresponds to a single enumerated contracted graph, and $\kappa_{lq} = 2$ for the asymmetric isomer. After setting $\lambda_{\text{lb}}(\mathbf{r}_i - \mathbf{r}_j) = \delta(\mathbf{r}_i - \mathbf{r}_j)$, the evaluation of the integral in (5.3) is reduced to the problem considered in the preceding, namely, to the evaluation of the configurational integral Q_{ms} for the labeled s-type graph with m vertices, which has the corresponding Kirchhoff matrix $\mathbf{K}(m, s)$ after all its loops are erased,

$$\frac{Z_{lq}}{\kappa_{lq}} = Z_{ms} = L^b Q_{ms} = L^b \left(\frac{\pi}{\gamma}\right)^{3(m-1)/2} |\mathbf{K}_p(m, s)|^{-3/2}$$
$$= \left(\frac{\pi}{\gamma}\right)^{3(l-1)/2} \left[L\left(\frac{\gamma}{\pi}\right)^{3/2}\right]^b |\mathbf{K}_p(m, s)|^{-3/2}. \tag{5.4}$$

The probability of an enumerated graph, that is, the ratio of its statistical weight to the sum Z of the statistical weights Z_{ms} for all similar graphs, is given, according to (5.4), by a factor $[L(\gamma/\pi)^{3/2}]^b$, in addition to the topological factor $|\mathbf{K}_p(m, s)|^{-3/2}$. For large values of the dimensionless parameter $L(\gamma/\pi)^{3/2}$, isomers with a maximum possible number b of labile bonds dominate among products of the cyclization reaction.

If the initial linear molecule contained, say, six units, then after the reaction process is over, the equilibrium ensemble of cyclized molecules (presented in Fig. 5.3) is a set of 15 enumerated graphs.[201] Since all of them contain the same number of bonds, according to (5.4), their relative probabilities $\mathscr{P}(\mathscr{G}_{ms})$ are determined solely by the topological structure of the graphs $\mathscr{G}_{ms}$. To calculate the probabilities, it is sufficient to evaluate the number of spanning trees that equals the principal minor of the Kirchhoff matrix $\mathbf{K}(m, s)$ for the graph with removed loops. For instance, for graph 8 in Fig. 5.3, the matrix $\mathbf{K}(3, 8)$ and its principal minors are

$$\mathbf{K}(3, 8) = \begin{Vmatrix} 2 & -1 & -1 \\ -1 & 3 & -2 \\ -1 & -2 & 3 \end{Vmatrix}, \qquad |\mathbf{K}_1(3, 8)| = |\mathbf{K}_2(3, 8)| = |\mathbf{K}_3(3, 8)| = 5. \tag{5.5}$$

The minors for other graphs are evaluated similarly, so one is able to determine their relative probabilities (see Table IV).

The approach described here is extended easily to the cyclization of molecules having an arbitrary initial topology specified by the Kirchhoff

TABLE IV Equilibrium Proportions of Isomers Depicted in Figure 5

	Isomer Number														
	1	2	3	4	5	6	7	8	9	10	11	12	13	14	15
Proportion (%)	34.4	12.2	12.2	6.6	4.3	6.6	6.6	3.1	3.1	3.1	1.9	1.5	1.5	1.5	1.5

matrix $\mathbf{K}_{\text{in}}$. An additional fact that must be taken into account is that the stoichiometric factor κ_{lq} in Eqs. (5.3) and (5.4) equals the number of enumerated contracted graphs representing the (l, q) isomer. In the general case, it is expressed in terms of the orders of the automorphism groups $\mathscr{S}(\mathscr{G}_{\text{in}})$ and $\mathscr{S}(\mathscr{G}_{lq})$ for the initial graph and the (l, q)-isomer, respectively,

$$\kappa_{lq} = \mathscr{S}(\mathscr{G}_{\text{in}})/\mathscr{S}(\mathscr{G}_{lq}). \tag{5.6}$$

B. Using Kirchhoff Matrix to Calculate Distributions in Radius of Gyration and Other Conformational Characteristics of Macromolecules

Consider a polymer molecule having a given topological structure specified by the Kirchhoff matrix $\mathbf{K}$ that is in a diluted solution at the θ temperature. In this case, the spatial interactions of units can be neglected, and one can calculate the nonperturbed dimensions of the molecules or their other conformational characteristics based on the Gaussian model, where the probabilities of different conformations for macromolecules are given by Eq. (5.1). Each conformation has a specific radius of gyration R_g, and its square is simply related to the matrix $\mathbf{R}$ considered in Section I.G, namely, $R_g^2 = \operatorname{Tr} \mathbf{R}\mathbf{R}^{\mathrm{T}}/l$. The distribution in the random quantity R_g^2 is obtained by integration of the probability measure (5.1) over all the conformations corresponding to a given R_g^2,

$$\mathscr{P}(R_g^2) = \int \delta(R_g^2 - \operatorname{Tr} \mathbf{R}\mathbf{R}^{\mathrm{T}}/l)\mathscr{P}(\mathbf{R})\, d\mathbf{R}. \tag{5.7}$$

This multidimensional integral can be reduced to a simple one,[75]

$$\mathscr{P}(R_g^2) = \frac{1}{2\pi} \int_{-\infty}^{\infty} D(\beta) \exp(i\beta R_g^2)\, d\beta, \qquad D(\beta) = \prod_{j=1}^{l-1} \left(1 + \frac{i\beta}{\lambda_j \gamma l}\right), \tag{5.8}$$

substituting the Fourier representation for the Dirac delta function and calculating the known Gaussian integral in $\mathbf{R}$. The characteristic function

$D(\beta)$, which is the Fourier transform of the distribution $\mathscr{P}(R_g^2)$, is a similar way of describing the system. The behavior of the distribution is determined by singularities of $D(\beta)$ in the complex plane, which are given, according to Eq. (5.8), by the spectrum of the Kirchhoff matrix, λ_j $(j=1,2,\ldots,l-1)$, for the graph of the macromolecule. In some simple cases, where the macromolecule is linear or a simple cycle, the integral in (5.8) can be calculated at once by means of the theorem of residue.[75,202–204] Numerical methods are necessary for the calculation of the distribution $\mathscr{P}(R_g^2)$ for more complicated topological structures; sometimes, the results permit an analytical approximation.[75,81]

When only the large-scale behavior of the distribution $\mathscr{P}(R_g^2)$ is interesting, incomplete information on the spectrum of the Kirchhoff matrix is sufficient. The asymptotics of $\mathscr{P}(R_g^2)$ for $R_g^2 \gg a^2$ is determined by the lower eigenvalues λ_j and the multiplicities of their degeneracy.[81]

The calculation of statistical moments for the random quantity R_g^2 can be performed by means of the generating function, which is the Laplace transform of the distribution $\mathscr{P}(R_g^2)$ and is expressed simply in terms of the characteristic function given in (5.8),

$$\int_0^\infty \exp(-\xi R_g^2)\mathscr{P}(R_g^2)\,dR_g^2 = D(-i\xi) = \exp\left\{-\frac{3}{2}\sum_{j=1}^{l-1} \ln\left(1+\frac{\xi}{\lambda_j \gamma l}\right)\right\}. \qquad (5.9)$$

To calculate the semi-invariants (cumulants) for the distribution $\mathscr{P}(R_g^2)$, one has to introduce their generating function,

$$\ln D(-i\xi) = \frac{3}{2}\sum_{k=1}^{\infty}\frac{1}{k}\left(-\frac{\xi}{\gamma l}\right)^k \sum_{j=1}^{l-1}\frac{1}{\lambda_j^k}. \qquad (5.10)$$

The linear term in the expansion in powers of $-\xi$ yields the mean square of the molecular radius of gyration, and the coefficient at ξ^2 is related to the fourth-order statistical moment,

$$\langle R_g^2\rangle = \frac{2a^2}{l}\sum_{j=1}^{l-1}\frac{1}{\lambda_j}, \qquad \langle R_g^4\rangle - \langle R_g^2\rangle^2 = \frac{8a^4}{3l^2}\sum_{j=1}^{l-1}\frac{1}{\lambda_j^2}. \qquad (5.11)$$

As an example, consider a macromolecule having the simplest cyclic topology (Fig. 5.4). The Kirchhoff matrix for the graph is cyclic, since every element $\mathbf{K}_{ij}$ depends only on the difference between the number of its row and column, $i-j$. Hence, one easily gets the spectrum[205]

$$\lambda_j = 2[1-\cos(2\pi j/l)] = 4\sin^2(\pi j/l), \qquad j=0,1,\ldots,l-1. \qquad (5.12)$$

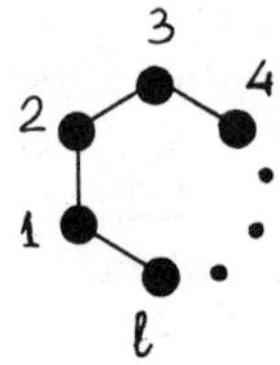

$$K = \begin{pmatrix} 2 & -1 & 0 & \cdots & 0 & -1 \\ -1 & 2 & -1 & \cdots & 0 & 0 \\ 0 & -1 & 2 & \cdots & 0 & 0 \\ \cdots & \cdots & \cdots & \cdots & \cdots & \cdots \\ -1 & 0 & 0 & \cdots & -1 & 2 \end{pmatrix}$$

Figure 5.4. Cyclic molecule with its Kirchhoff matrix.

Now we can sum exactly the first sum in (5.11),[204]

$$\langle R_g^2 \rangle = \frac{a^2}{2l} \sum_{j=1}^{l-1} \operatorname{cosec}^2 \frac{\pi j}{l} = \frac{a^2(l^2-1)}{6l}. \tag{5.13}$$

A typical problem in the statistical theory of polymers is the evaluation of the joint distribution in vectors $(\mathbf{x}_1, \ldots, \mathbf{x}_k, \ldots, \mathbf{x}_m)$, each being a linear combination of vectors $\{\mathbf{r}_i\}$, which are coordinates of the units,

$$\mathbf{x}_k = \sum_{i=1}^{l} \mathbf{r}_i f_{ik}, \qquad x_{\alpha k} = \sum_{i=1}^{l} r_{\alpha i} f_{ik}, \qquad \mathbf{X} = \mathbf{RF} \qquad (\alpha = 1, 2, 3; k = 1, \ldots, m). \tag{5.14}$$

Here the second and third equalities are matrix representations of the first equality, whereas the choice of concrete values of elements f_{ik} of the $l \times m$ matrix $\mathbf{F}$, corresponding to a linear transformation of the coordinates, depends on the particular problem to be solved. The integration of the probability measure (5.1) over the hypersurface in the conformation phase space $\{\mathbf{r}_i\}$, which is given by fixed values of the linear combinations of the coordinate vectors $\{\mathbf{r}_i\}$, Eqs. (5.14), results in[75]

$$\mathscr{P}(\mathbf{X}) = (\gamma/\pi)^{3m/2} |\mathbf{g}|^{-3/2} \exp[-\gamma \operatorname{Tr}(\mathbf{X}\mathbf{g}^{-1}\mathbf{X}^{\mathrm{T}})], \qquad \mathbf{g} = \mathbf{F}^{\mathrm{T}}\mathbf{K}^{+}\mathbf{F}. \tag{5.15}$$

This is the Wang–Uhlenbeck theorem.[206] The distribution in the random $3 \times m$ matrix $\mathbf{X}$ is determined only by a $m \times m$ matrix $\mathbf{g}$ that depends on the generalized inverse matrix $\mathbf{K}^+$ to the Kirchhoff matrix $\mathbf{K}$, according to (5.15).

As the $\mathbf{K}$ matrix spectrum contains zero eigenvalue, its determinant, $|\mathbf{K}|$, vanishes, so there is no inverse matrix $\mathbf{K}^{-1}$. Nevertheless, there is the generalized inverse[80] matrix $\mathbf{K}^+$, and its nature can be explained easily. The matrix $\mathbf{K}$ can be diagonalized with an orthogonal $l \times l$ matrix $\mathbf{U}$,

$$\mathbf{K} = \mathbf{U}^{\mathrm{T}}\mathbf{\Lambda}\mathbf{U} = (\mathbf{U}')^{\mathrm{T}}\mathbf{\Lambda}'\mathbf{U}', \tag{5.16}$$

where we have introduced an $(l-1) \times l$ submatrix $\mathbf{U}'$. The latter is obtained from U by omitting the first row, the elements of which are components of the eigenvector, corresponding to the zero eigenvalue. All other eigenvalues λ_j are elements of the diagonal $(l-1) \times (l-1)$ matrix $\mathbf{\Lambda}'$. Now the generalized inverse matrix to $\mathbf{K}$ is the rotated inverse matrix to $\mathbf{\Lambda}'$, namely,

$$\mathbf{K}^{+} = (\mathbf{U}')^{\mathrm{T}}(\mathbf{\Lambda}')^{-1}(\mathbf{U}'). \tag{5.17}$$

Various statistical characteristics of the Gaussian molecules are written in terms of elements k_{ij} of the matrix $\mathbf{K}^{+}$. For instance, the sum in the expression for $\langle R_g^2 \rangle$ in (5.11) is the trace of the matrix $\mathbf{\Lambda}'$, which equals $\mathrm{Tr}\,\mathbf{K}^{+}$, since the orthogonal transformation in (5.17) leaves the trace invariant. Thus, the mean square of the radius of gyration of macromolecules is given by

$$\langle R_g^2 \rangle = (2a^2/l)\,\mathrm{Tr}(\mathbf{K}^{+}). \tag{5.18}$$

The second expression in (5.15) can be written similarly, since the sum is just the trace of $(\mathbf{K}^{+})^2$.

The general Wang–Uhlenbeck formula, Eq. (5.15), enables one to get easily the distribution $\mathscr{P}_{ij}(\mathbf{x})$ in the distances $\mathbf{x} \equiv \mathbf{x}_1 = \mathbf{r}_i - \mathbf{r}_j$ between any pair of units with labels i and j. In this problem we have $m = 1$, and the matrix $\mathbf{F}$ is a single column vector with two nonzero elements $f_{i1} = 1$ and $f_{j1} = -1$. According to (5.15), the $m \times m$ matrix $\mathbf{g}$ is just a number,

$$g_{ij} = k_{ii}^{+} + k_{jj}^{+} - 2k_{ij}^{+}, \tag{5.19}$$

which equals, up to a factor of $2\gamma/3$, the dispersion of the Gaussian distribution,

$$\mathscr{P}_{ij}(\mathbf{x}) = (\gamma/\pi g_{ij})^{3/2} \exp(-x^2\gamma/g_{ij}). \tag{5.20}$$

Averaging over this distribution, one can calculate the specific contribution from a pair of units i and j to various conformational characteristics of the macromolecule, for example, $\langle R_g^2 \rangle$ or the mean hydrodynamical Stokes radius $\langle R_{\mathrm{St}}^{-1} \rangle^{-1}$. To calculate quantities of this kind, which can be evaluated experimentally, one has to sum the specific contributions from all pairs of molecule units,[164] namely,

$$\langle R_g^2 \rangle = l^{-2} \sum_{i<j} \langle (\mathbf{r}_i - \mathbf{r}_j)^2 \rangle = l^{-2} \sum_{i<j} g_{ij} \frac{3}{2\gamma}, \tag{5.21}$$

$$\langle R_{\mathrm{St}}^{-1} \rangle = \sum_{i<j} \langle |\mathbf{r}_i - \mathbf{r}_j|^{-1} \rangle = 2 \sum_{i<j} \left(\frac{\gamma}{\pi g_{ij}} \right)^{1/2}. \tag{5.22}$$

To get the generalized inverse Kirchhoff matrix, according to (5.17), one needs the total set of eigenvectors of **K** that correspond to nonzero eigenvalues. For a cyclic molecule, the result is[83]

$$k_{ij}^{+} = \frac{l^2 - 1}{12l} - \frac{|j-i|(l-|j-i|)}{2l}, \qquad g_{ij} = \frac{|j-i|(l-|j-i|)}{l}. \tag{5.23}$$

Setting these expressions in (5.18) and (5.21), one gets Eq. (5.13) for the mean square radius of gyration.

Starting from a definition of the two-point correlator of the molecular density of units of the (l, q)-isomer, one can get simple formulas for the Gaussian molecules, expressing the correlator and its Fourier transform in terms of elements of $\mathbf{K}^{+}$,

$$\chi_{lq}^{uu}(\mathbf{r}) = \frac{1}{V}\sum_{j}\sum_{i}\mathscr{P}_{ij}(\mathbf{r}) = \frac{1}{V}\left[\, l\delta(\mathbf{r}) + \sum_{j=1}^{l}\sum_{i\neq j}\mathscr{P}_{ij}(\mathbf{r})\right], \tag{5.24}$$

$$\tilde{\chi}_{lq}^{uu}(\mathbf{q}) = \frac{1}{V}\sum_{j}\sum_{i}\exp\left(-\frac{q^2 g_{ij}}{4\gamma}\right). \tag{5.25}$$

C. Spectrum of Kirchhoff Matrix for Regular Networks

Constructive analytical calculation of spectra of the **K** matrices for molecules having large numbers of units is possible only in the presence of a sufficient regularity in their topological structure. The limiting case of a regular graph is a lattice[207] (Fig. 5.5). As already noted, properties of the Gaussian molecules are determined solely by their topological structure, so one can neglect the position of the molecular graph vertices in the real three-dimensional space, dealing instead with the isomorphic graphs. When the topological structure of a molecular graph is regular, it can be represented as a lattice of

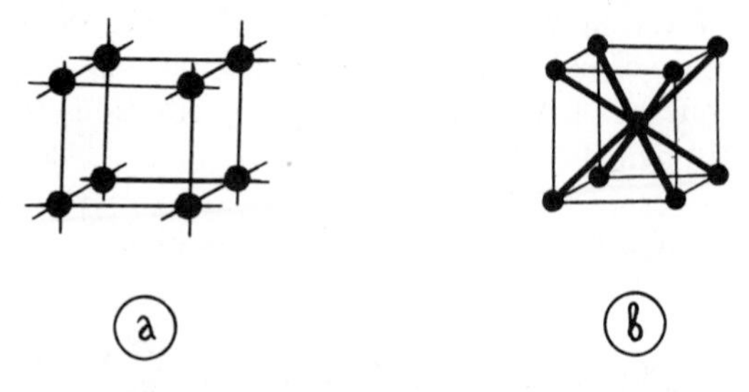

Figure 5.5. Fragments of (*a*) simple and (*b*) body-centered cubic lattices.

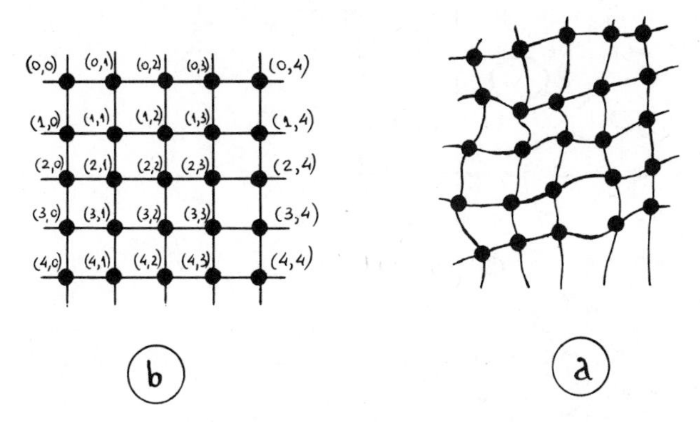

Figure 5.6. Topological structure of (*a*) real network and (*b*) equivalent regular two-dimensional lattice.

dimensionality d. For a linear molecule, $d = 1$, and for a network molecule similar to that shown in Fig. 5.6, $d = 2$. The position of a site in the d-dimensional lattice is parameterized with a vector $\mathbf{m}$, having integer components $m_1, \ldots, m_d$, which indicate the number of steps leading from the origin to the site in each of d basic directions (Fig. 5.6).

In the calculation of asymptotic dependences of network properties, as the number of their units l goes to infinity, the boundary effects become inessential,[207–209] so one can replace the spectrum of the Kirchhoff matrix for a real molecule (Fig. 5.6*a*) by that for a hypothetical molecule, which is obtained from the initial one by compactification into a torus (Fig. 5.7*a*).

The rows and columns of the Kirchhoff matrix $\mathbf{K}$ for the resulting graph, as well as the sites of a lattice in the torus, are given by d-dimensional vectors $\mathbf{m}$, integer components of which vary from 1 to an integer n. The position of a point in the torus can also be parameterized with a vector $\boldsymbol{\psi}$, the components of which are angles, $\psi_1, \ldots, \psi_d$, specifying different cross sections of the torus and acquiring discrete values, $\psi_\nu = 2\pi m_\nu / n$ $(m_\nu = 0, \ldots, n-1)$.

The matrix $\mathbf{K}$ has a block structure (Fig. 5.7*b*), corresponding to a distribution of all vertices between subsets, the cross sections. In each subset the vectors $\mathbf{m}$ and $\boldsymbol{\psi}$ are different only in their last components. Elements of the principal diagonal of $\mathbf{K}$ are equal to the coordination number of the lattice, an element $K(\mathbf{m}_1, \mathbf{m}_2) = -1$ if the nodes $\mathbf{m}_1$ and $\mathbf{m}_2$ are adjacent and zero otherwise. Because of the periodic boundary condition, the elements depend only on the difference of the lattice sites coordinates, $\mathbf{m} = \mathbf{m}_1 - \mathbf{m}_2$, so $K(\mathbf{m}_1, \mathbf{m}_2) = K(\mathbf{m})$. Matrices whose elements in each block row are the elements in the preceding row shifted one block to the right are called "circular" matrices. The characteristic matrix $\lambda\mathbf{E} - \mathbf{K}$, whose determinant

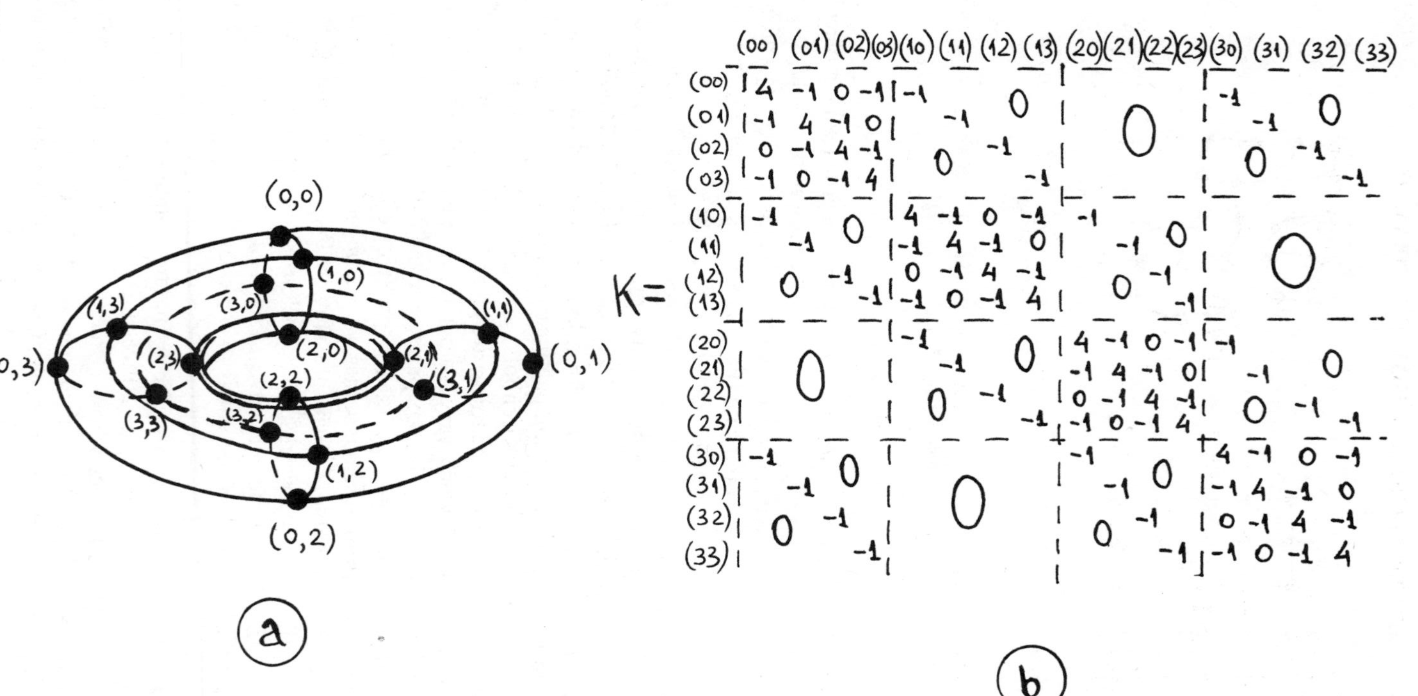

Figure 5.7. (*a*) Fragment of square lattice of Fig. 6*b* bended into torus by gluing respective vertices on opposite edges and (*b*) Kirchhoff matrix of this graph.

vanishes at the eigenvalues λ_j, has this circular property. Calculation of the determinants of circular matrices is a standard procedure[207] performed in terms of a connection function for the lattice $\mu_\lambda(\boldsymbol{\varphi})$,

$$|\lambda\mathbf{E}-\mathbf{K}|=\prod_{\varphi_1}\cdots\prod_{\varphi_d}\mu_\lambda(\boldsymbol{\varphi}). \tag{5.26}$$

In order to get the function $\mu_\lambda(\boldsymbol{\varphi})$, one must subtract from the diagonal elements of $\lambda\mathbf{E}-\mathbf{K}$, which are $\lambda-f$, other elements $K(\mathbf{m})$ multiplied by $\exp(i\boldsymbol{\varphi}\mathbf{m})$. As soon as these elements $K(\mathbf{m})$ are nonzero $[K(\mathbf{m})=-1]$, only for vectors $\mathbf{m}$ connecting adjacent sites of the lattice, the function μ_λ is the sum of $(\lambda-f)$ and exactly f terms,

$$\mu_\lambda(\boldsymbol{\varphi})=\lambda-f+\sum_{\mathbf{m}}\exp[i(\boldsymbol{\varphi}_1\mathbf{m}_1+\cdots+\boldsymbol{\varphi}_d\mathbf{m}_d)]. \tag{5.27}$$

Thus, the spectrum of the Kirchhoff matrix consists of n^d points $\lambda_{\mathbf{j}}$, which are given by, $\mu_\lambda(\boldsymbol{\varphi}(\mathbf{j}))=0$,

$$\lambda_{\mathbf{j}}=f\left\{1-\sum_{\mathbf{m}}f^{-1}\exp[i\mathbf{m}\boldsymbol{\varphi}(\mathbf{j})]\right\},\qquad 0\leqslant j_\nu\leqslant n-1. \tag{5.28}$$

The sum in braces is the characteristic function for a particle translation by one step in the process of its random walk in the lattice[207] if each of f possible directions $\mathbf{m}$ of the elementary step have equal probabilities.

Consider, for example, the random walk in a simple two-dimensional square lattice (Fig. 5.6). The possible directions $\mathbf{m}$ are four vectors, (0, 1), (0, − 1), (1, 0), and (− 1, 0), so that

$$\mu_\lambda(\boldsymbol{\varphi})=\lambda-4+\exp(i\varphi_1)+\exp(-i\varphi_1)+\exp(i\varphi_2)+\exp(-i\varphi_2),$$
$$\lambda_{j_1j_2}=4-2\cos(2\pi j_1/n)-2\cos(2\pi j_2/n). \tag{5.29}$$

Likewise, for three-dimensional cubic lattices—simple (SCL), face-centered (FCCL), and body-centered (BCCL)—the eigenvalues are, respectively,

$$\lambda_{\mathbf{j}}=6-2(\cos\varphi_1+\cos\varphi_2+\cos\varphi_3), \tag{5.30}$$

$$\lambda_{\mathbf{j}}=12-4(\cos\varphi_1\cos\varphi_2+\cos\varphi_1\cos\varphi_3+\cos\varphi_2\cos\varphi_3), \tag{5.31}$$

$$\lambda_{\mathbf{j}}=8-8\cos\varphi_1\cos\varphi_2\cos\varphi_3;\qquad \varphi_\nu=2\pi j_\nu/n. \tag{5.32}$$

The calculation of the mean-square radius, according to (5.11), at large n

is performed by integration over the d-dimensional hypercube,

$$\frac{\langle R_g^2 \rangle}{2a^2} = \frac{1}{n^d} \sum_{j \neq 0} \lambda_j^{-1} \simeq \left(\frac{1}{2\pi} \right)^d \int_0^{2\pi} \cdots \int_0^{2\pi} \frac{d\varphi_1 \cdots d\varphi_d}{f(1 - \sum_{\mathbf{m}} \exp(i\boldsymbol{\varphi}\mathbf{m}))}. \tag{5.33}$$

Integrals of this type have appeared in the problem of random walks, and they were calculated for some simple lattices.[207] If the particle returns inevitably, sooner or later, to the initial site, as it actually does in one- and two-dimensional lattices,[207,210] the integral is diverging. The integrals are finite for $d = 3$,[207,210] so the mean radius of gyration for an infinite three-dimensional network is comparable with the root-mean-square distance between adjacent sites. This result, which is known as a collapse of phantom networks, has been discovered earlier for SCL by means of an analogy between the polymer network and the resistance of an electric circuit.[211] It is known that the Kirchhoff matrix is exploited intensively in the calculations of electric circuits.[212] Equation (5.30), leading to the integral in (5.33), has been obtained by Eichinger[213] using the representation of the Kirchhoff matrix for SCL as a linear combination of matrix direct products,

$$\mathbf{K} = \mathbf{K}_c \otimes \mathbf{E} \otimes \mathbf{E} + \mathbf{E} \otimes \mathbf{K}_c \otimes \mathbf{E} + \mathbf{E} \otimes \mathbf{E} \otimes \mathbf{K}_c, \tag{5.34}$$

where $\mathbf{K}_c$ is the Kirchhoff matrix for a cyclic molecule (Fig. 5.4). Similarly, one can represent the Kirchhoff matrices for BCCL and FCCL and then evaluate the eigenvalues in (5.31) and (5.32). An advantage of using (5.26) and (5.28) for the evaluation of the matrix spectrum is a universal and simple approach that does not require the Kirchhoff matrices explicitly.

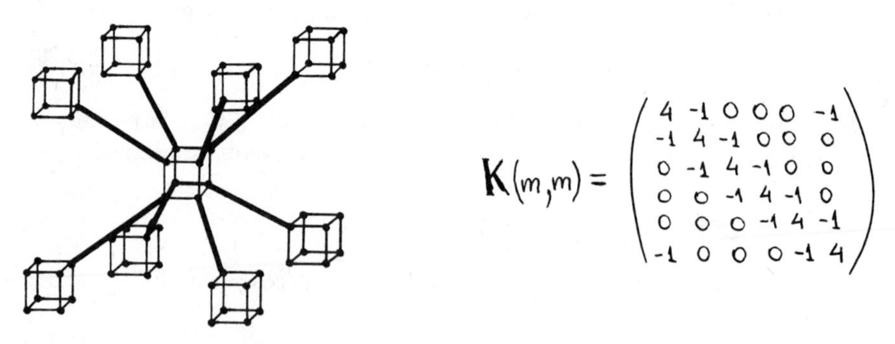

Figure 5.8. (a) Example of combined lattice obtained by substituting elementary simple cubic cell for vertices of body-centered cubic lattice of Fig. 5.5b and (b) matrix obtained from Kirchhoff matrix for cube by adding unity to each diagonal element.

Besides, the method we use is extended easily to complicated crystal lattices, the nodes of which are not single vertices but groups of q vertices, that is, elementary cells (Fig. 5.8). In the latter case, the Kirchhoff matrix for the lattice, $\mathbf{K}_q$, is constructed in a standard manner from $\mathbf{K}_1$ for the lattice with single vertices in every cell. To this end, each element $K_1(\mathbf{m}_1, \mathbf{m}_2)$ of the matrix $\mathbf{K}_1$ should be replaced by a $q \times q$ matrix $\mathbf{Q}(\mathbf{m}_1, \mathbf{m}_2)$, according to the following rules. Every zero element $K_1(\mathbf{m}_1, \mathbf{m}_2)$ is replaced by the zero matrix $\mathbf{Q}(\mathbf{m}_1, \mathbf{m}_2)$. Element $K_1(\mathbf{m}_1, \mathbf{m}_2) = -1$ is replaced by a matrix having elements $Q_{pr}(\mathbf{m}_1, \mathbf{m}_2)$, equal to -1 or 0, depending on whether the pth vertex of the $\mathbf{m}_1$ cell is connected with the rth vertex of the $\mathbf{m}_2$ cell. Any diagonal element $K_1(\mathbf{m}_i, \mathbf{m}_i)$ should be substituted for the Kirchhoff matrix of an elementary cell, with the number of edges connecting the corresponding vertex with vertices of other cells added to each diagonal element. As $\mathbf{K}_1$ is a circular matrix, the matrix $\mathbf{Q}$ depends, in fact, only on the difference $\mathbf{m} = \mathbf{m}_1 - \mathbf{m}_2$. In Fig. 5.9, we present an example of the Kirchhoff matrix $\mathbf{K}_2$ for a complex cyclic molecule constructed by means of that for a simple cycle, which is shown in Fig. 5.4.

The characteristic polynomial for the circular Kirchhoff matrix of a complex lattice is given by Eq. (5.26),[207] but in this case the function $\mu_\lambda(\boldsymbol{\varphi})$ is expressed in terms of the determinant for a $q \times q$ matrix,

$$\mu_\lambda(\boldsymbol{\varphi}) = |\lambda \mathbf{E} - \mathbf{Q}(\mathbf{0}) - \sum_{\mathbf{m}} \exp(i\mathbf{m}\boldsymbol{\varphi})\mathbf{Q}(\mathbf{m})|, \tag{5.35}$$

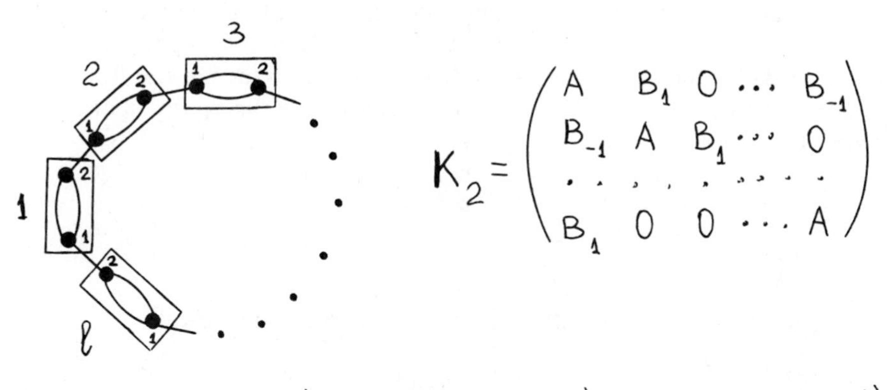

Figure 5.9. (*a*) Cyclic molecule built of elementary cells each involving two units and (*b*) its Kirchhoff matrix.

where the sum is over all the vectors **m** connecting the cell with adjacent ones.

The function $\mu_\lambda(\boldsymbol{\varphi})$ vanishes for $\lambda = \lambda_{\mathbf{j}}^{(r)}$ $(r = 1, \ldots, q)$ for each of n^d values of the vector $\boldsymbol{\varphi}(\mathbf{j})$. Thus, the spectrum of the Kirchhoff matrix consists of q separate branches; this is a feature specific for multiatomic lattices.[209] One branch corresponds to the so-called acoustic modes;[209] it vanishes for $\boldsymbol{\varphi} = \mathbf{0}$. The total number of eigenvalues is $l = qn^d$, which is the number of lattice sites.

As an example, consider a molecule obtained from the cyclic one of Fig. 5.4 by the substitution of each monomer unit for an elementary cell, consisting of two units (Fig. 5.9). According to Eq. (5.35), the factors in the characteristic polynomial (5.26) of its Kirchhoff matrix are

$$\mu_\lambda(\varphi) = \begin{vmatrix} \lambda - 3 & \exp(-i\varphi) \\ \exp(i\varphi) & \lambda - 3 \end{vmatrix} = \lambda^2 - 6\lambda + 4(1 - \cos\varphi), \tag{5.36}$$

and the eigenvalues belong to two branches, the first describing acoustic oscillations and the second describing optical oscillations,

$$\lambda_j^{(1)} = 3 - [5 + 4\cos\varphi(j)]^{1/2}, \qquad \lambda_j^{(2)} = 3 + [5 + 4\cos\varphi(j)]^{1/2}. \tag{5.37}$$

These two branches are separated; the corresponding eigenvalues belong to the intervals $(0, 2)$ and $(4, 6)$, respectively, whereas the interval $(2, 4)$ is empty.

D. Spectral Density and Dynamic Properties of Molecule

With an increase in the number of network units l, the eigenvalues λ, which are confined to a finite interval, become more dense and form the continuum in the limit $l \to \infty$. The spectral density $w(\lambda)$ is suitable for its description, where $w(\lambda)d\lambda$ is the fraction of eigenvalues belonging to the interval from λ to $\lambda + d\lambda$. Evidently, for finite networks, the density is a superposition of the Dirac δ functions,

$$w(\lambda) = \frac{\sum_j \delta(\lambda - \lambda_j)}{l}. \tag{5.38}$$

Equalities (5.11) can be rewritten in terms of the spectral density,[75]

$$\frac{\langle R_g^2 \rangle}{2a^2} = \int \lambda^{-1} w(\lambda)\, d\lambda, \qquad \frac{\langle R_g^4 \rangle - (\langle R_g^2 \rangle)^2}{2a^2} = \frac{2}{3l} \int \lambda^{-2} w(\lambda)\, d\lambda, \tag{5.39}$$

where the integration in λ is from a minimum nonzero eigenvalue $\lambda_{\min}$ to a maximum eigenvalue $\lambda_{\max}$. The generating function (5.9) for moments of the radius of gyration is represented similarly,[75]

$$\exp\left[-(3l/2)\int \ln(1+\xi/\gamma l\lambda)w(\lambda)\,d\lambda\right]. \tag{5.40}$$

The spectral density $w(\lambda)$ also characterizes the viscous-elastic properties of the polymer. In particular, the theories by Rouse[76] and Zimm[77] describe the relaxation behavior of linear molecules in terms of eigenvalues of some matrices. The matrix used by Zimm is the Kirchhoff matrix $\mathbf{K} = \mathbf{B}\mathbf{B}^{\mathrm{T}}$, whereas that used by Rouse, as well as its extension to tree molecules,[78] is $\mathbf{B}^{\mathrm{T}}\mathbf{B}$. Nonzero eigenvalues of these two matrices coincide for such molecules, so the methods are equivalent in this case.[75]

The **K**-matrix eigenvalues λ_j are related[75] to the molecule relaxation times τ_j by a simple relation, $\tau_j = \alpha/\lambda_j$, where α is a constant. Therefore, the stress relaxation module $G(t)$[214] is a sum of terms proportional to $\exp(-t/\tau_j)$. In terms of the spectral density, it is [75]

$$G(t) = lG_0\int \exp(-\lambda t/\alpha)w(\lambda)\,d\lambda = \int_{-\infty}^{\infty} \exp(-t/\tau)H(\tau)\,d(\ln\tau),$$
$$H(\tau) = H(\alpha/\lambda) = lG_0\lambda w(\lambda). \tag{5.41}$$

To calculate the spectral density $w(\lambda)$, one has to get a dispersion relation[209] between the eigenvalues λ and the "wave vector" $\boldsymbol{\varphi}/2\pi$. To perform the limit $l\to\infty$ in (5.38), which leads to the continuous spectrum and the integral instead of the sum, one should take a representation of the δ function. The result obtained in this way[209] is

$$w(\lambda) = \frac{1}{q}\sum_{r=1}^{q}\frac{1}{(2\pi)^d}\int\cdots\int\prod_{v=1}^{d} d\varphi_v\frac{1}{|\operatorname{grad}\lambda^{(r)}(\boldsymbol{\varphi})|}, \tag{5.42}$$

where the integration domain $\Omega_r(\lambda)$ is given by the dispersion relation $\lambda = \lambda^{(r)}(\boldsymbol{\varphi})$, and the index r enumerates q branches of the function $\lambda(\boldsymbol{\varphi})$. For instance, for a large cyclic molecule, according to (5.12), the spectrum has a single branch, $\lambda(\varphi) = 2(1-\cos\varphi)$. For such a molecule one has $m = 1$, so $\operatorname{grad}\lambda(\varphi)$ has a single component, $2\sin\varphi$. In this case (5.42) leads to

$$w(\lambda) = \frac{1}{2\pi}\int_0^{2\pi}\frac{\delta(\lambda-2+2\cos\varphi)}{2|\sin\varphi|}\,d\varphi = \frac{1}{\pi\sqrt{\lambda(4-\lambda)}}, \qquad 0<\lambda<4. \tag{5.43}$$

Since the periodic conditions are inessential for $l\to\infty$, as mentioned above, the eigenfunctions for a large linear molecule have the same limiting distribution. According to (5.41), the spectrum of the relaxation times $H(\tau)$ is proportional to $\tau^{-1/2}$ in the low-frequency region, $\tau \gg \alpha$, as expected from the Rouse theory.[76]

For a molecule that is a set of connected rings (Fig. 5.9), the spectral density in the $l \to \infty$ limit is calculated analogously by means of expressions (5.37) and (5.42),

$$w(\lambda) = \frac{1}{2}\sum_{r=1}^{2}\int_0^{2\pi} \frac{\delta(\lambda - 3 - (-1)^r\sqrt{5 + 4\cos\varphi})}{2\pi|\sin\varphi|}\sqrt{5 + 4\cos\varphi}\, d\varphi$$

$$= \frac{|\lambda - 3|}{\pi\sqrt{\lambda(2-\lambda)(4-\lambda)(6-\lambda)}}, \qquad (0 < \lambda < 2 \quad \text{or} \quad 4 < \lambda < 6). \qquad (5.44)$$

The methods exposed here for the calculation of spectra for networks are constructive only for regular structures, which are rather different from real polymer molecules. Networks having random topological structures have been investigated by computer simulation.[82,201,215–217] Analytical results[75,218] were based on the works of McKay,[219,220] who calculated the spectrum for a random network with identical degrees of all the vertices. All the possible enumerated graphs with fixed numbers of vertices were taken with equal probabilities; in other words, the differences in conformational entropy for the formation of cycles having various sizes and complexities were neglected completely. The result of that approach was a nonphysical conclusion that the number of cycles of any fixed size has a finite limit as the network size increases.[219,221] Evidently, the mutual spatial disposition of fragments of a growing molecule must be taken into account in the construction of the probability measure for networks.

CONCLUSION

The purpose of this chapter is to expose the power of the theoretical approaches and the variety of calculational methods that have been used recently for the description of the statistical properties of branched and network polymers. These method employ, more or less, the representation of polymer molecules as graphs, which makes it possible to formalize a number of problems in physics and chemistry of high polymer compounds. Their common feature is the fact that all experimentally observable characteristics of polymers are quantities obtained as averages over a set of configurations and conformations of molecules of the polymer sample. Therefore, one meets here certain problems of averaging in an ensemble of stochastic three-dimensional graphs. For equilibrium systems, the probability measure in the set of such graphs is given by the Gibbs distribution and is determined unambiguously by a chosen physicochemical model. The modern versions of the theory, taking into account intramolecular cycles and spatial physical interactions, require a new approach to the calculation of statistical properties of polymers. In our opinion the most effective approach is the use of field-theoretic methods; their potential power was shown in Section IV. The chemical physics of polymers must employ graphs here, since the diagram technique is the working language of field theory. One can predict with certainty that the concepts and methods of the theory of graphs will play a more important role in the development of the molecular theory of branched and network polymers, which involves more complicated polymer models.

APPENDIX: ELEMENTARY CONCEPTS OF THEORY OF GRAPHS

To date, no universal terminology has been adopted in the theory of graphs, so in those cases where standard textbooks in the field provide somewhat different definitions of some concepts, we choose those used more frequently in applications to physics and chemistry. Actually, we prefer more simple and spectacular definitions.

A *graph* is a figure consisting of points, called *vertices*, and segments, called *edges*, which connect some of the vertices. If the edges are directed, they are called *arcs*, and the graph is called *oriented* (*orgraph*). It is assumed usually that any two vertices can be connected by at most one edge. Otherwise, if multiple edges exist, the term *multigraph* is applied. If *loops* are possible besides, which are edges having both their endpoints at the same vertex, the graph is called a *pseudograph* (Fig. A.1). We do not exclude any such generalization in the following, providing special notes if necessary.

The accurate definition of a graph is as follows. *Graph G* is a combination of a set V of *vertices* v_i and a set X of nonordered pairs of vertices (v_i, v_j). Each pair $x = (v_i, v_j) = (v_j, v_i)$ is called an *edge* of the graph G, *connecting*

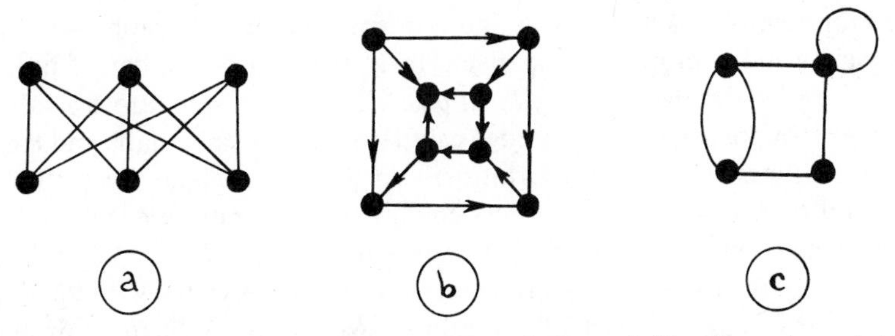

Figure A.1. (*a*) Graph; (*b*) orgraph; (*c*) pseudograph. Orgraph (*b*) and pseudograph (*c*) are planar, whereas graph (*a*) is nonplanar.

vertices v_i and v_j. If the pairs are ordered, that is, $(v_i, v_j) \neq (v_j, v_i)$, we have an *orgraph.* Every orgraph corresponds to a nonoriented graph, where vertices v_i and v_j are connected by an edge if and only if there is an arc (v_i, v_j) or (v_j, v_i). Thus, all the concepts defined for nonoriented graphs are valid also for orgraphs if their arcs are treated as edges. If the orientation of an arc is essential, we give special comments.

An edge $x = (u, v)$ is called *incident* to vertices u and v. An arc (u, v) *goes out* of the *initial* vertex u and *enters* the *final* vertex v. The vertices u and v, connected with an edge x, are called *adjacent.* A graph whose vertices are all adjacent is called *complete.* Two edges are *adjacent* if they are incident to the same vertex. *Degree* f of a vertex is defined as the number of edges incident to it. If an edge is a loop, it contributes twice to the vertex multiplicity. Some numbers (having, possibly, different signs) can be assigned to verices, edges, and arcs; the numbers are called *weights* of the elements. Such graphs are called *weighted.* Vertices of degree 1 are called pendant; those of degree 3 or higher are called *nodes.*

A course within a graph, that is, a sequence of transitions between adjacent vertices, is called a *walk.* A walk can be specified as a sequence of vertices $v_0, v_1, \ldots, v_n$ or as a sequence of edges $x_1 = (v_0, v_1)$, $x_2 = (v_1, v_2), \ldots, x_n = (v_{n-1}, v_n)$. The *walk length* is the number of edges n. If all the edges x_i are different, the walk is called a trail; if all the vertices v_i are different besides, it is a path. A close trail is called a *cycle.*

A graph is *connected* if any pair of its vertices is connected by a trail. An arbitrary nonconnected graph is a combination of its constituent connected graphs, which are called its connection *components.* The *distance* between two connected vertices is the length of the shortest walk connecting them. A graph is called *Eulerian* if it contains a Eulerian cycle with all its vertices and edges. Consequently, a Eulerian graph can be drawn with a continuous line, never passing an edge twice and returning to the start vertex. A cycle is called *Hamiltonian* if it passes its every vertex only once; respectively, a graph is called Hamiltonian if it contains a Hamiltonian cycle.

There is a number of representations of the same graph. For example, a cube graph can be sketched as a figure in space. Usually, graphs are drawn in a plane, and a graph can be given, in general, in a number of different pictures, called its *geometric realizations.* Edges of graphs are not always shown as straight segments; any edge can have an arbitrary shape and length. Besides, vertices can be posed arbitrarily in the figure plane. Distortions of edges and rearrangements of vertices change the graph outline (Fig. A.2) but conserve its topological structure, which is determined by the adjacency of its vertices. In the theory of graphs no information is associated with the shapes and lengths of edges. Clearly, the graph shown in Fig. A.2 can be drawn in such a manner that all intercepts of its edges are at its vertices. A graph that

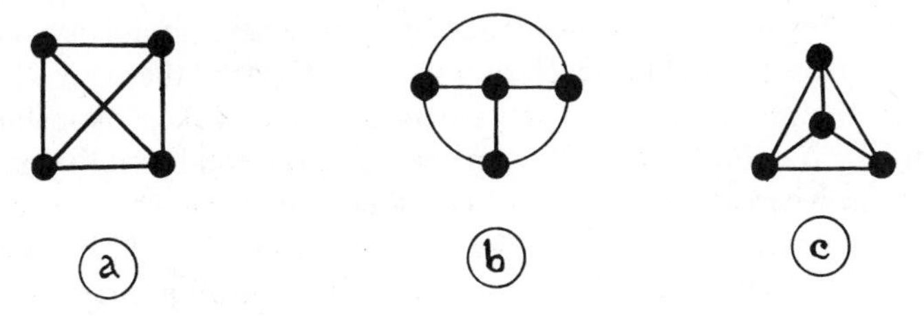

Figure A.2. Various geometric realizations of graph.

can be drawn in this way is called *planar*, and the corresponding geometric realization is called a *plane graph*.

Connected graphs with no cycles are called *trees*; they are especially important in application. A graph having trees as its connection components is called a *forest*. In the theory of graphs, a forest may be a single tree, in contrast to the common meaning of the word. A tree with n vertices has always $m = n - 1$ edges, that is, the minimum. This is true, since any two vertices are coupled by one edge, and each subsequent vertex is connected to a preceding vertex by one edge. Therefore, $n - 1$ is the necessary and sufficient number of edges connecting n vertices. If an edge is added to a tree, a cycle appears in the latter; if an edge is removed, the tree becomes a disconnected graph consisting of two trees.

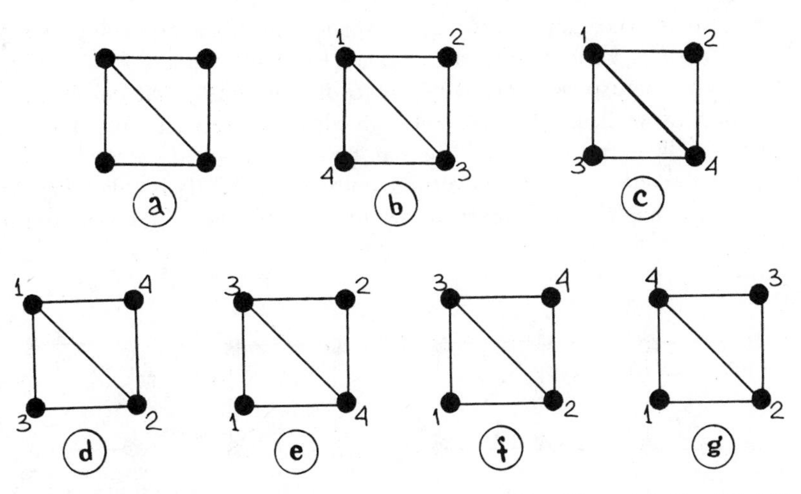

Figure A.3. Nonlabelled graph [(a)] and complete set of its different labelling [(b)–(g)].

Now we turn to *labeled* (*enumerated*) graphs, where nonidentical integers (from 1 to n) are assigned to each of n vertices. Clearly, there are $n!$ ways to label a graph, yet some of the ways may coincide. Take, for instance, the graph in Fig. A.3; only 6 enumerations (among the total number of $4! = 24$) are different, whereas each of the other 18 coincides with one of them. For example, the enumeration $(1, 4, 3, 2)$ is the same as $(1, 2, 3, 4)$ (the first one in Fig. A.3), since these two graphs coincide after a rotation around the edge connecting the vertices 1 and 3. Thus, the graphs are identical. Two (labeled) graphs are considered identical and are called *isomorphic* if there is a one-to-one mapping of one of them to the other, conserving the adjacency of the vertices and their labels. All the graphs isomorphic to that labeled with $(1, 2, 3, 4)$ (its *isomorphs*) are shown in Fig. A.4. Each of five other labeled graphs in Fig. A.3 has also $\mathscr{S} = 4$ isomorphs. So the total number of enumerations, $n! = 24$, is the product of the number of isomorphs of any particular enumeration, $\mathscr{S} = 4$, by the number of different enumerations, $W = 6$. Hence, one gets a known formula of the theory of graphs, $W = n!/\mathscr{S}$. Even though we have defined the number $\mathscr{S}$ for a set of enumerated graphs, it is meaningful also for nonenumerated graphs, since its value is independent of the enumeration. For each nonenumerated graph, the number $\mathscr{S}$, which is the *order* of its *automorphism group*, equals the number of isomorphisms for any corresponding enumerated graph. This number is a characteristic of the symmetry of nonenumerated graphs; the higher it is, the larger is the number of coinciding ways to label the dispositions. The value of $\mathscr{S}$ equals the number of automorphisms of the nonenumerated graph, that is, the number of such one-to-one mappings of the graph to itself, which do not change the adjacency relations.

When it is necessary to distinguish vertices or edges according to some internal features, various *colors* are assigned to the elements. For such colored graphs, an isomorphism means a one-to-one mapping that conserves the adjacency of vertices, the incidence of edges to vertices, and the color assignment. A graph whose vertices can be colored with two colors in such a way that each edge connects vertices colored differently is called *bipartite*.

If a graph has k vertices marked in some way, it is called the *k-root* graph,

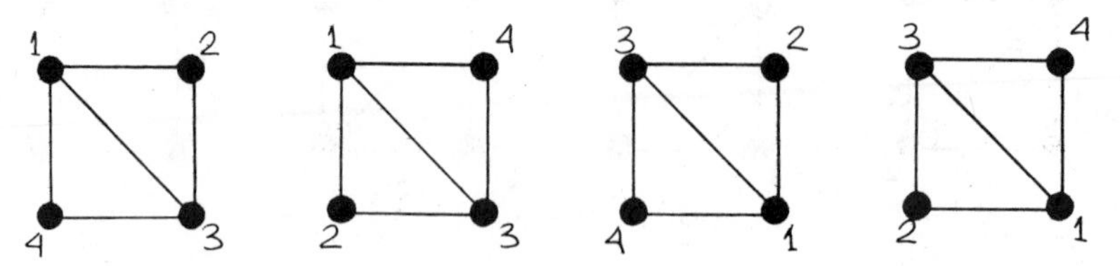

Figure A.4. Complete set of isomorphs for graph (b) in Fig. A.3.

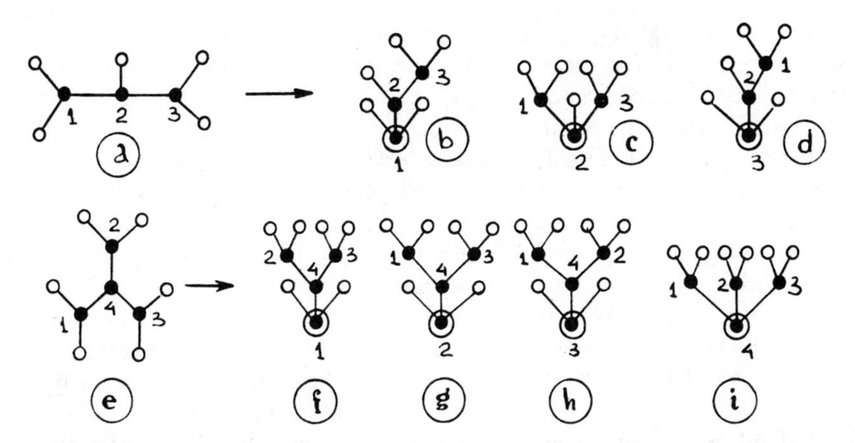

Figure A.5. Complete set of ordered rooted trees associated with nonordered rooted colored tree.

and the distinguished vertices are called *roots*. One-root graphs (or, simply, rooted graphs), trees in particular, are used most frequently. Each vertex is characterized by a number that is the length of the path connecting it to the graph root. This number is called the *number* of *generation* to which the vertex belongs. Any rooted tree is represented suitably as a genealogical tree, where all vertices belonging to a given generation are set at the same horizontal level. Any vertex of nonzero generations is the descendant of its adjacent node that belongs to the preceding generation. As seen in Fig. A.5, any rooted tree corresponds to a number of genealogical trees that differ in the orders of vertices originating from the same progenitor nodes. Such trees are called *ordered* rooted trees, since the disposition of descendants from a node can be interpreted as the order in which they have been born. Trees of this type are planar graphs; they are considered as identical if they can be transformed into each other with a continuous movement of their vertices in the graph plane with no intercepts of their edges and without passing through the horizontal line at the root level.

Besides the graphical picture of graphs as combinations of points (vertices) connected by lines (edges), there is also a matrix representation. Any graph corresponds to a matrix; given the matrix, the graph can be reconstructed unambiguously. There are various rules according to which such topological matrices can be compiled; two of them are used most frequently.

In the first method, all the n vertices of a graph are enumerated by integers from 1 to n. The corresponding $n \times n$ *adjacency matrix* has zeros on its diagonal, and its other elements a_{ij} are 0 or 1, depending on whether the ith vertex is connected to the jth vertex. For orgraphs, the arc directions are taken into account; $a_{ij} = 1$ only if the arc (v_i, v_j) belongs to the graph, so the adjacency matrix can be nonsymmetric. For multigraphs, a_{ij} equals the multiplicity of the edge connecting v_i and v_j; for pseudographs, the diagonal element a_{ii} equals the number of loops incident to the vertex v_i.

In the other method we deal with the *incidence matrix*, and n graph vertices and its m edges must be enumerated independently. The matrix has n rows and m columns; its element b_{ij} equals 1 if the edge x_j is incident to the vertex v_j; otherwise, it equals zero. For orgraphs, $b_{ij} = 1$ if the arc x_j goes out of the vertex v_i, and $b_{ij} = -1$ if the arc enters the vertex.

If an adjacency matrix (or an incidence matrix) is given, the corresponding nonlabeled graph is reconstructed directly. The inverse procedure is not unique since the graph vertices and edges can be enumerated arbitrarily.

We shall define now some concepts relevant to graph anatomy. If some edges and/or vertices (together with edges incident to the latter) of a graph are removed, one gets its *subgraph.* For instance, the graph of Fig. A.3 is a subgraph of that in Fig. A.2. If only edges are removed, one gets a *spanning subgraph. Spanning trees* are most important among such subgraphs. Any arbitrary connected cyclic graph can be transformed into a spanning tree by removing a number of edges. This number, r, is quite definite for each graph; it is called a *cyclomatic number* (*cyclic rank*) and is simply related to the numbers of the graph edges, m, and vertices, n; namely, $r = m - n + 1$. Clearly, $r = 0$ for trees, and $r > 0$ for other connected graphs. The graphs given in Figs. A.2 and A.3 have $r = 3$ and $r = 2$, respectively.

For a given graph, its spanning tree can be constructed by a consecutive inspection of all edges in two ways: (1) The edge considered belongs to the spanning tree if it is not in a cycle with other remaining edges. The procedure is continued until $n - 1$ edges remain, constituting the spanning tree. (2) The edge considered is removed from the graph if it constitutes a cycle with other edges until r edges are removed. These edges, called *chords*, together with $n - 1$ edges of the spanning tree, constitute the complete set of all the graph edges. In general, an arbitrary graph can have several spanning trees, each of them corresponding to a set of r chords.

There are some connected graphs that can be made disconnected by removing an element. A vertex, the removal of which (together with its incident edges) increases the number of graph components, is called an *cutpoint.* An edge having the same property is called a *bridge.* Other graph edges are called *cyclic* (belonging to a cycle); such also are loops in pseudographs. A connected graph having cutpoint is called *separable.* An arbitrary graph can be divided in *blocks*, each of them being a maximal *nonseparable* subgraph. Blocks are connected only at cut points. The graphs shown in Figs. A.1–A.3 are blocks; an example of a separable graph and the set of its blocks is shown in Fig. A.6. *Leaves* of that graph are also shown, which appear after all the

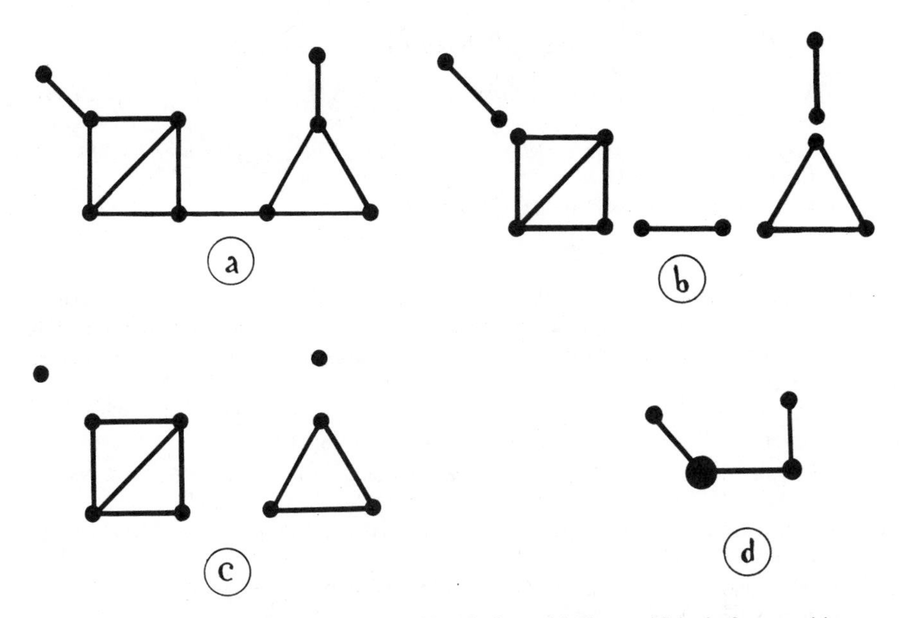

Figure A.6. (*a*) Separable graph; (*b*) set of its blocks and (*c*) leaves; (*d*) its leaf composition.

bridges are removed. Evidently, every nontrivial leaf is a maximal subgraph, that is, a connected set of cyclic edges. The trivial leaf is a single vertex. A graph obtained from the original one by *contracting* all vertices of every leaf to a single vertex is called the *leaf composition.* Thus, vertices of any leaf composition represent the leaves, and its edges are bridges of the initial graph. Clearly, for any graph, its leaf composition is a tree (Fig. A.6).

Note Added in Proof: In the case of small density of gel units in the domain of applicability of the equations (4.118–4.120) spinodal transition takes place at $\pi > 0$. The mentioned above of changing a sign of osmotic pressure π in the scope of the model, which has been under discussion in section IV, is possible only in a dense gel. As it has been shown by our consideration of gel obtained by crosslinking of linear chains in their functional groups (the result of which will be published elsewhere) the effect mentioned above can be observed also in the domain of small density of network.

Acknowledgments

We like to express our deep gratitude to Yu. L. Klimontovich and M. V. Volkenstein for their support and encouragement of this research. We are indebted to A. R. Khokhlov for his critical reading of the manuscript and to M. Gordon and K. Dušek for valuable discussions.

References

1. M. Gordon and W. B. Temple, in *Chemical Applications of Graph Theory*, Academic Press, New York, 1976, p. 299.
2. S. I. Kuchanov, *Methods of Kinetic Calculations in Polymer Chemistry* (in Russian), Chemistry, Moscow, 1978.
3. K. F. Freed, *Adv. Chem. Phys.* **22**, 1 (1972).
4. Z. Slanina, *Theoretical Aspects of Isomerism in Chemistry* (in Czech), Academia, Praha, 1981.
5. V. M. Tatevskii, V. A. Benderskii, and S. S. Yarovoi, *Rules and Methods for Calculation the Physicochemical Properties of Paraffinic Hydrocarbons*, Pergamon Press, Oxford, 1961.
6. S. S. Yarovoi, *Methods for the Calculation of Physical and Chemical Properties of Polymers* (in Russian), Chemistry, Moscow, 1978.
7. M. V. Volkenstein, *Configurational Statistics of Polymer Chains*, Interscience, N.Y., 1963.
8. P. J. Flory, *Statistical Mechanics of Chain Molecules*, Interscience, New York, 1969.
9. V. V. Korschak and S. V. Vinogradova, *Equilibrium Polycondensation* (in Russian), Nauka, Moscow, 1968.
10. L. B. Sokolov, *Principles of the Polycondensation Method of Polymer Synthesis* (in Russian), Chemistry, Moscow, 1979.
11. G. Schill, *Catenanes, Rotaxanes and Knots*, Academic Press, New York, 1971.
12. R. B. King (ed.), *Chemical Applications of Topology and Graph Theory*, Elsevier, Amsterdam, 1983.
13. J. W. Essam and J. W. Fischer, *Rev. Mod. Phys.* **42**, 272 (1970)
14. P. J. Flory, *Principles of Polymer Chemistry*, Cornell University Press, Ithaca, New York, 1953.
15. M. Gordon and G. R. Scantlebury, *Trans. Faraday Soc.* **60**, 604 (1964).
16. W. J. Stockmayer, *J. Chem. Phys.* **11**, 45 (1943).
17. M. Gordon and W. B. Temple, *J. Chem. Soc. A* **N5**, 729 (1970).
18. I. J. Good, *Proc. Cambridge Philos. Soc.* **61**, 499 (1965).
19. I. J. Good, *Proc. Cambridge Philos. Soc.* **56**, 367 (1960).
20. F. Harary and E. M. Palmer, *Graphical Enumeration*, Academic Press, New York, 1973.
21. T. E. Harris, *The Theory of Branching Processes*, Springer-Verlag, Berlin, 1963.
22. M. Gordon, *Proc. Royal Soc. (London), A* **268**, 240 (1962).
23. K. Dušek, *Makromol. Chem. Suppl.* **2**, 35 (1979).
24. S. V. Korolev, S. I. Kuchanov, and M. G. Slinko, *Dokl. Akad. Nauk USSR* **262**, 1422 (1982).
25. S. I. Kuchanov, S. V. Korolev, and M. G. Slinko, *Polymer J.* **15**, 775 (1983).
26. S. V. Korolev, S. I. Kuchanov, and M. G. Slinko, *Polymer J.* **15**, 785 (1983).
27. E. B. Brun and S. I. Kuchanov, *Vysokomol. Soedin., A* **21**, 1393 (1979).
28. E. B. Brun and S. I. Kuchanov, *Vysokomol. Soedin., A* **21**, 691 (1979).
29. D. W. Matula, L. C. D. Goenweghe, and J. R. Van Waser, *J. Chem. Phys.* **41**, 3105 (1964).
30. D. W. Matula, *Annals N. Y. Acad. Sci.* **159**, Part. 1, 314 (1969).
31. P. Luby, *J. Chem. Phys.* **78**, 1083 (1974).
32. D. Durand and C.-M. Bruneau, *Discr. Appl. Math.*, **3**, 79 (1981).
33. D. Durand and C.-M. Bruneau, *Macromolecules* **12**, 1216 (1979).

34. M. Gordon and J. A. Torkington, *Discr. Appl. Math.* **2**, 207 (1980).
35. M. Gordon and G. N. Malcolm, *Proc. Royal Soc. (London), A* **295**, 29 (1966).
36. P. Whittle, *Proc. Cambridge Philos. Soc.* **61**, 475 (1965).
37. P. Whittle, *Proc. Royal Soc. (London)* **A285**, 501 (1965).
38. P. Whittle, *Adv. Appl. Prob.* **12**, 94 (1980).
39. P. Whittle, *Adv. Appl. Prob.* **12**, 116 (1980).
40. P. Whittle, *Adv. Appl. Prob.* **12**, 135 (1980).
41. P. Whittle, *Theory Prob. Appl.* **26**, 350 (1981).
42. S. V. Korolev, S. I. Kuchanov, and M. G. Sliňko, *Dokl. Akad. Nauk USSR* **263**, 633 (1982).
43. M. Gordon and W. B. Temple, *J. Chem. Soc.* (Faraday II), **69**, 282 (1973).
44. G. Pólya, *Acta Math.* **68**, 145 (1937).
45. F. H. Harris, *J. Chem. Phys.* **23**, 1518 (1955).
46. H. Jacobson and W. J. Stockmayer, *J. Chem. Phys.* **18**, 1600 (1950).
47. F. Harary, *Graph Theory*, Addison-Wesley, London, 1969.
48. S. I. Kuchanov, S. V. Korolev, and M. G. Sliňko, *Vysokomolek. Soedin., A* **24**, 2160 (1982).
49. S. V. Korolev, S. I. Kuchanov, and M. G. Sliňko, *Vysokomolek. Soedin., A* **24**, 2170 (1982).
50. C. A. J. Hoeve, *J. Polym. Sci.* **21**, 11 (1956).
51. V. E. Stepanov, in *On Cybernetics* (in Russian), USSR Academy of Sciences, Moscow, 1973, p. 164.
52. C.-M. Bruneau, *Ann. Chim.* **1**, 271 (1966).
53. C.-M. Bruneau and P. Burgand, *Compt. Rend. C* **265**, 1422 (1967).
54. C.-M. Bruneau and P. Burgand, *Compt. Rend., C* **266**, 191 (1968).
55. M. Gordon and G. R. Scantlebury, *J. Chem. Soc., B* **N1**, 1 (1967).
56. M. Gordon and G. R. Scantlebury, *Proc. Royal Soc. (London), A* **292**, 380 (1966).
57. M. Gordon and G. R. Scantlebury, *J. Polym. Sci., C* **N16**, 3933 (1968).
58. V. I. Irzak, L. I. Kuzub, and N. S. Enikolopyan, *Dokl. Acad. Nauk USSR* **201**, 1382 (1971).
59. V. I. Irzak, *Vysokomolek. Soedin., A* **17**, 529 (1975).
60. B. G. Ozol' -Kalnin and J. A. Gravitis, *Vysokomolek. Soedin., B* **24**, 329 (1982).
61. I. Ya. Yerukhimovich, *Vysokomolek. Soed., A* **20**, 114 (1978).
62. I. Ya. Yerukhimovich, in *Mathematical Methods for Polymer Investigations*, Proc. 11 Nat. Conf., USSR Academy of Sciences, Pushchino, 1982, p. 52.
63. S. I. Kuchanov, S. V. Korolev, and M. G. Sliňko, *Dokl. Akad. Nauk USSR*, **267**, 122 (1982).
64. I. J. Good, *Proc. Royal Soc. (London), A* **272**, 54 (1963).
65. G. R. Dobson and M. Gordon, *J. Chem. Phys.* **41**, 2389 (1964).
66. P. J. Flory, *Proc. Royal Soc. (London), A* **351**, 351 (1976).
67. P. J. Flory, *Polymer* **20**, 1317 (1979).
68. P. J. Flory, *Macromolecules* **15**, 99 (1982).
69. J. Scanlan, *J. Polym. Sci.* **43**, 501 (1960).
70. L. C. Case, *J. Polym. Sci.* **45**, 397 (1960).
71. G. R. Dobson and M. Gordon, *J. Chem. Phys.* **43**, 705 (1965).
72. B. E. Eichinger, *Ann. Rev. Phys. Chem.* **34**, 359 (1983).

73. L. S. Priss, Preprint, Biolog. Data Science Center, Pushchino, 1981.
74. I. V. Irzak, B. A. Rozenberg, and N. S. Enikolopyan, *Network Polymers*, Nauka, Moscow, 1979.
75. B. E. Eichinger, *Macromolecules* **13**, 1 (1980).
76. P. E. Rouse, *J. Chem. Phys.* **21**, 1272 (1953).
77. B. N. Zimm, *J. Chem. Phys.* **24**, 269 (1965).
78. W. C. Forsman, *J. Chem. Phys.* **65**, 4111 (1976).
79. W. C. Forsman, *Macromolecules* **1**, 343 (1968).
80. A. Ben-Israel and T. N. E. Greville, *Generalized Inverses: Theory and Applications*, Wiley-Interscience, New York, 1974.
81. J. E. Martin and B. E. Eichinger, *J. Chem. Phys.* **69**, 4588 (1978).
82. B. E. Eichinger, *J. Chem. Phys.* **75**, 1964 (1981).
83. J. E. Martin and B. E. Eichinger, *Macromolecules* **16**, 1345 (1983).
84. R. J. Baxter, *Exactly Solved Models in Statistical Mechanics*, Academic Press, London, 1982.
85. A. Z. Patashinsky and V. A. Pokrovsky, *Fluctuation Theory of Phase Transitions* (in Russian), Nauka, Moscow, 1982.
86. S.-K. Ma, *Modern Theory of Critical Phenomena*, W. A. Benjamin, London, 1976.
87. P. G. De Gennes, *Scaling Concepts in Polymer Physics*, Cornell University Press, Ithaca, London, 1979.
88. D. Stauffer, A. Coniglio, and M. Adam, *Adv. Polym. Sci.* **44**, 103 (1982).
89. D. Stauffer, *Phys. Rep.* **54**, 1 (1979).
90. S. G. Whittington, in *The Mathematics and Physics of Disordered Media*, A. Donald and B. Eckman (eds.), Springer-Verlag, Berlin, 1983, p. 283.
91. D. J. Klein, *J. Chem. Phys.* **75**, 5186 (1981).
92. J. W. Essam, *Repts. Progr. Phys.* **43**, 833 (1980).
93. N. Boccaza, M. Daoud (eds.), *Physics of Finely Divided Matter*, Springer Verlag, N.Y., 1985.
94. M. E. Fisher and J. W. Essam, *J. Math. Phys.* **2**, 609 (1961).
95. D. Stauffer, *J. Chem. Soc. Faraday 11* **72**, 1354 (1976).
96. P. G. De Gennes, *J. Physique* **36**, 1049 (1975).
97. D. Stauffer, *Physica, A* **106**, 177 (1981).
98. H. E. Stanley, P. J. Reynolds, S. Render, and F. Family, in *Real Space Renormalization*, T. W. Burkhardt and J. M. J. van Leeuwen (eds.), Springer-Verlag, Heidelberg, 1982, p. 169.
99. D. J. Klein and W. A. Seitz, in *Chemical Applications of Topology and Graph Theory*, R. B. King (ed.), Elsevier, Amsterdam, 1983, p. 430.
100. A. M. Elyaschevich, Preprint, Chernogolovka, Publ. Inst. Chem. Phys. USSR Acad. Sci., 1985.
101. S. P. Obukhov, Preprint, Publ. USSR Academy of Sciences, Pushchino, 1985.
102. J. W. Essam, in *Phase Transitions and Critical Phenomena*, Vol. 2, C. Domb and M. S. Green (eds.), Academic Press, New York, 1972, p. 197.
103. P. Agrawal, S. Render, P. J. Reynolds, and M. E. Stanley, *J. Phys. A* **12**, 2073 (1979).
104. R. Nakanishi and P. J. Reynolds, *Phys. Lett. A* **71**, 252 (1979).
105. B. Shapiro, *J. Phys. C* **12**, 3185 (1979).
106. A. Coniglio, M. E. Stanley, and W. Klein, *Phys. Rev. Lett.* **42**, 518 (1979).

107. A. Coniglio and T. C. Lubensky, *J. Phys. A* **13**, 1783 (1980).
108. A. Coniglio, M. E. Stanley, and W. Klein, *Phys. Rev. B* **25**, 6805 (1982).
109. D. W. Beerman and D. Stauffer, *J. Phys. B* **44**, 339 (1981).
110. A. Coniglio and W. Klein, *J. Phys. A* **13**, 2775 (1980).
111. T. C. Lubensky and J. Isaacson, *Phys. Rev. Lett.* **41**, 829 (1978).
112. T. C. Lubensky and J. Isaacson, Errata, *Phys. Rev. Lett.* **42**, 410 (1979).
113. T. C. Lubensky, J. Isaacson, *Phys. Rev. A* **20**, 2130 (1979).
114. T. C. Lubensky, J. Isaacson, *J. Physique* **42**, 175 (1981).
115. A. Coniglio, *J. Phys. A* **16**, 187 (1983).
116. G. Parisi and N. Soulras, *Phys. Rev. Lett.* **46**, 871 (1981).
117. F. Family and A. Coniglio, *J. Phys. A* **13**, 1403 (1980).
118. R. W. Kasteleyn and C. M. Fortuin, *J. Phys. Soc. Japan*, Suppl., **26**, 11 (1969).
119. C. M. Fortuin and R. W. Kasteleyn, *Physica* **57**, 536 (1972).
120. F. Y. Wu, *J. Stat. Phys.* **18**, 115 (1978).
121. P. G. De Gennes, *J. Physique* **38**, L355 (1977).
122. D. Stauffer and A. Coniglio, *Z. Phys. B* **38**, 267 (1980).
123. M. Daoud, *J. Physique Lett.* **40**, L201 (1979).
124. A. Coniglio and M. Daoud, *J. Phys. A* **12**, L259 (1979).
125. G. Ord and S. G. Wittington, *J. Phys. A* **13**, L307 (1980).
126. A. J. Guttman and S. G. Wittington, *J. Phys. A* **15**, 2267 (1982).
127. S. I. Kuchanov and S. V. Korolev, in Transactions of Intern. Conf. Rubber 84, Sec. A, Vol. 3, No. 84, Moscow, 1984.
128. D. Stauffer, *Pure Appl. Chem.* **53**, 1479 (1981).
129. M. Gordon and J. A. Torkington, *Pure Appl. Chem.* **53**, 1461 (1981).
130. A. B. Harris and T. C. Lubensky, *Phys. Rev. B* **23**, 3591 (1981).
131. A. B. Harris and T. C. Lubensky, *Phys. Rev. B* **24**, 2656 (1981).
132. S. F. Edwards and K. F. Freed, *J. Phys. C* **3**, 739, 750, 760 (1970).
133. I. Ya. Yerukhimovich, Candidate's Dissertation, Char'kov, 1979.
134. S. V. Panyukov, *Zh. Eksp. Teor. Fiz.* **88**, 1795 (1985); *Sov. Phys. JETP* **61**, 1065 (1985).
135. S. I. Kuchanov, Dokl. *Akad. Nauk USSR*, (1985).
136. G. G. Lowry (ed.), *Markov Chains and Monto-Carlo Calculation in Polymer Science*, Marcel Dekker, New York, 1970.
137. B. D. Coleman and T. G. Fox, *J. Polym. Sci. Part A* **1**, 3183 (1963).
138. H. L. Frish, C. L. Mallows, and F. A. Bovey, *J. Chem. Phys.* **45**, 1565 (1966).
139. A. Ziabicky and J. Walasek, *Macromolecules* **11**, 471 (1978).
140. S. I. Kuchanov, S. V. Korolev, and M. G. Sliňko, *Vysokomol. Soedin. A* **26**, 263 (1984).
141. M. Gordon and J. W. Kennedy, *J. Chem. Soc. Faraday 11* **69**, 484 (1973).
142. J. W. Essam, J. W. Kennedy, and M. Gordon, *J. Chem. Soc. Faraday II* **73**, 1289 (1977).
143. M. Gordon and J. W. Kennedy, *SIAM J. Appl. Math.* **28**, 376 (1975).
144. J. W. Kennedy and M. Gordon, *Ann. N.Y. Acad. Sci.* **319**, 331 (1979).
145. R. Becker and G. Neumann, *Plast. Kautsch. B* **20**, 809 (1973).
146. R. Becker, *Plast. Kautsch. B* **22**, 790 (1975).
147. T. I. Ponomareva, V. I. Irzak, and B. A. Rozenberg, *Vysokomol. Soedin. A* **20**, 597 (1978).

215. A. M. Elyashevich, *Polymer* **20**, 1382 (1979).
216. I. S. Remeev and A. M. Elyashevich, *Vysokomol. Soedin. A* **27**, 629 (1985).
217. Y.-K. Leung and B. E. Eichinger, *J. Chem. Phys.* **80**, 3885 (1984).
218. B. E. Eichinger, J. E. Martin, and B. D. McKay, *ACS Polymer Prepr.* **22**, 155 (1981).
219. B. D. McKay, Mathematical Research Report N 9, University of Melbourne, 1979.
220. B. D. McKay, *Linear Algebra Appl.* **40**, 203 (1981).
221. N. C. Wormald, Ph.D. Thesis, Department of Mathematics, University of Newcastle, 1978.

AUTHOR INDEX

Numbers in parentheses are reference numbers and indicate that the author's work is referred to although his name is not mentioned in the text. Numbers in *italics* show the pages on which the complete references are listed.

Adam, M., 155(88), 156(88), 162(88), 165(88), *322*
Agrawal, P., 161(103), 163(103), 196(103), *322*
Aksel'rod, B. Ya., 173(151), *323*
Alekseeva, S. G., 173(151), *323*
Alexander, M., 74(63), 77(65), *112*
Alexander, M. H., 82(92), *113*
Alkamade, C. Th. J., 48(43), *112*
Allan, R. J., 39(20), 57(54), 59(54), 77(20), *111, 112*
Allegra, G., 308(211), *325*
Allegrini, M., 38(6), *110*
Andersen, N., 40(32, 33), 45(34), 77(33), *111, 112*
Andersen, T., 54(50), *112*
Anderson, N., 44(32), 54(47–50), *111, 112*
Andrew, K. L., 21(32), 22(32), *35*
Archirel, P., 101(119), *114*
Arshava, B. M., 173(151), *323*

Bähring, A., 39(13), 54(46), 55(13, 51–53), 58(52), 64(46), 66(13), 67(46, 52), 68(46), 69(46, 60), 82(60), 83(60), 84(52), 85(60), *111, 112*
Baig, M. A., 31(40), 33(43), *35, 36*
Balko, B. A., 39(16), *111*
Barbier, L., 39(27, 28), 77(27, 28), *111*
Barg, M. A., 29
Barker, J. R., 81(80), 97(80), 101(80), 102(80), *113*
Batson, C. H., 1(5), *34*
Baxter, R. J., 153(84), *321*
Becker, R., 173(145, 146), *323*
Beerman, D. W., 164(109), *322*
Beijerinck, H. C. W., 39(26), 77(26), *111*
Bender, Ch., 48(44), *112*
Benderskii, V. A., 123(5), 124(5), 173(5), *319*
Ben-Israel, A., 153(80), 302(80), *321*
Berge, K., 221(176), *324*
Berkowitz, J., 1(1–6, 8, 9), 2(10), 3(3, 4, 6, 8, 9, 11), 5(1), 6(3, 4), 7(3, 4), 8(8, 9), 16(1), 18(10), 19(8, 9, 26, 27), 20(26–28), 22(28), 24(28), *34, 35*
Berry, H. G., 4(14), *35*
Beyer, W., 48(44), *112*
Bijl, P., 82(91), *113*
Blais, N. C., 97(105), 101(117, 118), 102(117), *113, 114*
Blum, K., 40(31), *111*
Bochner, B., 193(173), *324*
Bonacić-Koutecký, V., 97(108), 100(114), 104(114), 106(114), 109(114), *114*
Bonanno, R., 77(76), *113*
Botschwina, P., 100(112), 101(112), *114*
Boulmer, J., 77(75), 78(75), *113*
Bovey, F. A., 171(138), 174(138), 178(138), *323*
Breckenridge, W. H., 97(100), *113*
Bregel, T., 39(20, 22), 77(20, 22), *111*
Briggs, J. S., 45(36), *111*
Brochu, M., 4(15), *35*
Brout, R., 254(186), 255(186), 256(186), 275(186), *324*
Brown, C. M., 25(35), 26(33), 27(34, 35), *35*
Brun, E. B., 138(27, 28), *320*
Bruneau, C. M., 140(32, 33), 146(52–54), *320, 321*
Burchard, W., 184(155, 159, 161–164), 185(155, 159), 211(164), 303(164), *323, 324*
Burgand, P., 146(53, 54), *321*
Burke, P. G., 29(38), 30(38), *35*
Burnett, K., 62(59), *112*
Bussert, W., 4(12), *35,* 39(20–23), 70(62), 76(62b), 77(20–23, 62b), *111, 112*

Campbell, E. E. B., 45(35), 61(35), 91(35), 92(35), 93(35), 95(35), 96(35), *111*
Campbell, M. L., 82(93), *113*
Cardon, B. L., 29
Carter, S. L., 27(36), 28(36), *35*
Case, L. C., 150(70), *321*
Celotta, R. J., 82(84, 85), *113*
Chaves, C. M., 290(194), 291(194), *325*
Cheret, M., 39(27, 28), 77(27, 28), *111*
Cho, H., 19(27), 20(27, 28), 22(28), 24(28), *35*
Chupka, W. A., 1(6, 7), 3(6, 7), 19(26), 20(26), *35*
Clark, C. W., 33(41), *36*
Coleman, B. D., 171(137), *323*
Condon, E. U., 9(17), *35*
Coniglio, A., 155(88), 156(88), 162(88, 106), 163(106–108), 164(107, 110,115), 165(88), 167(115), 169(122, 124), 290(106–108, 110, 115, 117, 124), 291(107, 196), *322, 323*
Connerade, J. P., 29, 31(40), *35*
Cooper, J. W., 11(20), *35*
Covinsky, M. H., 39(16), *111*
Cowan, E. E. B., 101(116), *114*
Croxton, C. A., 254(185), 255(185), 275(185), *324*

Dagdigian, P. J., 82(93), *113*
Dahler, J. S., 77(69, 78), 81(78), *112, 113*
Dalby, Z., 54(50), *112*
Dasgupta, C., 290(194), 291(194), *325*
Daud, M., 169(123, 124), 290(124), *323*
Dawbarn, M., 173(152), *323*
De Gennes, P. G., 155(87), 161(87, 96), 169(121), 196(87), *322, 323*
Dehmer, P. M., 1(6, 7), 3(6, 7), 19(26), 20(26), *35*
Devdariani, A. Z., 74(64), *112*
de Vries, M. S., 77(71, 74, 76), 78(71), 81(71, 74), *112, 113*
Ditkin, V. A., 216(179), 231(179), *324*
Dobson, G. R., 149(65), 150(71), 184(71), *321*
Donavan, R. J., 38(2), *110*
Drouin, R., 4(15), *35*
Durand, D., 140(32, 33), *320*
Düren, R., 39(12), 46(37), 47(12, 42), 82(12, 42, 90), 85(12, 42), 86(42, 95, 96), 87(42, 96), 88(42), 89(42), *111, 112, 113*
Dušek, K., 135(23), 149(23), *320*
Dyke, J. M., 20(30), *35*

Ebdon, J. R., 173(152), *323*
Edlen, B., 17(22), 20(22), 21(22), *35*
Edwards, S. F., 170(132), *323*
Eichinger, B. E., 150(72), 151(75), 152(75), 153(75, 81–83), 300(75), 301(75, 81), 302(75), 304(83), 308(213), 311(75), 312(75, 82, 217, 218), *321, 325*
Elbel, S., 20(30), *35*
Elyaschevich, A. M., 161(100), 297(201), 299(201), 312(201, 215, 216), *312, 325*
Eminyan, M., 40(30), *111*
Enikolopyan, N. S., 147(58), 150(74), 192(171), *321, 324*
Essam, J. W., 127(13), 158(92), 161(92, 102), 171(13), 173(13, 142), 174(142), 198(92), 201(13), 279(92), *320, 322*
Estera, J. M., 27(37), 29(37), 31(37), 32(37), *35*

Family, F., 161(98), 164(98), 290(117), *322*
Fano, U., 4(16), 5(16), 11(20), *35*, 61(57), 89(97), *112, 113*
Feynberg, V. Z., 181(154), *323*
Feynman, R. P., 235(181), 238(181), 240(181), *324*
Fink, M., 4(12), *35*
Fixman, M., 310(202), *325*
Flory, P. J., 124(8), 128(14), 150(66–68), 151(8), 261(187), *320, 321, 324*
Fluendy, M. A. D., 101(116), *114*
Flynn, G. W., 97(98), *113*
Ford, G. W., 191(168, 169), 192(169), 205(168, 169), *324*
Forsman, W. C., 152(78, 79), *321*
Fortuin, C. M., 166(118), 167(118, 119), 279(118, 119), *322*
Fox, T. G., 171(137), *323*
Freed, K. F., 118(3), 170(132), *319, 323*
Frish, H. L., 171(138), 178(138), *323*
Froese-Fischer, C., 12(21), 17(23), *35*
Fujita, H., 301(203), *325*

Gadea, F. X., 39(18), *111*
Gallagher, J. W., 40(33), 45(33), 77(33), *111, 112*
Ganz, J., 4(12), *35*, 39(22), 77(22), *111*
Garret, B. L., 101(118), *114*

Garton, W. R. S., 31(40), 33(43), *35, 36*
Geiger, J., 4(12), *35*
Gibson, S. T., 1(8, 9), 3(8, 9), 8(8, 9), 19(8, 9), *35*
Ginter, M. L., 25(35), 26(33), 27(34, 35), *35*
Goenweghe, L. C. D., 138(29), 139(29), 178(29), *320*
Good, I. J., 132(18), 133(19), 140(18, 19), 145(19), 149(64), *320, 321*
Goodman, G. L., 1(5), 19(27), 20(27), *34, 35*
Gordon, M., 117(1), 128(15), 129(1, 17), 132(1, 17), 135(22), 138(15), 140(1, 17, 34), 141(35), 144(1, 43), 146(55–57), 148(55), 149(65), 150(71), 170(129), 173(141–144), 174(141–143), 179(153), 182(153), 184(65, 155, 158–160), 185(155, 159), 210(15), 240(1), 295(200), *319, 320, 321, 323, 324, 325*
Gravitis, J. A., 148(60), 149(60), *321*
Greene, C. H., 39(7, 8), 82(8), *110*
Greene, J. P., 3(4, 8, 9, 11), 6(3, 4), 7(3, 4), 8(8, 9), 19(8, 9, 27), 20(27, 28), 22(28), 24(28), *34, 35*
Greville, T. N. E., 153(80), 302(80), *321*
Grosberg, A. M., 297(198), *325*
Grosser, J., 62(58), *112*
Guidotti, C., 38(6), *110*
Gurman, I. M., 173(151), *323*
Guttman, A. J., 169(126), *323*

Haberland, H., 48(44), *112*
Habitz, P., 100(113), 101(119), *114*
Haensel, R., 33(42), *36*
Hale, M. O., 70(61, 62), 73(62a), *112*
Hanne, G. F., 82(82), *113*
Harary, F., 133(20), 145(47), 152(47), 173(47), 201(47), 202(20), 221(47), 225(47), *320, 321*
Harris, A. B., 170(130, 131), 290(130), 291(130, 131), 294(130), *323*
Harris, T. E., 134(21), 208(21), 209(21), 212(21), *320*
Harth, K., 4(12), *35*
Hartmann, B., 173(148), *323*
Hasselbrink, E., 39(12), 47(12, 42), 82(12, 42), 85(12, 42), 86(42, 95, 96), 87(42, 96), 88(42), 89(42), *111, 112, 113*
Hausamann, D., 48(44), *112*
Helbing, R. K. B., 46(38), *111*
Hermann, H. W., 40(32), 44(32), *111*
Hershbach, D. R., 52(45), *112*
Hertel, I. V., 39(10, 11, 13, 14), 40(32, 33), 44(32), 45(14, 34–36), 54(46), 55(13, 52, 53), 58(52), 61(35), 64(46), 66(13), 67(46, 52), 68(46), 69(46), 70(61, 62), 73(62a), 74(62a), 77(33, 72), 78(72), 82(83), 84(52), 91(35), 92(35), 93(35), 95(35), 96(35), 97(14, 99, 101, 104, 106–111), 98(14), 99(14), 100(14, 112), 101(101, 112), 102(14), 103(14, 120), 104(120), 105(14), *110, 111, 112, 113, 114*
Hewitt, S. J., 173(152), *323*
Hill, W. T., III, 33(41), *36*
Hillrichs, G., 86(96), 87(96), *113*
Hoeve, C. A. J., 146(50), *321*
Hoffmann, H., 97(104), *113*
Hoppe, H. O., 46(37), *111*
Horiguchi, H., 97(102), *113*
Hormes, J., 33(43), *36*
Hotop, H., 4(12), *35*, 39(19, 20, 22, 23, 24), 77(19, 20, 22, 23, 24), *111*
Houston, P. L., 39(9), *110*
Howard, L. E., 21(32), 22(32), 23(32), *35*
Hsu, D. S. Y., 97(103), *113*
Huffman, R. E., 4(13), *35*
Hunt, I. E., 173(152), *323*
Hürvel, L., 46(38, 39), 47(39), *111*
Hüwel, L., 46(38, 39), 47(39), *111*

Irzak, V. I., 147(58, 59), 150(74), 173(147), 192(171), *321, 323, 324*
Isaacson, J., 164(111–114), 290(113), 292(113, 114), 293(114), 294(111–114, 199), *322, 323*

Jacobson, H., 145(46), 296(46), *320*
Jaecks, D. H., 40(29), *111*
Jamieson, G., 39(14), 45(14), 97(14, 109), 98(14), 99(14), 100(14), 102(14), 103(14), 105(14), *111, 114*
Jones, D. M., 77(69, 78), 81(78), *112, 113*
Joshi, Y. N., 22(31), 24(31), *35*

Kajiwara, K., 184(155–160, 163), 185(155–157, 159), 186(157), *323, 324*
Kalal, J., 184(159), 185(159), *324*
Karamatskos, N., 29(39), 30(39), *35*
Kasteleyn, R. W., 166(118), 167(118, 119), 279(118, 119), *322*

Kaufman, V., 33(41), *36*
Keller, J., 77(75, 76), 78(75), *113*
Kelley, M. H., 82(83–85), *113*
Kelly, H. P., 27(36), 28(36), *35*
Kempter, V., 40(32), 44(32), *111*
Kennedy, J. W., 173(141–144), 174(141–143), 184(159), 185(159), *323, 324*
Kertesz, J., 291(196), *325*
Khnokhlov, A. R., 297(198), *325*
King, R. B., 127(12), 196(12), *320*
Kircz, J. G., 77(67), 78(67), 81(67), *112*
Klein, D. J., 158(91), 161(99), 164(99), *322*
Klein, W., 163(106, 108), 164(110), 290(106, 108, 110), *322*
Kleinpoppen, H., 40(30, 31), 45(36), *111*
Kleyn, A. W., 101(115), *114*
Klimontovich, Yu. L., 193(172), *324*
Knystautas, E. J., 4(15), *35*
Kohl, J. L., 29
Kohmohto, M., 89(97), *113*
Korolev, S. V., 137(24–26), 142(24, 42), 146(48, 49), 170(127), 171(25, 26, 140), 186(26), 233(24), *320, 321, 323*
Korsch, H. J., 57(54), 59(54), *112*
Korschak, V. V., 126(9), *320*
Kozlov, M. G., 29
Kryloo, B. E., 29
Kuchanov, S. I., 117(2), 128(2), 129(2), 133(2), 135(2), 137(24–26), 138(2, 27, 28), 142(24, 42), 146(48, 49), 170(127, 135), 171(2, 25, 26, 140), 182(26), 185(2), 186(2), 210(2), 211(2), 233(24), 235(2), 274(190), 279(135), 218(2, 135), *319, 320, 321, 323*
Kurucz, R. L., 29
Kuzub, L. I., 147(58), *321*
Kwei, G. H., 97(105), *113*

Lackschewitz, U., 46(40, 41), *112*
Landau, L. D., 188(165), 193(165), *324*
Larrabee, J. C., 4(13), *35*
Lawley, K. P., 38(4), *110*
Ledermann, W., 305(208), *325*
Le Dourneuf, M., 29(38), 30(38), *35*
Lee, G., 173(148), *323*
Lee, S. T., 20(29), *35*
Lee, Y. T., 39(16), *111*
Lemont, S., 97(98), *113*
Leone, S. R., 39(17), 70(61, 62), 73(62a), 77(62b), *111, 112*
Letuchy, B. A., 262(188), *324*
Leung, Y. K., 312(217), *325*
Levine, R. D., 52(45), *112*
Lifshitz, E. M., 188(165), 193(165), 235(180), 238(180), 240(180), 253(184), 260(184), 261(184), 274(180), 297(198), *320, 324, 325*
Lin, M. C., 97(103), *113*
Lombardi, G. G., 29
Lorentz, D. C., 45(34), *111*
Lorenzen, J., 39(19), 77(19), *111*
Los, J., 101(115), *114*
Lowry, G. G., 171(136), *323*
Lubensky, T. C., 163(107), 164(107, 111–114), 170(130, 131), 290(107, 113, 130, 194), 291(107, 130, 131, 194), 292(113, 114), 293(114), 294(111–114, 130, 199), *322, 323, 325*
Luby, P., 139(31), 140(31), *320*
Ludescher, H. P., 48(44), *112*
Luken, W. L., 1(7), 3(7), *35*
Lutz, H. O., 45(36), *111*

Ma, S. K., 155(86), 167(86), 169(86), *322*
Mac Adam, K. B., 40(30), *111*
McClelland, J. J., 82(83–85), *113*
Macek, J., 40(29), *111*
Macek, J. H., 61(57), *112*
McGowan, J. W., 38(3), *110*
McKay, B. D., 312(218–220), *325*
Maier, J., 46(38–41), 47(39), *111, 112*
Malcolm, G. N., 141(35), *320*
Mallows, C. L., 171(138), 178(138), *323*
Mansfield, M. W. D., 33(43), *36*
Manson, S. T., 19(24), *35*
Maradudin, A. A., 305(209), 310(209), 311(209), *325*
Martin, J. E., 153(81, 83), 301(81), 304(83), 312(218), *321, 325*
Martlengiewicz, M., 150(173), *323*
Matula, D. W., 138(29, 30), 139(29, 30), 174(29, 30), 178(29, 30), *320*
Mayer, J. E., 188(166), 190(166), 192(166), *324*
Mayer, M. G., 188(166), 190(166), 192(166), *324*
Mehlman-Ballofet, G., 27(37), 29(37), 31(37), 32(37), *35*
Meijer, H. A. J., 77(72, 79), 78(72, 79), 79(79), 80(79), 81(79), *112, 113*
Mestdagh, J. M., 39(16), *111*

Meulen, H. P. v. d., 77(72, 79), 78(72, 79), 79(79), 80(79), 81(79), *112, 113*
Meyer, E., 39(13), 55(13, 51–53), 54(46), 58(52), 64(46), 66(13), 67(46, 52), 68(46), 69(46), 77(72), 78(72), 84(52), 100(112), 101(112), *111, 112, 114*
Meyer, W., 39(24), 55(52), 58(52), 67(52), 77(24), 84(52), *111, 112*
Meyerhoff, W. E., 45(34), *111*
Miller, B., 55(51), *112*
Montroll, E. W., 304(207), 305(207, 209), 307(207), 308(207), 309(207), 310(209), 311(209), *325*
Moore, C. E., 19(25), 20(25), *35*
Morgenstern, R., 48(43), 77(67, 72, 79), 78(67, 72, 79), 79(79), 80(79), 81(67, 79), 82(91), *112, 113*
Moritz, G., 86(95), *113*
Morris, A., 20(30), *35*
Moutinho, A. M. C., 101(116), *114*
Msezane, A., 19(24), *35*
Muller, M., 29(39), 30(39), *35*
Murata, K. K., 291(195), *325*

Nakanishi, R., 161(104), 163(104), *322*
Neumann, G., 173(145), *323*
Neuschäfer, D., 70(62), 73(62a), 74(62a), 76(62b), 77(62b), *112*
Newson, G. H., 31(40), *35*
Nielsen, S. E., 54(47, 48, 49), *112*
Nienhuis, G., 77(67, 68, 77), 78(67), 79(68), 81(67), *112, 113*
Nikitin, E. E., 82(94), *113*
Norisuye, T., 301(203), *325*

Obukhov, S. P., 161(101), 294(199), *322, 325*
Ord, G., 169(125), *323*
Ore, O., 200(174), *324*
Orlikowski, T., 82(92), *113*
Ozol-Kalnin, B. G., 148(60), 149(60), *321*

Palenius, H. P., 4(13), *35*
Palmer, E. M., 133(20), *320*
Panev, G. S., 54(50), *112*
Panyukov, S. V., 170(134), 266(134), 270(134), 274(134), 275(191), 278(134, 191), 284(191), 287(191), 292(134), 294(191), *323, 325*
Papierowska-Kaminski, D., 100(114), 104(114), 106(114), 109(114), *114*
Parisi, G., 165(116), 290(116), *322*
Parisi, J., 266(189), 267(189), *324*
Parker, T. G., 179(153), 182(153), *323*
Parkinson, W. H., 29
Patashinsky, A. Z., 155(85), 167(85), 169(85), 271(85), *322*
Pauley, H., 46(37–41), 47(39), *111, 112*
Pelissier, M., 39(18), *111*
Peniche-Covas, C. A. L., 184(160), *324*
Penning, F. M., 77(66), *112*
Percus, J. K., 190(167), *324*
Perdrix, M., 39(25), 77(25), *111*
Persico, M., 97(108), *114*
Pesnelle, A., 39(25, 27), 77(25, 27), *111*
Peterson, J. R., 45(34), *111*
Picque, J. L., 38(5), *110*
Pitaevsky, L. P., 235(180), 238(180), 240(180), 274(180), *324*
Pokrovsky, V. A., 155(85), 167(85), 169(85), 271(85), *322*
Polya, G., 145(45), 308(210), *320, 325*
Ponomareva, T. I., 173(147), *323*
Popov, V. N., 235(182), 237(182), 238(182), 240(182), *324*
Poppe, D., 100(114), 104(114), 106(114, 121), 109(114), *114*
Pouilly, B., 74(63), *112*
Praxedes, A. J. F., 101(116), *114*
Priss, L. S., 150(73), *321*
Prudnikov, A. P., 216(179), 231(179), *324*

Rachman, N. K., 38(6), *110*
Radler, K., 2(10), 18(10), 33(42), *35, 36*
Rahmat, G., 39(18), *111*
Ray, W. J., 4(14), *35*
Reck, G. P., 77(70, 73), 78(70), *112, 113*
Reed, M. B., 308(212), *325*
Reiland, W., 39(14), 45(14), 97(14, 106–110), 98(14), 99(14), 100(14, 112), 101(112), 102(14, 120), 104(120), 105(14), *111, 113, 114*
Remeev, I. S., 312(216), *325*
Render, S., 161(98, 103), 163(103), 164(98), 196(103), *322*
Rettner, C. T., 39(15), 48(15), 49(15), 50(15), 51(15), 52(15), *111*
Reynolds, P. J., 161(98, 103, 104), 163(103, 104), 164(98), 196(103), *322*
Richter, C., 45(35), 61(35), 91(35), 92(35), 93(35), 95(35), 96(35), *111*
Roig, R. A., 29

Romand, J., 27(37), 29(37), 31(37), 32(37), *35*
Ronca, G., 308(211), *325*
Ross-Murphy, S. B., 184(160), *324*
Rost, K., 97(104), *113*
Rothe, E. W., 77(70, 73), 78(70), *112, 113*
Rouse, P. E., 152(76), 312(76), *321*
Rozenberg, B. A., 150(74), 173(147), 192(171), *321, 323, 324*
Ruf, M. W., 4(12), *35*, 39(19, 20, 22–24), 77(19, 20, 22–24), *111*
Runge, S., 39(25), 77(25), *111*
Ruscic, B., 1(2–4, 8, 9), 3(3, 4, 8, 9, 11), 6(3, 4), 7(3, 4), 8(8, 9), 19(8, 9), *34, 35*

Saakyan, L. L., 297(201), 299(201), *325*
Sarma, V. N., 22(31), 24(31), *35*
Scanlan, J., 150(69), *321*
Scantlebury, G. R., 128(15), 138(15), 146(55–57), 148(55), 210(15), *320, 321*
Schill, G., 127(11), *320*
Schmidt, 29(39), 30(39), *35*
Schmidt, H., 39(13, 16), 45(35), 54(46), 55(13, 51–53), 58(52), 61(35), 64(46), 66(13), 67(46, 52), 68(46), 69(46, 60), 77(72), 78(72), 82(60), 83(60), 84(52), 85(60), 91(35), 92(35), 93(35), 95(35), 96(35), *111, 112*
Schmidt, M., 184(161, 162), *324*
Schulz, C. P., 97(106, 107, 109, 111), 103(120), 104(120), *113, 114*
Seaton, M. J., 10(19), *35*
Seitz, W. A., 161(99), 164(99), *322*
Seshu, S., 308(212), *325*
Sevastyanov, B. A., 209(175), 212(175), *324*
Shahabi, S., 19(24), *35*
Shakeshaft, R., 61(55), *112*
Shapiro, B., 161(105), 163(105), *322*
Shirley, D. A., 20(29), *35*
Shortley, G. H., 9(17), *35*
Siegel, A., 4(12), *35*
Silver, J. A., 97(105), *113*
Slanina, Z., 120(4), 173(4), 193(4), *319, 324*
Slevin, J., 40(30, 31), *111*
Slinko, M. G., 137(24–26), 142(24, 42), 171(25, 26, 140), 182(26), 233(24), *320*
Slonim, I. Ya., 173(149, 151), *323*
Smitt, R., 10(18), *35*
Sokolov, L. B., 126(10), *320*
Šolc, K., 301(204), 302(204), *325*
Somerville, L. P., 4(14), *35*
Sommer, K., 33(43), *36*
Sonntag, B., 33(42), *36*
Soulras, N., 165(116), 290(116), *322*
Spiegelmann, F., 39(18), *111*
Spies, N., 55(52), 58(52), 67(52), 84(52), *112*
Spiess, G., 38(5), *110*
Standage, M. C., 40(30), *111*
Stanley, H. E., 161(98), 164(98), *322*
Stanley, M. E., 161(103), 163(103, 106, 108), 196(103), 290(106, 108), *322*
Starace, A. F., 19(24), *35*
Startsev, G. P., 29
Stauffer, D., 155(88, 89), 156(88, 89), 157(89), 158(93), 161(88, 93, 95, 97), 162(88), 164(109), 165(88), 169(122), 170(128), 279(93), 284(128), 219(196), *322, 323*
Stepanov, V. E., 146(51), *321*
Stevens, J. C. H., 20(30), *35*
Stockmayer, W. J., 129(16), 130(16), 145(46), 184(162), 296(46), *320, 324*
Stoll, W., 39(10, 11), *110, 111*
Storey, P. J., 10(19), *35*
Suzer, S., 20(29), *35*
Sugar, J., 33(41), *36*
Szmytkowski, Cz., 82(82), *113*

Tanaka, Y., 4(13), *35*
Tatevskii, V. M., 123(5), 124(5), 173(5), *319*
Taylor, K. T., 29(38), 30(38), *35*
Temple, W. B., 117(1), 129(1, 17), 132(1, 17), 140(1, 17), 144(1, 43), 179(153), 182(153), 240(1), *319, 320, 323*
Theyunni, R., 77(70, 73), 78(70), *112, 113*
Tilford, S. G., 25(35), 26(33), 27(34, 35), *35*
Tischner, H., 39(12), 47(12), 82(12), 85(12), *111*
Tittes, H. U., 39(14), 45(14), 97(14, 106–109, 111), 98(14), 99(14), 100(14), 102(14), 103(14, 120), 104(120), 105(14), *111, 113, 114*
Torkington, J. A., 140(34), 170(129), *320, 323*
Tousey, R., 27(34), *35*

Truhlar, D. G., 101(117, 118), 102(117, *114*
Tsuchiya, S., 97(102), *113*
Tung, C. C., 77(70, 73), 78(70), *112*

Uhlenbeck, G. E., 191(168, 169), 192(169), 205(168, 169), 302(206), *324, 325*
Umanskii, S. Ya., 82(94), *113*
Umemoto, H., 97(100), *113*
Urman, Ya. G., 173(149, 151), *323*

Van den Berg, F., 48(43), 82(91), *112, 113*
Van der Weil, M., 82(82), *113*
Van Kleef, Th. A. M., 22(31), 24(31), *35*
Van Waser, J. R., 138(29), 139(29), 174(29), 178(29), *320*
Vasil'ev, A. N., 235(183), 237(183), 253(183), *324*
Verges, J., 39(18), *111*
Verheijen, M. J., 39(26), 77(26), *111*
Vernon, F., 39(16), *111*
Vetter, R., 39(18), *111*
Vinogradova, S. V., 126(9), *320*
Vo Ky Lan, 29(38), 30(38), *35*
Volkenstein, M. V., 124(7), 151(7), *319*

Waibel, H., 39(23, 24), 77(23, 24), *111*
Walasek, J., 171(139), *323*
Wallace, D. J., 291(197), *325*
Wang, M. C., 302(206), *325*
Wang, M. X., 77(71, 74–76), 78(71, 75), 81(71, 74), *112, 113*
Watel, G., 39(25), 77(25), *111*
Weiner, J., 77(71, 74–76), 78(71, 75), 81(71, 74), *112, 113*
Weiss, J. H., 305(209), 310(209), 311(209), *325*
Weiss, P. S., 39(16), *111*
Weston, R. E., 81(80), 97(80), 101(80), 102(80), *113*
Westwood, A. R., 173(152), *323*
Whittington, S. G., 156(90), 157(90), 161(90), 165(90), *322*
Whittle, P., 141(36, 41), 146(36–41), *320*
Williams, I. E., 173(152), *323*
Witte, R., 45(35), 61(35), 69(60), 77(72), 78(72), 82(60), 83(60), 85(60), 91(35), 92(35), 93(35), 95(35), 96(35), *111, 112*
Wittington, S. G., 169(125, 126), *323*
Wolff, H. W., 33(42), *36*
Wood, R. W., 38(1), *110*
Wormald, N. C., 312(221), *325*
Wu, F. Y., 167(120), 280(120), *322*
Wuilleumier, F. J., 38(5), *110*

Yang, Y., 301(205), *325*
Yarovoi, S. S., 123(5, 6), 124(5, 6), 173(5), *319*
Yerukhimovich, I. Ya., 149(61, 62), 170(133), 253(133), 258(133), 262(133, 188), 266(133), 292(133), *321, 323*
Young, L., 4(14), *35*
Yu, T., 301(205), *325*

Zagrebin, A. L., 74(64), *112*
Zare, R. N., 39(7, 8, 15), 48(15), 49(15), 50(15), 51(15), 52(15), 82(8), *110, 111*
Zia, R. K. P., 291(197), *325*
Ziabicky, A., 171(139), *323*
Zimm, B. N., 152(77), *321*
Zimmermann, P., 29(39), 30(39), *35*

SUBJECT INDEX

Absorbing sphere model:
of collision velocity, 81
of energy transfer, 101–102
Acoustic modes, of multiatomic lattices, 310
Acyclic units, in cyclation topologies, 224
Adjacency matrix, defined, 317
Adjacent edge, defined, 314
Alignment angle (γ):
cylindrical geometry, 73
in p-state distribution, 44
in scattering experiments, 62–65
trajectory schematic, 105
Alignment parameters, in linear polarization, 62
Alkaline earth-rare gas interactions, typical potentials, 46
Alkaline earths, 30–31
autoionization experiments, 2
Alkali-rare gas systems:
orientation effects on spin-orbit interaction, 85–89
typical potentials, 46, 48
Alkalis, 31–33
autoionization experiments, 2, 31–33
fluorescence in cells, 38
Aluminum:
photoabsorption spectrum of, 30
reduced autoionization widths for, 29
Angular momentum:
in circular polarization, 67–68
and collisional energy, 66
in collisions with laser-excited polarized atoms, 37–114
at large internuclear distances, 68
in p-state distribution, 44
quenching process, 109
rotational alignment of sodium atoms, 81
spin-orbit interaction, 83–84
transfer and charge exchange, 96
Antimony, autoionization experiments, 19–20
Arbitrary fields, generating functionals and, 212
Argon:
autoionization experiments, 2
autoionization lineshapes, 5–7
photoabsorption studies, 30
Arsenic:
autoionization experiments, 20
centroid diagrams, 21, 23
quasi-discrete and continuum states, 24
Associative ionization:
alignment angles and, 81
collisional cross sections for, 79–81
cylindrical geometry, 77–82
of laser-excited polarized atoms, 77–82
Morgenstern experiment, 78
polarization effects, 77–79
schematic diagram, 78
Atom–atom collisions, at suprathermal energies, 110
Atomic frames, and molecular frames, 51–52
Atomic polarizability, 15–17
Atomic resonance radiation, collisional depolarization of, 62
Atom–ion collisions, at suprathermal energies, 110
Atom–molecule dynamics, 39
Atoms, systematic autoionization features, 1–36
Autoionization:
characteristics, of discrete-continuum interaction, 9–17
core function integral, 13–14
features in atoms, 1–36
in first–row atoms, 4
improved wave functions, 14–17
limiting values of narrow resonances, 4
lineshapes, 4–8
of pnicogen atoms, 19–24
Rydberg and continuum wave function integrals, 11–13
between spin-orbit components, 25–27
trends inferred from experiments, 2–8
widths and shapes, 2

Autoionization (*Continued*)
 Z-dependence of resonance width in isoelectronic sequence, 17
Automorphic groups:
 defined, 316
 four-function monomers and, 202–205
 of polymeric molecules, 129

Barium, Z-dependence of resonance width, 17
Basis states:
 atomic or molecular, 42
 for p-state atoms, 42, 43
Beryllium:
 autoionization experiments, 31
 photoabsorption spectrum of, 32
Bethe lattice model, 154–155
 enumeration problem in, 160
 random percolation on, 162
Binary collision experiment, schematic view, 41
Bipartite graph, defined, 316
Bismuth, autoionization experiments, 19–20
Black particles, in polymer branching, 135–136, 137
Body-fixed electronic motion, 62, 63
Body-fixed frames, center-of-mass frames and, 51–52
Bonds. *See also* Chemical bonds
 problem in percolation models, 158
 stretching attraction in energy transfer, 101
 symbols used in diagram technique, 246–247
Born–Oppenheimer approximation, in collisions of laser-excited atoms, 40
Boron, photoabsorption spectrum of, 29
Branched polymers:
 bond substitution in, 218–219
 continual models, 292–293
 cyclation models, 144–150
 ensembles and random graphs, 124–129
 Feynman diagrams, 292–293
 field theory descriptions, 289–294
 Flory models, 129–136, 293
 graph theory, 116–119
 Hilhorst model, 290
 invariance principle, 142–144
 kinetic equations, 128
 Kirchhoff matrix and molecular conformation, 150–152
 lattice and scaling graphs, 153–158
 lattice models, 290
 Lubensky–Isaakson model, 292–294
 macromolecular fragments and subgraphs, 171
 macroscopic features, 118
 models and approaches for description of, 120–170
 molecular graphs, 120–124
 multicomponent systems, 142
 percolation and lattice models, 158–170
 physicochemical models of, 128, 129–136
 process diagram, 135
 random graphs, 124–129
 statistical properties, 290–291
 substitution effect models, 137–142
Branching process:
 diagram technique, 234, 242
 displacement of particles in, 208, 210
 Lifshitz–Erukhimovich model, 258
Bridge, graph, defined, 318
Bromine, autoionization experiments, 2–3, 5–7

Cactus graphs, of molecular cyclation, 145–146
Calcium atom-rare gas collisions:
 experimental geometry, 71
 intersystem crossings in, 70–77
 molecular potentials, 74–75
 polarization effects, 75–76
Calcium atoms:
 alignment studies, 70
 autoionization experiments, 31
 collision studies, 41
 energy level diagram, 70
Calcium-hydrochloric acid system:
 chemiluminescence spectra, 49, 50
 laser polarization studies, 48–52
Canonical ensemble, of polymeric molecules, 137
Carbon, photoabsorption measurements, 28
Cathenans, from topological linkages, 127
Center-of-mass frames, and body-fixed frames, 51–52
Cesium:
 photoabsorption spectrum of, 33
 Z-dependence of resonance width, 17
Chalcogen atoms, 18–19
 autoionization experiments, 3
 core function integral, 13–14

photoionization of, 1
reduced autoionization widths for, 4
Charge cloud, alignment and orientation effects of, 97
Charge cloud height, 65
in p-state distribution, 44
and quenching process, 107
Charge exchange:
alignment and orientation parameters, 92–94
apparatus diagram, 90
in laser-excited polarized atoms, 90–97
scattering geometry, 90–91
in sodium atom–ion system, 90–97
Chemical bonds:
analytical methods of field theory, 265–267
self-consistent field approximation, 268–269
spatial physical interactions, 251
system description, 253
Chemical physics, of polymers (graphs), 115–325
Chemiluminescence, laser-induced spectra, 49, 50
Chlorine:
autoionization experiments, 2–3, 5–7
photoabsorption studies, 30
Chords, defined, 318
Chord trees, and gel formation, 275
Circular polarization:
in collisions of laser-excited atoms, 54–55
orientation parameter in, 66–68
Clebsch–Gordan coefficients, for collisional energy transfer, 77
Clusters, in percolation models, 159–160
Cocycles, of spanning trees, 221–222
Collision:
binary experiment, 40–41
optical excitation prior to, 42–44
planar symmetry model, 55–69
symmetry and angular momentum in laser-excited polarized atoms, 37–114
velocity cross sections, 80–81
Combined lattice, Kirchhoff matrix, 308
Components, graph, defined, 314
Condensation reaction, in polymer formation, 122
Connected graphs:
defined, 314
transition to, 191–192
Continuum wave functions:
autoionization integral involving, 11–13
L–S notation, 13
Contracting graph, defined, 319
Coordinate graphs:
correlation functions and generating functionals, 193–215
contributions from cycles of different topologies, 218–227
edge-chain substitution in, 202
molecules as, 186–233
one-point correlators and molecular weight distribution, 215–217
perturbation theory, 228–230
possible model generalizations, 230–233
statistics of cyclic fragments, 230–233
Copolymers, 120
macromolecular fragments and subgraphs, 171
molecular graphs, 178–179
Core functions, autoionization integral involving, 13–14
Coriolis coupling, 51
Correlated percolation, through sites and bonds, 162–166
Correlation functions:
and generating functionals, 193–199
symbols and equation, 256
Correlators, diagram technique, 249
Coulomb–Bethe approximation, for dielectronic recombination, 10
Critical exponents, in lattice models, 154
Crystal lattices, Kirchhoff matrix, 309
Cubic lattices:
Kirchhoff matrix, 308
simple and body-centered, 304, 307
Cyclation:
contributions of different topologies, 218–227
and coordinate graphs, 186–233
diagram technique, 242–246
graphs of topologies, 226, 227
molecular configuration, 144–150
perturbation theory, 228–230
possible model generalizations, 230–233
statistics of cyclic fragments, 230–233
topological orientations, 204
Cycle, defined, 314
Cyclic edge, defined, 318
Cyclic molecule, Kirchhoff matrix, 302, 309

Cyclic rank, defined, 318
Cyclomatic number, defined, 318
Cylindrical geometry, of laser-excited polarized atoms, 69–82

Deflection functions, for elastic scattering, 92
Degeneration probability, in cyclation models, 149
Density correlators:
  diagram technique, 262–263
  for sol molecules, 272
  variational differential technique, 254
Density matrix (σ), 42
  of collisionally excited atom, 45
  for electronic-vibrational energy transfer, 106
  for intersystem crossings, 76
Diagram technique, 237–251
  correlator notations, 249
  for cyclic density, 242–246
  defined, 233
  for density correlation functions, 262–263
  density equations, 239
  in graph theory, 116
  Lifshitz–Erukhimovich model, 258–265
  mathematical and graphical symbols, 237–238
  for monomer units, 241–242
  one-trunk model, 259
  in polymer chemistry, 233–294
  for polymer system with physical interactions, 258
  semishaded and shaded bonds, 246–247
  spatial physical interactions, 251–258
  thermodynamic relations, 260–261
Diatomic molecules, 97
Dilute solutions, branching molecules in, 294
Dilution factor, in percolation models, 161
Dimensionless variables, in coordinate graphs, 223–224
Dimer-monomer graph system, 186–187
Direct correlation function, in spatial physical interactions, 255
Discrete-continuum interaction, autoionization characteristics of, 9–17
Distance, graph, defined, 314
Dumbell orbital description:
  for large internuclear distances, 53
  in molecular collision studies, 42, 51
Dyads, 171, 172
  coefficient values, 176
  nonrooted, 178
  ordered, 178
Dyad trees, 179–181
Dye lasers, in collision studies, 39

Edge, defined, 313, 314
Elastic free energy, in gel calculations, 150
Elastic scattering:
  deflection functions for, 92
  and fine-structure changing collisions, 85
  in molecular collisions, 46
Electric circuits, Kirchhoff matrix calculations, 308
Electronic-vibrational energy transfer, 97–109
  apparatus diagram, 98
  experimental geometry, 98
  molecular orbital schematic, 103
  orientation effects, 108–109
  polarization effects, 100
Electrostatic interaction, in iodine autoionization, 3
Energy transfer, electronic-vibrational, 97–109
Eulerian graph, defined, 314
Excited rare-gas atoms, elastic collisions of, 48

Feynman diagrams, of many-particle systems, 235
Field theory:
  analytical methods, 265–268
  basic relations, 235
  and branching polymers, 289–294
  Mayer diagram technique, 254–257
  multicomponent systems, 277–282
  in polymer chemistry, 233–294
  Potts model extension, 277–282
  spatial physical interactions, 251–258
  steepest descent method, 237
Final vertex, defined, 314
Fine-structure components:
  autoionization between, 20–24
  splitting of pnicogen atoms, 19–23
First-row atoms:
  limiting values of narrow resonances, 4
  polarizability of, 15
Flexible monomers, 143

Flory model, of polycondensation, 129–136
Fluorescence intensity:
  and alignment angle, 74
  collisional energy transfer and, 76–77
Fluorine:
  autoionization experiments, 3
  narrow resonances in, 4
  photoabsorption studies, 30
Forest graph, defined, 315
Forward scattering, kinetic energy and, 107–108
Functional integration, stochastic fields and, 233–237

Gallium, reduced autoionization widths for, 29
Gaussian molecules:
  automorphic groups, 298–300
  conformations and spectral properties, 294–312
  graphs of, 294–312
  Kirchhoff matrix, 300–304
  radius of gyration distribution calculations, 300–304
  space metric and graph metric models, 294–300
  spectral density and dynamic properties, 310–312
  spectrum for regular networks, 304–310
Gel (macromolecule), 124
  molecular graphs, 149–150
Gelation:
  of branched polymers, 124
  diagram technique, 263–265
  Flory theory, 270
  lattice models, 154–155
  and percolation, 160, 164
  perturbation theory, 228–229
  physicochemical model of, 170
  in polymer branching process, 135
  self-consistent field approximation, 270–276
  swelling theory, 284–285
Generating functionals:
  correlation functions and, 193–199
  equations for, 199–206
  Flory model, 206–207
  in rooted graphs of polymeric molecules, 132
  studies and models, 206–215
Geometric realizations, defined, 314, 315
Germanium, photoabsorption measurements, 26
Gibbs distribution, and probability measure on space graphs, 188–191
Ginzburg number, in percolation models, 169
Graph metric models, of Gaussian molecules, 294–300
Graphs:
  in chemical physics of polymers, 115–325
  defined, 313
Graph theory:
  elementary concepts, 313–319
  in polymer chemistry, 115–325
Group I atoms, *see* Alkalis
Group II atoms, *see* Alkaline earths
Group III atoms, 27–30
  autoionization experiments, 27–30
  reduced autoionization widths for, 29
Group IV atoms, 25–27
  autoionization between spin-orbit components, 25–27
  core function integral, 13–14
  shell excitation, 27
Group V atoms, *see* Pnicogen atoms
Gyration, *see* Radius of gyration, distribution calculations using Kirchhoff matrix

Halogen atoms, 18
  autoionization experiments, 2–3
  core function integral, 13–14
  fine-structure components, 18
  photoionization of, 1
  reduced autoionization widths for, 4
Hamiltonian graph, defined, 314
Hartree–Slater wave functions, 11
Heavy-particle motion, electron spin coupling and, 82
Hilhorst model, of branching polymers, 290
Homeomorphic graphs, of cyclation topologies, 218
Homeomorphism *vs*. spatial topology, 127
Homopolycondensation, 232–233
Homopolymers, 120
  molecular graphs, 179
Hydrogenic wave functions:
  calculated value of radial matrix elements using, 11, 12
  matrix elements using, 19

Incidence matrix, defined, 318
Incident edge, defined, 314
Indium, reduced autoionization widths for, 29
Inelastic scattering, 57–59
Initial vertex, defined, 314
Intersystem crossings, in calcium atom-rare gas collisions, 70–77
Invariance principle, in molecular configuration, 142–144
Inverse collision, of laser-excited polarized atoms, 44–54
Iodine atoms, autoionization experiments, 3
Ion–atom dynamics, 39
Isoelectronic sequence, Z-dependence of resonance width in, 17
Isomers:
  "assemblage" methods, 131
  conformations in graphs, 124
  cyclic formation and linear memory, 296–297
  equilibrium proportions of, 296, 300
  labeled combinations, 131
  molecular graphs of, 120, 122
  polymerization degree, 122
  topological, 127
Isomery, in molecular graphs, 120
Isomorphic graphs, of polymeric molecules, 131–132

J_c K coupling, *see* Pair coupling

Kernel, integral operator, 276
Kirchhoff matrix:
  block structure of, 305–307
  for Gaussian molecules, 295–299
  in macromolecule distribution, 119
  and molecular conformation, 150–152
  in radius of gyration distribution calculations, 300–304
  spectrum for regular networks, 304–310
K-root graph, defined, 316
Krypton, autoionization experiments, 2, 5–7

Labeled graph, defined, 315, 316
Landau–Zener curve crossing model, 74
Landau–Zener probability, for intercrossing energy transfer, 102
Laser-excited polarized atoms:
  alignment and orientation, 38, 40
  associative ionization, 77–82
  binary collision experiment, 40–41
  charge exchange in, 90–97
  cylindrical geometry examples, 69–82
  electronic-vibrational energy transfer, 97–109
  inverse collision, 44–54
  optical excitation prior to collision, 42–44
  planar symmetry case, 55–69
  spin-orbit interaction, 82–89
  symmetry and angular momentum in collisions, 37–114
  three-vector correlation experiments, 38–39
Laser polarization, dynamics of, 46–52
Lattice animals, 156–157
  of branching polymers, 290
  Potts model, 291
  random model, 164
  spatial types of, 157
  thermodynamic features, 164–167
Lattice gas, 153
  Potts models, 290
Lattice models:
  cross-linking theories, 169
  field theory descriptions, 289
  macromolecules as, 153–158
  of pair interaction potential, 153–154
  percolation, 158–170
  Potts variants, 289
  spatial physical interactions, 256
Leaf composition:
  defined, 319
  of molecular graph, 200–201
Leaves, graph, defined, 318
Lifshitz–Erukhimovich models:
  of branching polymers, 292
  diagram technique, 258–265
  factorization in, 261–262
  of linear polymer molecules, 253
  for thermodynamic relations, 261
Linear memory, and cyclic isomer formation, 296–297
Linear molecules, cyclation models, 145–146
Linear polarization:
  alignment parameters, 61
  and charge cloud length, 44
  and scattering intensity, 59–63
Lineshapes, autoionization, 4–8
Locking radius:
  and associative ionization, 81

in charge transfer, 94
and intersystem crossings, 74
in linear polarization, 62
Poppe model, 106
Loops:
defined, 313
molecules, with and without, 295
Lubensky–Isaakson model, of branching polymers, 292–294

Macromolecules:
conformational characteristics of, 300–304
fragment subgraphs, 171–175
as lattice and scaling graphs, 153–158
Magnesium:
autoionization experiments, 31
photoabsorption spectrum of, 32
Many-particle systems, field theory, 235
Matrix, definition of, 317
Matrix elements:
autoionization width, 9
monopole and quadrupole, 11
for Rydberg and continuum wave functions, 11–13
Mayer diagram technique:
in field theory, 255–257
total density symbol, 257
Mean (self-consistent) field, of Bethe lattice model, 154
Microcanonical ensemble, of polymeric molecules, 129
Microscopic density, correlation function and, 193–196
Molar mass, of polymer molecule, 123
Molecular collisions, dynamics of, 46–54
Molecular fragments, external field effects, 213
Molecular frames, atomic frames and, 51–52
Molecular graphs:
of branched and network polymers, 120–124
cacti with cycle vectors, 145–146
functional groups in, 120
invariance principle, 142–144
particularization levels in, 120
with quasi-monomers, 200
and subgraphs of polymer microstructures, 171–186
Molecular structure distribution (MSD):
diagram technique, 263
in random graphs, 126–127
Molecular weight distribution (MWD):
one-point correlators, 215–217
in random graphs, 126
Molecules:
as coordinate graphs, 186–233
spectral density and dynamic properties, 310–312
Monads, 171, 172
NMR measurements, 174
Monomers:
diagram technique, 241–242
displacement of particles corresponding to, 210
flexible and rigid, 143
polycondensation, 294
of polymer molecules, 120
starlike, 189
Multiatomic lattices, Kirchhoff matrix, 310
Multichannel quantum defect theory (MQDT), 34
Multicomponent systems:
field theory, 277–282
matrix elements, 281
Potts model extension, 277–282
self-consistent field approximation, 280
spinodal transition in, 282, 288
Multigraph, defined, 313

Narrow resonances, limiting values in first-row atoms, 4
Natural frame:
charge cloud distribution in, 45
in polarization studies, 44
scattering geometry in, 91
in spin-orbit interaction studies, 89
Neon:
narrow resonances in, 4
photoabsorption studies, 30
Z-dependence of resonance width, 17
Network molecules:
Kirchhoff matrix spectrum for, 304–310
with random topologies, 312
topographical structure, 305
Network polymers:
cyclation models, 144–150
ensembles and random graphs, 124–129
Flory model of photocondensation, 129–136
graph theory, 116–119

Network polymers (*Continued*)
inhomogeneities in, 148–149
invariance principle, 142–144
Kirchhoff matrix and molecular conformation, 150–152
lattice and scaling graphs, 153–158
models and approaches for description of, 120–170
molecular graphs, 120–124
percolation and lattice models, 158–170
random graphs, 124–129
ring defects in, 150
substitution effect models, 137–142
swelling theory, 283–289
topological limitations, 127
Newton binomial formula, graphical representation, 132, 133
Nitrogen, autoionization experiments, 19–20
Noble gases, 18
core function integral, 13–14
photoionization of, 1
reduced autoionization widths for, 4
Nodes, defined, 314
Nonadiabatic transitions:
alignment parameters, 84
angular momentum transfer, 84
apparatus diagram, 55
collision-induced, 55–69
at large internuclear distances, 82–83
spin-orbit interaction in sodium atom–ion system, 82–85
Nonlabeled graph, defined, 315, 316
Nonrooted graphs, of polymeric molecules, 124–125
Nonseparable graph, defined, 318
Nuclear magnetic resonance (NMR), 173–174
Number, root, defined, 317

One-point correlators, and molecular weight distribution, 215–217
One-root graphs, diagram technique, 259–260
Optical excitation, prior to collision, 42–44
Optical pumping, laser methods, 48
Orbital angular momentum (ℓ), 9, 13
Orbital following:
in associative ionization, 82
in linear polarization, 62
in molecular collisions, 51, 53
and polarization dependence, 74
Orbital steering, or stereospecificity, 52, 69, 77
Ordered rooted trees, defined, 317
Orgraphs, 150–151
defined, 313, 314
and momentum conservation, 221
Orientation. *See also* Circular polarization
in collisional studies, 38, 54
spin-orbit effects, 85–89
Ornstein–Zernike equation, for direct correlation function, 255
Osmotic pressure, 286
Oxygen:
autoionization experiments, 3, 4
photoabsorption studies, 30

Pair coupling:
at large internuclear distances, 82–83
notation for fine-structure components, 21–22
Particle-photon coincidence studies, 40, 45
Partition function, in field theory, 266
Path, defined, 314
Pendant vertices, defined, 314
Percolation models, 158–170
by bonds and sites, 158–164
cluster ensemble, 291
correlated, 291
cross-linking theories, 169
geometric and thermal phenomena, 167
polymer descriptions, 266
postulated and mathematical, 170
Potts type, 279–280
random, 158–162
Perturbation theory, and coordinate graphs, 228–230
Phantom networks, 308
Phosphorous, autoionization experiments, 19–20
Photoabsorption, spectra of group III atoms, 28–30
Photochemistry, and photophysics, 38
Physicochemical properties of polymeric molecules, 125
Planar graph, defined, 315
Planar symmetry, model collisional case, 55–69
Pnicogen atoms, 19–24
fine-structure components, 19–23
Point, graph, defined, 318

Polarized atoms, laser-excited collisions, 37–114
Polyatomic molecules, 97
Polycondensation:
conversion factor in, 130
equilibrium system, 163–164
Flory model, 129–136
polymer synthesis, 128, 129–136
self-consistent field approximation, 273
spatial physical interactions, 251
swelling theory, 283
Polyethylene, molecular graph, 121
Polymer chemistry:
branching processes in, 135
diagram technique and field theory, 233–294
graph theory, 116–119
physics graphs, 115–325
stochastic fields and functional integration, 233–237
Polymerization:
topology, 123
weight-average degree of, 134
Polymer microstructures:
macromolecular fragments and subgraphs, 171–175
molecular graphs and subgraphs, 171–186
NMR measurements, 173
probability measure on subgraphs, 178–183
subgraph stoichiometry, 175–178
trails and molecular conformations, 184–186
Polymer molecules:
averaging techniques, 125–126
colored ordered trees, 141
ensembles and random graphs, 124–129
kinetic equations, 127–128
Polymers. *See also* Branched polymers; Network polymers
chemical physics of, 115–325
diagram technique and field theory, 233–294
formation and bond reaction, 122
gelation in, 270–276
with high substance combinations, 259
processing conditions, 126–127
semidiluted and strong solutions, 290
stochastic fields and functional integration, 233–237
swelling theory, 283–289
Potassium, photoabsorption studies, 33
Potassium-argon collision pairs, rainbow oscillations in, 85–87
Potassium-rare gas system, orientation effects on spin-orbit interaction, 85–89
Potts model:
extension and field theory, 277–282
multicomponent systems extension, 277–282
Profile index (g), for autoionization lineshapes, 4–5
Proliferating superparticles, in polymer branching process, 140
Propensity rules:
for autoionization, 2
for matrix elements, 13
Pseudograph, defined, 313
p-state atoms, basis states for, 42, 43

Quasi-monomers, molecular graph with, 200
Quenching process, in electronic-vibrational energy transfer, 101–105

Radial coupling, internuclear distance and, 83
Radiative emission, *vs.* autoionization, 3, 4
Radius of gyration, distribution calculations using Kirchhoff matrix, 300–304
Rainbow oscillations:
in potassium–argon collision pairs, 85
of scattering signal, 46
Random graphs:
averaging techniques, 125–126
ensembles of polymeric molecules, 124–129
of polymeric molecules, 138–139
rooted *vs.* nonrooted, 124–125
and subgraph "continuations," 139
Rare-gas atoms, associative ionization, 77
Reactive scattering, in collisions of laser-excited atoms, 39
Reduced widths, in autoionization experiments, 2, 4
Reflection symmetry:
in charge exchange, 95–96
in collisions with polarized atoms, 52–53
conservation of, 61
in inverse collision, 44

Replics method:
in field theory, 267
in swelling theory, 284
Resonance widths:
autoionization, 2–3
Z-dependence in isoelectronic sequence, 17
Resonating atoms:
nonsymmetric disposition of, 175
in polymer microstructures, 174–175
Rigid monomers, 143
tetrahedron and cube models, 143
Rooted graphs:
defined, 317
of polymer molecules, 124–125, 131–132
Rotational coupling:
and charge exchange, 96
and intersystem crossings, 74
in nonadiabatic transitions, 64
transition probability, 58–59
Rotational-vibrational energy transfer, in laser-excited polarized atoms, 41
Rouse matrix:
and molecular conformation, 152
for spectral density, 311
Rydberg series, autoionizing members, 2
Rydberg wave function, 9
autoionization integral involving, 11–13
L-S notation, 13

Scaling graphs, macromolecules as, 153–158
Scattering intensity:
for collisional excitation, 57, 58
equation, 45
linear polarization and, 59–63
Scattering signals, in circular polarization, 54
Second-row atoms, polarizability of, 15
Selenium, autoionization experiments, 3, 8
Self-consistent field:
approximation, 268–277
diagram technique, 260–263
in field theory, 236
Sensitivity parameter, in scattering experiments, 72–73
Separable graph, defined, 318
Separate units, system description, 253
Shell excitation, in group IV atoms, 27
Six-function molecule, bond formation in, 297–299
Slater integral ($_Fk$), 9
Sodium:
alignment and orientation effects, 59
associative ionization, 77–80
collision studies, 40
photoabsorption studies, 33
Z-dependence of resonance width, 17
Sodium–argon collision pairs, rainbow oscillations in, 86
Sodium atom–ion system:
apparatus diagram, 90
charge exchange in, 90–97
collision-induced nonadiabatic transition experiment, 55–69
energy loss spectra from scattering, 56
potential energy curves, 58
spin-orbit interaction in nonadiabatic transitions, 82–85
Sodium–hydrogen system, potential energy surfaces, 100–101
Sodium iodide, isoelectronic sequence experiments, 17
Sodium–mercury system, scattering intensity of, 46–47
Sodium–nitrogen system:
contour map for, 102
electronic-vibrational energy transfer, 97
potential energy surfaces, 100–101
Sodium-rare gas system, orientation effects on spin-orbit interaction, 85–89
Sol fraction:
in cyclation models, 149
in polymer formation, 124
Space-fixed electronic motion, 62, 63
Space-fixed orbital model, 68, 69
charge exchange and, 94–95
following, 53
Space graphs:
Gibbs distribution and probability measure on, 188–191
thermodynamic parameters, 190–191
Space metric models, of Gaussian molecules, 294–300
Spanning trees:
defined, 318
of molecular cyclation, 147–148
Spatial physical interactions:
diagram technique and field theory, 251–258
Mayer diagram technique, 254–257
Spectral density, of molecules, 310–312
Spin-orbit components:
autoionization between, 25–27
of group IV atoms, 25–27

Spin-orbit interaction:
in chlorine autoionization, 3
incoherence due to, 66
of laser-excited polarized atoms, 41, 82–89
nonadiabatic transitions in sodium atom–ion system, 82–85
orientation effects, 85–89
polarization effects, 89
Spontaneous symmetry breaking, at gel point, 271
Square lattice model, 154–155, 157
torus sites, 305, 306
Starlike molecules, and monomers, 142, 189
Stereochemistry, graph theory, 116
Stereoisomers:
molecular graphs, 120
polymeric pattern, 144
Stochastic fields, and functional integration, 233–237
Stockmayer's method, of molecular cyclation, 144–145
Structure-additive properties, in molecular graphs, 123–124
Stueckelberg oscillations, in potassium–argon collision pairs, 87–88
Subgraph, defined, 318
Substitution effects:
first-shell model, 138
higher-order model, 140
kinetic or thermodynamic, 128
in molecular graphs, 128, 137–142
physicochemical models, 137–142
Sulfur, photoabsorption studies, 30
Sulfur atoms, autoionization experiments, 3, 8
Superelastic scattering, 57–59
charge transfer and, 94
Superparticles, proliferation in polymer branching process, 140
Suprathermal energies, atom-ion collisions at, 110
Surface hopping, trajectory calculations, 102, 106
Swelling theory:
diagram technique, 287
for stochastic polymer network, 283–289
Symmetry, in collisions with laser-excited polarized atoms, 37–114

Tellurium, autoionization experiments, 3, 8
Tetrahedron lattice model, 158
Thallium, reduced autoionization widths for, 29
Three-dimensional lattices:
random percolation on, 162
simple and body-centered, 304, 307
Three-dimensional space, graphs embedded in, 186–187
Three-vector correlation:
with laser-excited polarized atoms, 38–39
in polarization experiments, 69
Topological matrix, in molecular graphs, 123
Trails:
defined, 314
and molecular conformations, 184–186
Tree molecules, self-consistent field approximation, 273
Trees, defined, 317
Triads, 171, 172
coefficient values, 176
cyclic, 17
ordered, 183
Triatomic molecules, 97
Trimer concentration, one-trunk diagrams, 259
Tropological isomers, configuration *vs.* confirmation, 127
Two-dimensional lattice, topographical structure, 305
Two-vector correlation, in linear polarization experiments, 69

Vertices, defined, 313

Walk, defined, 314
Wang–Uhlenbeck theorem, for vector distribution, 302–303
Wave functions, improved, 14–17
Weighted graphs, defined, 314
White particles, in polymer branching, 135–136

Xenon:
autoionization experiments, 2, 5–7
Z-dependence of resonance width, 17

Z-dependence, of resonance width in isoelectronic sequence, 17
Zimm matrix:
and molecular conformation, 152
for spectral density, 311